Geothermal energy

GEOTHERMAL ENERGY

Its past, present and future contributions to the energy needs of man

Second edition

H. CHRISTOPHER H. ARMSTEAD

B.Sc., C. Eng., F.I.C.E., F.I.Mech.E., F.I.E.E., F.C.G.I.

LONDON NEW YORK

E. & F.N. SPON

First published 1978 by
E. & F.N. Spon Ltd
11 New Fetter Lane
London EC4P 4EE
Second edition 1983
Published in the USA by
E. & F.N. Spon
733 Third Avenue
New York NY 10017

Photoset by Enset Ltd, Midsomer Norton, Bath, Avon
Printed in Great Britain at J.W. Arrowsmith Ltd, Bristol.

ISBN 0 419 12220 6

British Library Cataloguing in Publication Data

Armstead, H.C.H.
  Geothermal energy.—2nd ed.
  1. Geothermal resources
  I. Title
  333.8′8   HD9682.A2

    ISBN 0-419-12220-6

Library of Congress Cataloging in Publication Data

Armstead, H. Christopher, H.
  Geothermal energy.

  Bibliography: p.
  Includes index.
  1. Geothermal engineering.   2. Geothermal resources.
  I. Title.
  TJ280.7.A74 1983   333.79′2   83-337

    ISBN 0-419-12220-6

Dedication

To all those who feel profound unease at the profligate squandering of the earth's dwindling resources and at the wanton fouling of our once beautiful planet this book is dedicated, in the hope that it may at least contribute to the solution of one of the grave crises confronting Mankind – that of finding abundant energy and of simultaneously reducing the terrifying degree of pollution associated with the combustion of huge quantities of fossil fuels on which we are now so dependent for our social needs.

Contents

List of figures

List of plates

List of tables

Unit conversion tables

Physicists, professors and engineers are usually accustomed to working in different units, and despite the tendency to encourage the use of the SI system, it is a fact that old customs die hard. Engineers, especially of the middle and older generations, are unaccustomed to thinking in terms of newtons, pascals or even joules: they prefer to talk of atmospheres (ata and atü), psia and psig, Btu, foot-pounds and watts. To most engineers a statement that the world's annual energy consumption is so many millions of terajoules is more or less meaningless until he has converted it into kilowatt-hours. Even amongst engineers of different nationalities customs vary. New Zealanders usually refer to geothermal bore yields in terms of kilopounds/hour, Americans in tons/hour, Continentals in kilograms/second. The author has generally tended to use the metric system and the Celsius temperature scale except where custom, or quotation from other authors' works makes the use of other unit systems more convenient. On many of his graphs he has used multiple scales to suit different readers. However, to avoid ambiguities and to facilitate comparisons, the following conversions should serve a useful purpose.

Class	To convert	into	Multiply by
Length	centimetres	inches	0.394
	metres	feet	3.281
Volume	cu centimetres	cu inches	0.061 02
	cu metres	cu feet	35.31
	cu metres	litres	1000
	cu metres	US gallons	264.2
	cu metres	Imp gallons	220.5
	US gallons	Imp gallons	0.833 [1]
	US gallons	cu feet	0.1337
	Imp gallons	cu feet	0.16

[1] 1 Imperial gallon of water at 4° C weighs 10 lb.
1 US gallon of water at 4° C weighs 8.33 lb.

Unit conversion tables

Class	To convert	into	Multiply by
Mass	kilograms	pounds	2.205
	kilograms	long tons	0.000 9842[2]
	kilograms	short tons	0.001 102[3]
	kilograms	metric tons, or tonnes (t)	0.001
Density	grams/cm^3	lb/in^3	0.036 13
	kg/m^3	lb/ft^3	0.062 43
Pressure	atmospheres	cm of Hg	76.0
	atmospheres	kg/cm^2	1.033
	atmospheres	lb/in^2, or psi	14.70[4]
	bars	atmospheres	0.987
	bars	newtons/m^2	100 000
	bars	psi	14.5
	bars	kg/cm^2	1.02
	atmospheres absolute (ata)	atmospheres gauge (atü)	subtract 1.0
Energy and Calorific value	British thermal units, or Btu	ergs	1.055×10^{10}
	Btu	ft lb	778.3
	Btu	joules	1054.8
	Btu	kg cal	0.252
	Btu	kWh	0.000 292 8
	Btu/lb	cal/g, or kcal/kg	0.555 56
	tonnes coal equivalent (tce)	kWh	8130
Concentration	parts/10^6, or ppm	grains per US gallon	0.0584
	ppm	grains per Imp gallon	0.070 16
	ppm	grams/litre	0.001
	ppm	% by weight	0.000 01
Enthalpy	joules/gram, or J/g	Btu/lb	0.43
	kcal/kg	Btu/lb	1.8
Heat flow	kilowatt (thermal)	Btu/h	3413
	kilowatt (thermal)	kcal/h	860

[2] 1 long ton = 2240 lb
[3] 1 short ton = 2000 lb
[4] Absolute pressure in psia = gauge pressure in psig + 14.7
Absolute pressure in ata = gauge pressure in atü + 1.

Temperature conversion $°F = (1.8 \times °C) + 32$
$°C = (°F - 32) \div 1.8$
Absolute Celsius temperature $= °C + 273 = K \text{ (Kelvin)}$
Absolute Fahrenheit temperature $= °F + 460$

Prefixes			
	Tera (T)	10^{12}	
	Giga (G)	10^9	
	Mega (M)	10^6	
	Kilo (k)	10^3	
	Hecto (h)	100	
	Deca (da)	10	
	Deci (d)	10^{-1}	
	Centi (c)	10^{-2}	SI system
	Milli (m)	10^{-3}	
	Micro (μ)	10^{-6}	
	Nano (n)	10^{-9}	
	Pico (p)	10^{-12}	
	Femto (f)	10^{-15}	
	Atto (a)	10^{-18}	

Reciprocals It is perhaps scarcely necessary to add that inverse relationships in the above table can be derived from the reciprocals of the last column of figures.

When dealing with immense quantities of energy in the context of world or national resources, the following units are sometimes used:

1 petrajoule $=$ 1000 TJ or 10^{15} J. (This unit is used by the World Energy Conference in their published energy statistics)

1 quad $=$ 10^{15} Btu (favoured in the USA).

Preface to second edition

The harnessing of earth heat to the service of man calls for the collaboration of a team of specialists, each versed in a different discipline. Geologists, hydrogeologists, geophysicists, geochemists, petrologists, engineers, metallurgists, economists, industrialists, environmentalists, financiers, lawyers – even medical men, politicians and others – will or may become involved. To embrace a full description of the activities of all these specialists and of the theories and hypotheses on which they work would virtually require an encyclopaedia, and would certainly be too great a task for any single author. This book makes no claim to being encyclopaedic: nor is it a 'popular' work for the man-in-the-street, although the mathematical content is both simple and minimal. It is addressed to the well-educated reader who has enjoyed the benefits of a scientific training, who may perhaps be one of the specialists mentioned above but whose activities have not hitherto been slanted towards the geothermal aspects of his own particular discipline, and who wants to have a fairly broad idea as to 'what geothermal energy is all about'. There is many a well-trained specialist who knows virtually nothing of geothermics but whose knowledge could well be channelled profitably into geothermal work did he but know how to apply it in that direction. In short, he lacks the background information that would illuminate the scene sufficiently to show him how to direct his talents most usefully towards the science of geothermics, which is of rapidly growing importance and which will soon urgently need all the qualified manpower that can possibly be recruited in its service. This book seeks to provide that necessary background.

As the author is an engineer, it is only to be expected that the engineering aspects of geothermal development will here receive perhaps rather more than their fair share of attention. For this he tenders his apologies, but at the same time he would point out that had someone of a different provenance written the book it seems probable that rather greater stress would have been placed upon some other facet of what is essentially a many-sided subject. He, nevertheless, hopes that he has done justice to those aspects of geothermics in which he cannot claim to be a specialist.

Since the early 1950s there has been a spate of technical, economic, legalistic and other writings on geothermal matters. Since, as is to be expected, such writings have been partly good, partly bad and partly indifferent, it follows that too much rather than too little 'literature' is available. However, this very plethora of writings is apt to be confusing to the would-be student of geothermics, who cannot see the wood for the trees and who (to change the metaphor) needs guidance in separating the chaff from the wheat. One of the purposes of this book is to provide that guidance by drawing the reader's attention to some of the more apposite writings by means of an extensive bibliography which will enable him to pursue any aspect of geothermics in which he may be particularly interested, as far as he wishes; for most of the writings to which the reader will be referred will in turn contain their own bibliographies for still more detailed study . . . and so *ad infinitum*. The author is very conscious of the fact that in selecting works of reference for inclusion in his bibliography he will undoubtedly have omitted many excellent writings that would equally, and in some cases perhaps even better, have illustrated or amplified the various points made in the text of the book. He offers his apologies to those authors whose works have been omitted – not by intent, but by his inability to be familiar with *all* the relevant writings and by the necessity of limiting the length of his reference list.

With a science that is developing as rapidly as geothermics, it is inevitable that a book of this type is overtaken by events between the time of submission of the manuscript and the date of publication. The information given here, except where specifically stated otherwise, was broadly valid at the end of 1980. The facts that this second edition has appeared only five years after the first, and that it contains many revisions and additions, are indicative of the speed at which the geothermal sciences and developments are advancing.

Finally, the author would like to thank those many co-operative people from several countries who have so helpfully responded to his persistent, and probably troublesome, requests for statistical and other information. He hesitates to mention them by name for fear of inadvertently omitting any.

Rock House,   H.C.H.A.
Ridge Hill,
Dartmouth,
South Devon

September 1982

Introduction

γη Earth   θερμη Heat

Within a few years earth heat – or to use its more scientific but ponderous title, geothermal energy – may well become a factor of supreme importance for the survival of industrial man. Its resources are gigantic, could we but gain easier access to them, and it is vitally necessary that the gospel of its study and exploitation be propagated as widely and as rapidly as possible. Hence this book.

It will be shown in Chapter 19 that the world is in serious danger of approaching an energy famine within a few decades, if not earlier. In fact it is maintained by some that the first harbingers of such a famine are already with us. However, public complacency is such that very few people can read the writing on the wall. Although alarmist preachings are generally to be deplored, a few well-placed jeremiads can be salutary. By recognizing in good time the potential dangers that lie ahead we can perhaps take timely action to avoid them: by ignoring them we may be inviting disaster.

Two of the most promising sources of long-term salvation, of greater ultimate importance even than that of fossil fuels and nuclear fission, are controlled thermo-nuclear fusion (which has so far eluded us) and geo-thermal energy which has progressed rapidly and for which the promise of impressive new developments can now be discerned. Active research into thermo-nuclear fusion will doubtless be pursued, and if successful that would be a 'consummation devoutly to be wished'. However we must not, we *dare* not, rely solely upon that problematic success. It is essential that we have an alternative and it is suggested that this should be the exploitation of earth heat – not in the relatively puny quantities hitherto won, but on an altogether vaster scale.

There are of course other resources available, of which hydro-power is perhaps the most important. At present we are using only about one-sixth of the estimated total world hydro resources that are technically usable, and in 1980 about 23% of the world's electricity needs were contributed from this source. Thus at first sight it might be thought that there should be enough hydro-power to satisfy all or most of our electrical needs for a reasonable time, even allowing for growth; but constraints of economics, of relative locations of resource and of market, and of seasonable availability would almost certainly impose a limit on hydro-development well below what is 'technically usable'. Moreover, at present about two-thirds of the world's total primary energy consumption is used for non-electric purposes. Hydro-power, even if developed as rapidly as possible to its fullest practicable extent, could therefore never replace more than a fraction – though a very valuable fraction – of the fossil and fissile fuels that are now necessary to support the world economy, and which must inevitably become scarcer in the foreseeable future.

Then there is solar energy – inexhaustible and abundant – but its diffuseness at present renders it difficult to harness on the required grand scale, though it can be of considerable local small-scale importance. Tidal, wind and wave power are all renewable resources that should be encouraged as far as possible; but none of them is likely to have more than a modest palliative influence on the world energy scene.

With the exception of hydro-power, earth heat would at present seem to be capable of giving quicker relief than any of the other available resources to the pressures on our dwindling fuel supplies; so we must lose no time in grasping the rewards it can offer.

Even today, geothermal energy is regarded by most people, if they think about it at all, merely as an interesting freak of nature that serves as a tourist attraction in certain parts of the world, either for its spectacular or for its alleged medicinal properties. It would scarcely be an exaggeration to say that tomorrow it may well become one of the best hopes for the survival of civilization.

There are promising signs of a belated, but rapidly growing awareness of the great potentialities of earth heat. Ever since the United Nations first convened in Rome an international conference on new sources of energy in 1961 [1], at which geothermal energy had to share the platform with wind and solar energy and attracted less than one-third of the total number of papers submitted, there has been a steady and rapid expansion of international and national geothermal activity. Universities, professional societies, the technical press, consultants, international organizations, national departments of energy and commercial laboratories are all devoting more and more time and effort to geothermal affairs; and a spate of papers, articles, printed reports on specialized studies, symposia, conferences and seminars bears witness to this fact. The membership of that international body, the Geothermal Resources Council, is growing rapidly. No longer can it be said that geothermal energy is a neglected subject, although in the popular press it is still the Cinderella of the 'alternative' energy rcsources. The sun, the wind, the waves and the tides still seem to appeal more to the imagination of the layman; but in terms of dedicated manpower it is probable that earth heat now leads.

The United Nations should be given due credit for the contributions they have made towards geothermal development, not merely by organizing international conventions [1, 2, 3] but by providing technical assistance and funds for exploration projects in several countries that lacked the resources to proceed without aid. UNESCO too have helped by sponsoring specialized training courses in the geothermal sciences and by publishing a *vademecum* in the form of a Review of geothermal research and development [4].

From all this it is clear that throughout the world there is a marked growth of interest in earth heat. The oil crisis of 1973 undoubtedly provided a great stimulus to this interest, which will no less certainly gather impetus in the years to come. Each successful geothermal project engenders confidence that encourages the development of further projects, and the science of geothermics, born at the turn of the twentieth century, will undoubtedly have grown to an impressive maturity before the century closes. New developments are now taking place that may well increase the rate of geothermal exploitation to a staggering degree, so that earth heat may soon assume an importance comparable with that accorded during the last hundred years to fossil fuels.

In spite of a handful or so of very impressive installations, it must be admitted that so far we have scarcely scratched at the surface of geothermal potentialities. At present earth heat is contributing only a small fraction (less than a quarter of 1%) of the world's electrical needs and a far smaller fraction of its non-electrical requirements. Clearly we are a very long way from being able to claim that geothermal energy is yet a major factor in the world economy. However, with the techniques now available we can exploit earth heat only in what are known as 'geothermal fields' and low-grade aquifers (see Chapter 5), which are confined by nature to certain favoured parts of the earth. At last, however, we seem to be within reach of *creating* fields where none exist in nature (see Chapter 18) and perhaps of ultimately tapping immense reserves of heat that have hitherto been regarded as inaccessible.

Wherever earth heat has been exploited, it has proved to be cheap by comparison with alternative sources of energy (see Chapter 14). Why then, it may reasonably be asked, has its development not been much more rapid since the 1930s when the Italians had so successfully exploited the Larderello field in Tuscany for power generation and the Icelanders had made such great progress in the heating of Reykjavík's homes by means of subterranean heat? The explanation lies partly in the comparative rarity of geothermal fields, some of which are topographically remote from potential energy markets. Others occur in countries well endowed with cheap alternative energy sources that can be developed without risk by well-established techniques involving no more than conventional engineering methods. New Zealand and Iceland are cases in point; for both are rich in hydro-power resources, and New Zealand also possesses coal and natural gas. It is true that New Zealand *has* developed geothermal power to some extent and that Iceland *has* used low-grade earth heat for space heating and domestic hot water supplies (and is now starting to develop geothermal power); but had either of these countries been poor in other energy resources it is highly probable that by now they would have exploited their earth heat to a far greater extent.

But these are not the only reasons for the relatively slow past development of geothermal energy. There has also been a philosophical conflict of ideas between 'risk capital' on the one hand and 'safe development' on the other hand. Although there are many places in the world where the existence of geothermal resources is suspected, there can be no advance certainty that the right geological, hydrogeological and chemical conditions occur, without which exploitation would be impossible. Geothermal development must be preceded by exploration to establish whether truly exploitable heat exists. Exploration requires a fairly large outlay of risk capital which, if successful, can be amply repaid by results; but which, if unsuccessful, will represent just so much money wasted (except, perhaps, for any purely scientific knowledge that may have been gained). Governments – particularly those of developing countries – tend to be reluctant to invest risk capital if an alternative energy source is clearly obtainable without risk, even at a higher price than that at which geothermal energy could *perhaps*, but by no means *certainly*, be won if the outcome of the risk were successful. In short, governments are seldom willing to take a gamble. Private enterprise, on the other hand is often very willing to invest risk capital (e.g. the oil companies) but, not unnaturally, it expects a fairly high return on its money if the results of the risk should be successful. Now the commonest application for geothermal energy has hitherto been electricity generation, with district heating second. But electricity and domestic heat supply are industries which, being essential public services, are apt to be intolerant of high profits; so there is less inducement to stake risk capital. Moreover, whereas a successful oil strike produces a commodity that is readily marketable and easily shipped all over the world, an enterprise that wins geothermal energy must find a local market, which may be more limited.

Reluctance to take financial risks has, moreover, been reinforced by a certain conservatism and unwillingness on the part of electricity supply undertakings and industries to take geothermal energy seriously: there has been a tendency to regard it as 'new fangled', and to cling to more conventional sources of energy that have for longer stood the test of time. Prejudicial though this attitude may have been, it has undoubtedly been an adverse psychological factor that has retarded the rate of geothermal development, and one which even now has not been fully eliminated.

Examples can of course be cited both of governments and of private enterprises that have been venturesome and willing to invest risk capital in geothermal exploration that has paid off handsomely. In many countries, however, the philosophical gap between risk and caution remains, despite the valuable work done by the United Nations in helping to bridge this gap by partly financing geothermal exploration programmes through the medium of their United Nations Development Programme Special Fund, thus reducing the element of risk that has to be taken by client nations.

Certain trends, however, are already doing much, and will undoubtedly do more in future, to remove these barriers of reluctance. First, many countries have by now 'skimmed the cream' off their readily available indigenous energy resources and must now perforce seek less conventional supplies. Secondly, the spectacular increases in fuel prices since 1973 are helping to create a greater willingness to face the required risks inherent in rather costly exploration work, in view of the great prizes that may be won in the form of relatively cheap energy and, equally or even more important, self-sufficiency and independence from the political whims and economic policies of other nations. Thirdly, there have been great advances in recent years in exploration techniques that have gone far in reducing the risk element. In the early days of geothermal development, before we knew much about what is happening below ground, exploration was a rather chancy hit-or-miss business largely based on 'wild-cat' drilling in the neighbourhood of visible surface thermal manifestations. Drilling is a very costly operation, and much money was sometimes spent before any positive results could be shown. Now, by using more refined techniques and by applying the accumulated knowledge gained during the last two or three decades, we are able to collect much more reliable information from relatively inexpensive measurements made at the surface, from which we can build up a reasonably probable 'model' of the structure of a field so that we may postpone expensive drilling operations until we can choose bore sites with a much better chance (though still without certainty) of success. Lastly, whereas at one time there was a tendency to regard earth heat as being suitable almost exclusively for electricity generation or district heating, there is now a growing awareness of its industrial and farming potentialities which can thus widen the choice of markets.

A useful rule-of-thumb which graphically illustrates the potential value of earth heat is that *every kilowatt of geothermal base load power is capable of saving about two tons of oil fuel per annum,* or its equivalent, in an integrated power system. Thus if the price of oil were, for example, US $220 per ton, one kilowatt of geothermal base load could save about US $440 p.a. which, if capitalized at 16% p.a., would be worth US $2750 per kilowatt! It is of interest to note that in California the capital cost of a 110 MW geothermal power plant (excluding the steam supply) in 1980 was US $254 per kW: with the cost of the steam supply included, the gross costs would probably be less than double this figure. As will be shown in Chapter 14, capital costs per kilowatt can vary greatly from country to country, from time to time and from plant capacity to plant capacity. Nevertheless, this example will suffice to show that in terms of fuel saved, investment in geothermal power can be very attractive. It can be still more attractive for low-grade heat.

The example also emphasizes the importance of *time* in an oil-importing country (or for that matter also in an oil-exporting country, for oil not burned at home is oil available for export). For example, with oil fuel at US $220 per ton, a year's delay in constructing a 100 MW geothermal power plant would effectively burden the national balance of payments by about US $44 million, even if the possible effects of inflation upon the price of the plant during the year's delay were disregarded. This underlines the importance with pressing ahead as rapidly as possible with geothermal development in countries possessing potentially exploitable earth heat, especially in those lacking cheap indigenous fuels.

It is of interest to study the past and present growth trends of geothermal development, and also the expected growth during the next few years. Only power generation will here be considered – not because other applications of earth heat are in any way to be despised, but simply because heat statistics are less readily available than kilowatt figures. It took more than half a

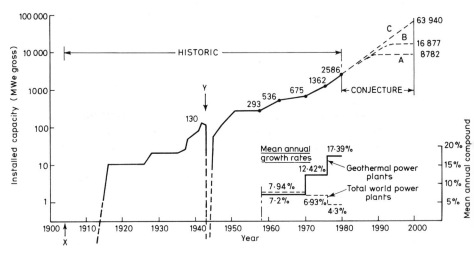

Figure 0.1 World total of gross installed geothermal power capacity and growth trends. X, Prince Conti starts to develop power at Larderello; Y, destruction of Italian plant by war damage; A, definitely planned in 1980; B, tentatively planned in 1980; C, continued growth at 1976/80 rate.

century, from 1904 to 1958, for 293 MW of geothermal electric power to be developed, but the growth of installed capacity of such plants since then has shown a marked upturn, as can be seen from Fig. 0.1, which is drawn to a logarithmic vertical scale.

It is of particular interest to note the mean annual growth rates, plotted in the lower right-hand part of Fig. 0.1. These show the growth trends in the capacities both of geothermal power plants and of world power plants of all kinds. Until about 1970 the growth rate for geothermal power plants had been failing to keep pace with that of the world total installed capacity; but in the 1970s the growth in the world total plant capacity declined conspicuously as a result of the 1973 Middle East crisis and of the subsequent world recession. By contrast, the growth in the installed *geothermal* capacity during the 1970s rose very markedly. In the last four years of the 1970s, geothermal power plant capacity grew more than four times as rapidly as the capacity of world power plants of all kinds, and was thus taking upon itself an increasing share of the world electricity market. Moreover, these growth figures are not entirely fair to geothermal power, for whereas the average annual load factor of all the world's power plants is about 47.5%, geothermal power plants generally operate at much higher load factors and therefore contribute a larger share of electrical *energy* than would appear from the installed capacities. This point, however, should not be over-stressed as much would depend upon the types of alternative plants that would have been installed if the geothermal plants had not been adopted. However, it would generally be true to say that a geothermal kilowatt is worth more in terms of energy than any other kind of kilowatt – except perhaps for run-of-river hydro-plants.

Extrapolation of geothermal growth into the future is bound to be speculative, but curves A and B of Fig. 0.1 show what the growth is expected to be on the basis of geothermal power plants definitely and tentatively planned, respectively. Curve A can therefore be regarded as an absolute minimum forecast, as much of its growth is accounted for by plants already under construction. Even curve B is likely to be an under-estimate, because new fields are continually being discovered and exploited, and also because the pressures on governments to become less dependent upon fuels are continually intensifying. Curve C shows extrapolation on the assumption that the growth trend of the late 1970s (17.39% p.a.) is maintained until the end of the century. From one point of view this could be regarded as an under-estimate, because it assumes an abrupt halt to the accelerating growth rate shown over the 22 years before 1980. But it is also necessary to recognize the fact that undiscovered and undeveloped fields of high quality, suitable for large-scale power development, are limited in number and will therefore ultimately become scarce; also that fields now being exploited are likely to run down in the course of time. Some kind of 'logistic' growth pattern would seem to be probable, with consecutive phases of hyper-exponential, simple exponential and sub-exponential characteristics until growth ultimately falls to zero, perhaps with subsequent *decline*. It would be rash to attempt to quantify such a pattern, but an order-of-magnitude perspective can nevertheless be obtained, as follows.

If, for the sake of argument, the installed geothermal plant capacity in the year 2000 were about midway between the forecasts shown by curves B and C – say 40 000 MWe – and if the total world power plant capacity should grow at the mean rates shown in the first column of Table 0.1, then by the end of the century geothermal power plant could account for the share of the world's total installed power plant capacity shown in the third column of the table. If it were further assumed that the geothermal power plants will operate at 85% average annual plant factor, while the average world power plant factor remains at its present level of 47½%, then the electrical *energy* contribution from the geothermal plants would be as shown in the fourth column of the table.

Table 0.1 Proportional contribution of geothermal power to the world total power demand, in terms of growth rates.

Mean compound growth rate in world total installed power capacity (%)	Total world installed power plant capacity in the year 2000 (MWe)	Geothermal	
		Proportion of power plant capacity (%)	Proportion of electrical energy contribution (%)
5	5 262 000	0.760	1.36
4	4 348 000	0.920	1.65
3	3 585 700	1.116	1.20
2	2 949 700	1.356	2.43
1	2 422 100	1.651	2.95
nil	1 981 400	2.019	3.61

Now although these figures are not innocent of guesswork, they do seem to suggest that *with present technology* it is improbable that geothermal energy could ever contribute more than quite a small percentage of the world's total electrical needs. In the absence of new technologies the growth rate of geothermal power capacity must inevitably decline in the course of time; and unless the growth rate of the total world electricity demand is extremely small, the two growth rates will sooner or later become equal and the proportion of the geothermal contribution must thereafter fall.

The direct application of earth heat for non-electrical purposes has hitherto developed very slowly, the lack of historical records makes it impossible to present a graphical representation of its growth in the manner of Fig. 0.1. As far as can be estimated it is unlikely that direct applications accounted for as much as one-tenth of 1% of the world's total primary energy requirements in 1980. There are, however, promising signs of a new awareness of the great potentialities of such applications (see Chapter 11). In the USA alone, the 1980 level of direct uses of geothermal heat will be more than trebled if all the projects now under development, planned, and the subject of feasibility studies come to fruition – and it is not improbable that this will be almost, if not totally, achieved. If all these new projects should become operational in five years, the mean growth rate would exceed 25% p.a. – a far higher rate than that at which geothermal power is now growing. This shows that in the USA at least efforts are being made to make up for lost time. If other countries show the same drive, a long period of sustained growth in direct applications may be expected; for even with existing technology it has been reckoned that the resources of low-grade heat suitable for direct use are probably far greater than those of high-grade heat suitable for power generation. Growth in direct applications should not therefore become inhibited by shrinking resources nearly so soon as in the case of geothermal power, even if no technological advances are made. Thus it may not be long before direct uses of earth heat will be making as great a contribution to the world's total primary energy needs as geothermal power; but it seems probable that in 1980 direct applications were extracting only about one-sixth as much heat from the earth's crust as geothermal power plants (taking into account both plant factors and efficiencies of the power plants and including hot water for bathing purposes in the direct applications).

From all this it might be thought that we are enjoying a fairly short-lived 'boom' that has a very minor influence upon the world energy pattern and which must later peter out as the twenty-first century advances. This would probably be true if technology remained static: but why should it? And if improved techniques for winning earth heat are to be developed, why should they lie in the remote future? Necessity is the mother of invention, and necessity is pressing hard upon us. Interesting and impressive developments are even now the subject of experiment in the USA and in other countries; and these hold out reasonable hope that within a decade or so we may come to possess the means of vastly extending the areas of the world where earth heat may be won commercially. It is even conceivable that one day it will become possible to extract that heat in large quantities at any convenient point on the earth's surface. There is now a concentration of human effort dedicated to these ends; and the driving force is the same as that which inspired Mallory to give as his simple reason for wishing to conquer Mt Everest – 'because it is there'.

There is a common tendency amongst writers and commentators to write off geothermal energy as little more than a resource which, though sometimes of great local importance, could certainly never make a major contribution to the solution of the world's energy problems. The author believes this to be an unwarranted defeatist attitude, and that we are, on the contrary, on the eve of a geothermal 'renaissance' that will ultimately assume a significance comparable with, and perhaps exceeding, that of fossil fuels at the present day. After all, less than half a century ago there was probably only a handful of dreamers who seriously believed it would ever be possible for man to reach the moon and planets, because, *with the techniques then available,* such an achievement could clearly never have been realized.

It is useful to summarize the 'virtues' of earth heat as an energy source:

(i) Its resources are immense – virtually infinite on the scale of history (see Chapter 4).

(ii) It is highly versatile (see Chapter 2).

(iii) It is far less guilty of pollution than fuel combustion or nuclear fission (see Chapter 16).

(iv) It has the merit of security.

This question of security requires amplification. Unlike hydro-power it is not at the mercy of seasonal rainfall. If indigenous geothermal resources can be developed they can liberate a country from political and economic dependence upon other countries and can help the national balance of payments. The continued exploitation of earth heat, once won, is not labour-intensive and is therefore less vulnerable to industrial disputes than, say, coal mining. A geothermal installation, unlike a marine oil rig, is relatively in little danger from sabotage which, in these lawless days, is a threat that cannot be ignored. For all these reasons earth heat may be regarded as having the merit of security.

In 1967 in Washington DC, the author publicly expressed the opinion that humanity might have reaped a far greater reward had the vast expenditure hitherto lavished upon space exploration been devoted instead to activities in a downward direction. He still stands by that opinion.

1 *Historical note*

Visual evidence of earth heat is provided by such natural phenomena as volcanoes (Plate 1), geysers (Plate 2), fumaroles, hot springs and pools of boiling mud (Plate 3). These manifestations are to be found in various parts of the world and are often associated with the occurrence of earthquakes. It is not surprising that the ancients should have regarded the depths of the earth with horror, as the seat of Hell and of malignant gods – as do some primitive peoples to this day. Fortunately we have now come to view them as the seat of hope. The transition from horror to hope will now be described in this historical note, which may conveniently be treated for the most part under headings corresponding with the various applications of earth heat. Some of these applications have developed collaterally with others.

1.1 Applications of earth heat

1.1.1 *Balneology.* Although the ancients who inhabited the thermal regions of the world tended to hold in fear the more violent manifestations of earth heat such as volcanoes, and therefore to shun them – and after the destruction of Pompeii and Herculaneum who can really blame them? – they were not slow to exploit for their comfort the gentler thermal phenomena such as hot springs, which have been used for centuries for bathing purposes by the Etruscans, Romans, Greeks, Turks, Mexicans, Japanese, Maoris and no doubt others. Sometimes the waters from these springs have a rather foul taste and smell, suggestive of rotten eggs or worse; and as there is a curiously perverse human tendency to think that that which is unpleasant must be good for you, a belief grew up that to drink one's bath water was commendable and beneficial to the health! Certainly it is true that some thermal waters act as powerful aperients. Furthermore, as hot water is undoubtedly soothing to tired limbs, the cult of 'taking the waters' was built up over the centuries. Thermal waters were alleged to possess healing and prophylactic properties when applied externally, as in bathing, or internally when taken

1

Plate 1 Mount Ngauruhoe volcanoe, North Island, New Zealand. (By courtesy of the New Zealand High Commissioner, London)

orally or used for douches. Thus was born the balneology industry – the oldest application of earth heat. It flourished in the days of Imperial Rome where the *thermae* became an institution, not only as centres for health and hygiene but also as foci for social intercourse – rather like the eighteenth century coffee houses in London. Later, in the eighteenth and nineteenth centuries the custom of taking the waters underwent a tremendous revival, enthusiastically supported by the jaded world of fashion whose members sought, for a few weeks annually, to recoup from the effects of excessive port-drinking during the rest of the year. 'Spas', 'hydros' and 'watering places' sprang up all over Europe and elsewhere, and were frequented by invalids, hypochondriacs and the merely affluent. Not all, but most of these spas were of thermal origin. The balneology industry thrives to this day, as will be shown in Chapter 11 (though without its erstwhile aristocratic over-tones), and millions of bottled 'mineral waters' from thermal springs are consumed every year all over the world. The *hamams* of Turkey, from which originated the westernized 'turkish bath', were patronized by the Ottomans,

Plate 2 Castle Geyser, Yellowstone National Park, USA. (Zeta)

and even today they are frequented by women who squat over jets of hot water in the hopes of inducing fertility or of curing unmentionable diseases.

1.1.2 *Domestic services.* The Maoris have adapted geothermal phenomena to their domestic needs ever since they settled in New Zealand in the fourteenth century, and their traditional rural way of life in the thermal areas may be witnessed today. At a village near Rotorua in North Island one may see a fisherman catch his trout in a cool river and drop it into

Plate 3 Boiling mud pool, Rotorua, New Zealand. (By courtesy of the New Zealand High Commissioner, London)

a nearby pool of boiling water to cook it. A few yards away his wife may be seen administering a geothermal bath to the baby, while his daughter is doing the household laundry in a hot spring and the vegetables are cooking over a fumarole. Each of these simple domestic chores is performed with natural fluids at appropriate temperatures, and it is important not to confuse the trout with the baby. Although the hidden hand of organized tourism may perhaps be detected behind this scene of rural bliss, the activities nevertheless have a factual historical basis.

1.1.3 *Mineral extraction.* Another very early application of natural thermal activity was for the procurement of minerals. The Etruscans extracted boric acid from the boiling springs – later known as *lagoni* – to the south of the ancient town of Velatri (modern Volterra), and used it for making the splendid enamels with which they decorated their vases. From the mid-thirteenth to the end of the sixteenth century these *lagoni* were exploited for the extraction of sulphur, vitriol and alum. In the early nineteenth century Francesco Larderel, Count of Montecerboli – a Frenchman who had emigrated to Italy – built up a prosperous boric acid industry in Tuscany which survived until well after the middle of the present century, by which time it was no longer profitable. The count gave his name to the now famous township of Larderello.

Another ingenious exploiter of geothermal minerals was none other than Hernan Cortez, who used them for military purposes. The Fates seem to have favoured this *conquistador* by spreading superstitious dismay among the Aztecs. Not only did these doomed Mexicans believe Cortez to be the reincarnation of the mythical Quetzalcoatl who, centuries earlier, had come out of the East and had later sailed away into the Atlantic, promising to return one day; not only did they identify the hitherto unknown horse with its rider so as to produce in their imagination a sort of synthetic centaur; but Mount Popocatépetl, after centuries of dormancy, decided to stage a miraculous eruption at the precise time when the small Spanish force was advancing upon the Emperor Moctezuma's capital city. With such supernaturally loaded dice, the conquest of a mighty empire by a small band of determined adventurers was perhaps not so surprising. Cortez was not only a soldier: he was a great opportunist. He decided to earn a double dividend from the eruption of Popocatépetl. Having burned his fleet to ensure that his comrades-in-arms could not defect, he was still able to maintain his supplies of ammunition by making his own gunpowder, using as an ingredient the sulphur deposited near the crater of the volcano after its obliging activity.

The extraction of chemicals from geothermal fluids continues. Elemental sulphur is recovered from fumaroles in Japan and Taiwan, sulphuric acid and ammonium salts are being extracted from thermal fluids in Japan and Italy; common salt and calcium chloride have been produced from hot brines beneath the Salton Sea area in Southern California; and the possibilities of winning valuable trace elements from geothermal fluids are being closely studied.

1.1.4 *Electric power generation.* The most spectacular advance to be made in the exploitation of earth heat was heralded when Prince Piero Ginori Conti first promoted electric power generation at Larderello in 1904. Attempts were first made to use reciprocating steam engines fed with natural steam, but these were short-lived owing to virulent chemical attack. Then, at first by using heat-exchangers with which to raise 'clean' steam from 'dirty' natural steam, and later by improving the quality of materials used in the manufacture of prime movers so that natural steam could be used directly without incurring the losses inherent in the use of heat-exchangers, the chemical problems were gradually overcome. A 250 kW power station was put into service in 1913; and thereafter a steady expansion in the sizes and numbers of generating units took place until, by the early 1940s, some 130 MW of geothermal power plant in Tuscany were feeding the electrified Italian railway system. This plant, then the only installation of its kind in the world, was totally destroyed by enemy action during the latter part of the

Second World War, but shortly after the return of peace to Italy reconstruction was begun – not only in Larderello, but also at several other places in the vicinity. Now, a complex of several geothermal power plants having a total installed capacity exceeding 400 MW and no longer confined to Tuscany alone is supplying the integrated power network of the Italian state organization *Ente Nazionale per l'Energia Elettrica*, commonly known as 'ENEL'. Although larger geothermal power plants are now in service in America and elsewhere, it is to the eternal credit of the Italians that the first really impressive break-through in geothermal power exploitation was achieved.

About half a century was to pass before any other country followed Prince Conti's early pioneering work in geothermal power generation. The New Zealand Government started serious exploration in the Wairakei area in North Island in about 1950, and in 1955 they decided in conjunction with the United Kingdom Government to embark upon a dual purpose chemical/power project which was to have produced heavy water and 47 MW of electric power. While the plant was under construction the market price of heavy water fell to a level at which its production would no longer have been economic, so the installation was modified to become a 'power only' project. Meanwhile, successful drilling operations had proved a lot more steam, and the project was stepped up in scale to a total installed capacity of 192 MW. The first power unit was commissioned in 1958 and by 1963 the entire project was completed. The discovery of natural gas in North Island and the construction of the 600 MVA Cook Strait cable which enabled large quantities of cheap hydro-power to be transmitted from South Island to the more populous North Island cause a postponement of the tentative plans that had been drawn up for additional geothermal power development in New Zealand. After the oil crisis of 1973, however, the New Zealand Government decided to proceed with further exploitation of natural steam. As the Wairakei field had meanwhile shown signs that its ultimate potential might fall rather short of earlier expectations, it was decided to develop other known fields in North island, from which it is hoped that power will start to flow by late 1986.

California was the next country to produce geothermal power. After a very cautious start – 12 MW in 1960 and a further 14 MW in 1963 – the pace of development accelerated rapidly as more and more steam was proved. By 1980 the total geothermal generating plant capacity installed in the Geysers field, some 60 miles or so north of San Francisco, had reached a figure of 943 MW (gross), and the rate of further expansion is now limited more by the need to comply with the stringent environmental standards than by the ability to win steam or to build the plant. According to the US Geological Survey Circular 790, 1978 [5], an ultimate potential of 48 300 MWye – or 1610 MWe for 30 years – has been estimated for the Geysers field alone.

In a historical note such as this it would be tedious to make a full catalogue of every geothermal power development. Suffice it to say that after the great economic success of geothermal power had been clearly demonstrated in Italy, New Zealand and California a spate of other such developments followed from about the mid-1960s. By the end of 1980 a total of 2586 MWe (gross) of geothermal power plants had been installed in 13 countries, and more than 6½ times that capacity had by then either been definitely or tentatively planned in a total of 24 countries. Active exploration is now being carried out in a further three countries for high-grade earth heat suitable for power generation. The production of electric power from geothermal energy is now a well-established activity.

It is of interest to note that Iceland, one of the most thermally active countries in the world (and whence the word 'geyser' is derived from an Icelandic word *geysir,* which can most nearly be translated as 'gusher') was a late starter in the development of geothermal power. It is true that Iceland pioneered large-scale geothermal district heating, and that a 17 MW power plant was planned in the early 1960s to exploit the thermal field of Hveragerdi [6], but this power plant was cancelled when a large hydro-power installation which could not at that time have competed economically with Hveragerdi for normal demand growth, suddenly became justified by the unforeseen emergence of a large new power-intensive industry which changed the economic perspective. It was only in 1969 that a small non-condensing 3 MW geothermal power unit was first installed at Námafjall. This was rather an anti-climax, but one which was quite natural in a country of so small a population. Iceland expected to become a fully qualified member of the community of geothermal power generating nations when the 60 MWe plant was commissioned in 1977 at Krafla; but unfortunately that event too was an anti-climax. For in 1975, when the plant was under construction and drilling was in progress, rift formations appeared about 2 km from the power station and volcanic eruptions occurred; thus providing unwelcome confirmation of the alleged behaviour of the Mid-Atlantic Rift (see Chapter 3). One bore was totally destroyed by an explosion, while others were damaged. As a result, by the end of 1980 the 60 MWe power station was capable of generating only 5 or 6 MWe through lack of steam. A new drilling programme in another part of the field is now in progress. The Krafla episode has been one of the very few temporary setbacks in the otherwise almost unblemished 'success story' of geothermal power.

1.1.5 *District heating, domestic hot water supply and air-conditioning.* More or less simultaneously with the major development of the Larderello field in Italy for power generation, the Icelanders were initiating large-scale exploitation of their earth heat for district heating and domestic hot water supplies, not on the small scale that had been practised for centuries in various parts of the world, but as a large-scale public supply system similar to the electricity and gas distribution networks of big cities. After some modest beginnings at the turn of the twentieth century a pilot district heating scheme was established in 1930 in the capital city of Reykjavík to supply about 70 houses, two public swimming pools, a school and a hospital. The scheme was so successful that the Municipality of Reykjavík started to drill for hot water about 15 km outside the city in 1933 with the intention of ultimately heating the entire capital. By 1943 no less than 2300 houses were thus supplied. New areas were then drilled, some within the city boundaries, and gradually the system grew until by 1975 all but 1% of the buildings in Reykjavík were supplied with domestic heat – an amenity now enjoyed by about 100000 people. While the Reykjavík system had been extended to outlying suburbs, other smaller public heating supply systems were springing up in various parts of Iceland. Almost two-thirds of the entire population were enjoying the benefits of geothermal heating by 1980 and the proportion was still rising. Few Icelandic houses of less than 40 years of age are provided with chimneys.

In the USA too there have been smaller public heating systems utilizing earth heat. At Boisé, Idaho, the Warm Springs residential area was geo-thermally heated as long ago as 1890, while at Klamath Falls, Oregon, geothermal space-heating has been practised since the turn of the century and is now in the process of rapid expansion. In Japan and New Zealand geothermal domestic heating has developed more on individualistic lines for single homesteads, though recently a group of public buildings in Rotorua, North Island, New Zealand, was collectively supplied with natural hot waters. In 1962 space heating began to be developed in Hungary, where a large low-grade thermal reservoir occurs below ground, and similar activi-ties'have been pursued in the USSR in the Caucasus and in Kamchatka. In the early 1970s public heat supplies were made available on a modest scale at Melun, in the Paris Basin, by circulating water at depth through bore-holes sunk into a deep aquifer in a non-thermal area. The success of the Melun scheme led to the initiation in 1976 of several other geothermal district heating projects in suburban areas around Paris: far more ambitious plans for district heating by means of earth heat in other parts of France are now being contemplated (see Section 11.2.4).

As a complementary development to space-heating, house-*cooling* (or air-conditioning) was first developed in Rotorua for a hotel building, by making use of a single geothermal bore [7]. Einarsson suggested in 1975 that the destruction of Managua, Nicaragua, in the devastating earthquake of 1972, offered an excellent opportunity for providing a geothermally activated public cooling system [8], but this proposal was not pursued. Geothermal air-conditioning is also being practised to a modest extent in Japan and elsewhere.

1.1.6 *Farming and aquaculture.* Another thriving application of earth heat was first practised in Iceland in the 1920s, when natural hot waters were used to heat greenhouses in which vegetables, fruit, mushrooms and flowers were raised in a country where such produce could not be farmed under natural conditions owing to the prevailing inhospitable climate (Plates 4 and 5). Before then, virtually all food other than fish, meat and potatoes had to be imported into Iceland, but by using geothermal heat under glass or plastic roofs, the country's economy was greatly improved and the population was enabled to enjoy foods that were formerly unobtainable except at exorbitant prices. By 1980 almost 110 000 m² were under cultivation in Iceland in geothermally heated greenhouses. More recently the USSR and Hungary have exploited their natural hot water fields to an even greater extent; and in these two countries alone the areas of geothermally heated greenhouses by the end of 1980 were 420 000 m² and 1 900 000 m² respectively, accounting together for about 645 MWt of heat. The USA, Italy, Japan and New

Plate 4 Grapes grown in geothermally heated greenhouse, Iceland. (Photo by Icelandic Photo and Press Service, Reykjavík)

Plate 5 Tropical fruit grown in geothermally heated greenhouse, Iceland. (Photo by Icelandic Photo and Press Service, Reykjavík)

Zealand have all developed this kind of cultivation to some extent in more recent years, and the farming application of earth heat has been extended to animal husbandry, soil heating, fish hatcheries, dairy farming, egg incubation, poultry raising and the biodegradation of farm wastes. At Agamemnon, near Ismir in Turkey, waste geothermal waters from *hamams* are used to irrigate the fields, where crop yields are alleged to be substantially increased as a result of the warmth. In the State of Oregon, USA, the application of geothermal soil heating is said to have raised the output of corn by 45%, tomatoes by 50% and soya beans by 66%, as well as showing improvement in the *quality* of certain crops [9]. Experiments in geothermal greenhouse heating are being conducted in Ladakh in the Himalayan districts of India where the exceptionally severe climate virtually precludes the possibility of natural farming.

Farming applications require only low grade heat, ranging from about 20° C for fisheries, 40° C for soil warming to 60° C, or perhaps as much as 80° C, for greenhouse heating.

1.1.7 *Industry*.   The word 'industry' of course embraces a very wide range of activities, including some of those already referred to in this chapter. With a growing realization that many industries are very heat-intensive – some requiring high grade, but many more requiring only low-grade heat – there has been a steady, but slow, expansion in the industrial applications of earth heat since the end of the Second World War, before which such application was more or less confined to the borax industry in Italy, referred to in Section 1.1.3. The first large scale industrial application of geothermal heat was initiated in the 1950s at Kawerau, North Island, New Zealand, where the Tasman Pulp and Paper Mills have for many years now been producing paper – particularly newsprint. At the Kawerau plant about 200 tons of natural steam are consumed hourly for processing purposes, and incidentally for small scale power generation also. Other examples of geo-thermally activated industries are referred to in Section 11.4, and the scope continues to widen.

1.2 Scientific precursors

No historical note of this description would be complete without mention of certain great scientists who, while not having directly contributed to the harnessing of earth heat in the service of man, have nevertheless influenced thought that has promoted geothermal development.

First there was *Goethe* (1749–1832) who, though principally renowned as a poet, was a man of great versatility. He might well be regarded as the Father of Geology, for it was he who developed the study of rocks into a serious science worthy of systematic treatment – a science that figures so largely in geothermal exploration. Under Goethe's initial impetus, geology developed rapidly during the nineteenth century, particularly after Charles Darwin endowed the whole of pre-history with an evolutionary perspective, and it advanced further in the twentieth century when isotopic studies became possible.

Lord Kelvin (1824–1907) was fascinated by theories to account for the origin and nature of the interior heat of the earth, and hoped that in the pursuit of these theories he would be able to determine the age of our planet. Although he outlived by about ten years the discovery of radioactivity by Becquerel in 1896, his theorizing was mainly based upon the classical nine-teenth century conceptions of heat. He was unable to reconcile the observed rates of outward heat flow from below ground with any calculable store of 'original' heat with which the earth could have been endowed at the time of its formation, without conflicting with well-established geological evidence as to the age of the earth. But he was largely responsible for directing much thought towards the search for an explanation of the mysteries of sub-terranean heat.

Sir Charles Parsons (1854–1931), the eminent British engineer whose name is more generally associated with the development of the steam turbine and with astronomical telescopes, also took an interest in subterranean heat. In 1904 he proposed that £5 million – a very considerable sum in those days – should be spent in sinking a shaft twelve miles deep into the earth. He estimated that the work would take about 85 years to complete, and on the basis of 'normal' temperature gradients he expected to encounter temperatures of the order of 500°C at the base of his shaft. The practical way in which this exposed heat was to be reached and used seems to have been rather left to the imagination, and the fact that he referred to his proposal as 'The Hellfire Exploration Project' suggests that Sir Charles's tongue was in his cheek. Nevertheless, he again referred to this proposal in 1919 in his presidential address to the British Association.

Another outstanding scientist who influenced geothermal thought was the German, *Albert Wegener,* who was a pioneer in developing the theory of 'floating continents' (see Chapter 3). It was in 1915 that Wegener first suggested that the earth's crust was behaving in a peculiar manner. His theories at first aroused fierce scepticism, but modern theory has apparently vindicated his ideas, which have had a profound influence upon geothermal thinking.

There can be little doubt that if Goethe, Kelvin, Parsons and Wegener were alive today they would all be geothermal enthusiasts.

1.3 Development of the environmental conscience

It is fortunate that the rapid acceleration of geothermal development in recent years has coincided with the growing public awareness of the dangers threatening our human environment. The fantastic and self-evident growth of pollution from fuel combustion has at last frightened the more responsible elements of society into a realization of the terrible things we are doing to our planet, and belated recognition of the awesome problems to be faced is now becoming evident. Geothermal energy was at first rather unjustifiably hailed as entirely 'clean', and altogether innocent of pollutive side effects: it was sometimes argued that its development should be encouraged even for that reason alone. But on closer examination it was found that the exploitation of earth heat has its pollutive problems too, but that these problems are far more manageable than with fuel combustion. A good cause can be harmed by exaggerated claims, but fortunately the pollutive aspects of geothermal development are being attacked with vigour – in fact (as will be shown in Chapter 16) with rather excessive vigour sometimes. Although earlier irresponsible claims of entirely non-pollutive earth heat were not well founded, the balance is now being redressed. So long as we remain alive to the pollutive risks, no serious difficulty need arise in combating them.

1.4 Conclusion

This brief historical note will serve to show that the exploitation of geo-thermal energy, apart from one or two earlier minor applications, is really a child of the twentieth century, and that it entered the period of adolescence after the Second World War. It is still quite juvenile, but it may be expected to attain adulthood early in the twenty-first century.

By definition, 'history' stops at the present moment. Nevertheless, the discerning eye can detect, embedded in the present, certain 'arrows' that point to exciting future possibilities – even probabilities. These future prospects will be dealt with more fully in Chapter 18, but it may here be conjectured with reasonable confidence that many more natural geothermal fields and low-grade aquifers will be discovered and exploited within the next few years, that it will before long become possible to *create* artificial geothermal fields (where none now exist) in places where relatively impermeable hot rock is to be found at moderate depths, and that ultimately we shall be able to win earth heat from very great depths *wherever we wish*.

2 *The versatility of Earth heat*

In Chapter 1 the principal applications of earth heat have been broadly classified in the process of their historical development. Moreover, in Chapters 10, 11 and 12 these applications will be dealt with in greater technical detail; but the purpose of this chapter is to emphasize the great *versatility* of geothermal energy.

There was until recently a strong tendency to think of electricity generation as by far the most important application of earth heat. This was understandable. In the first place electricity seemed to be one of the world's most marketable commodities, since the demand for it was growing at an accelerating pace and since it was easily transported over considerable distances. Furthermore, most of the geothermal fields that were then attracting attention all provided fluids at fairly high temperatures at which the steam turbine could operate at acceptable (though not high) efficiencies. Apart from these factors, electricity itself is so versatile that it can be adapted to almost any application – lighting, heating, motive power, etc. – so that its generation was analogous to the minting of coin, a commodity having the great merit of exchangeability into whatever energy form we may need. Hence the interest in discovering geothermal fields was chiefly concentrated on finding fairly high temperatures and pressures suitable for power generation. The occurrence of low-grade earth heat was then thought to be chiefly of interest to the balneologist or, if sufficiently hot, to the suppliers of district heating and of domestic hot water. But district heating and domestic hot water supply usually become commercially attractive only for sizeable communities which are often not sufficiently obliging as to be located close to a low-grade geothermal field (and the transmission of low-grade heat over long distances is not economically attractive); it is also suited to cold, or fairly cold, climates only, which somewhat restricts the choice of available fields. This is not intended to belittle the importance of district heating and domestic hot water supply: it is only intended to show that it lacks some of the ease of marketability possessed by electricity. In

14

point of fact it has been shown by Kunze *et al.* [10] that nearly 20% of the total energy consumed in the USA is accounted for by space-heating – but at rather low annual load factor.

This former bias towards electricity generation as the 'natural' application for earth heat tended to overlook the fact that power generation is essentially an inefficient operation owing to inescapable thermodynamic constraints; and the lower the temperature of the heat source the lower will be the efficiency of energy conversion. On the other hand, the direct use of geothermal heat for space heating, industrial processes or husbandry can be highly efficient, since the losses incurred are not imposed by the laws of thermodynamics but only by such imperfections as must inevitably arise from insulation losses, drains and terminal temperature differences in heat-exchangers. These imperfections can, moreover, be controlled within economic constraints: they are not dictated by natural laws.

Another important fact is that the sources of high enthalpy natural heat which are both accessible to man and suitable for power generation are believed to be far less abundant than those of lower enthalpy fluids that can be used for other purposes. Kunze *et al.* [11] underline the importance of moderate temperature geothermal fluids (e.g. 150°C) on the interesting proposition of a probable Poisson distribution curve, from which they draw the conclusion that what lower temperature fluids lack in enthalpy they make up for in quantity. This theory is supported by Lund [12] who suggests that in the USA about five times as much earth heat is recoverable with present technology in the 50–150°C range as that available at temperatures of more than 150°C. As utilization techniques improve, it is argued, lower temperature fluids will become of increasing economic importance. Moreover, low enthalpy fluids are likely to contain less dissolved solids and gases than high enthalpy fluids and are therefore less likely to cause corrosion troubles. On the other hand Kumze *et al.* admit that a low temperature geothermal field would require more land for a given thermal scale of development than a high-temperature field, and that subsidence (see Chapter 16) could perhaps present problems.

The advantages of low-grade heat for farming and animal husbandry were realized fairly early in this century though, as with district heating, most of these applications are of greater interest to cold than to hot countries. It was not until after the middle of the twentieth century that the enormous *industrial* potential of earth heat began to be appreciated – for industries cover a wide spectrum of uses that require heat of all grades from about 20°C upwards, and some industries are very heat-intensive.

Líndal [9] published an instructive chart (Fig. 2.1) which sets out a large number of uses of heat over a temperature range of 20–200°C. He would not claim to have shown every conceivable use on his chart, and he emphasizes that many of the applications shown are practised over a *range* of

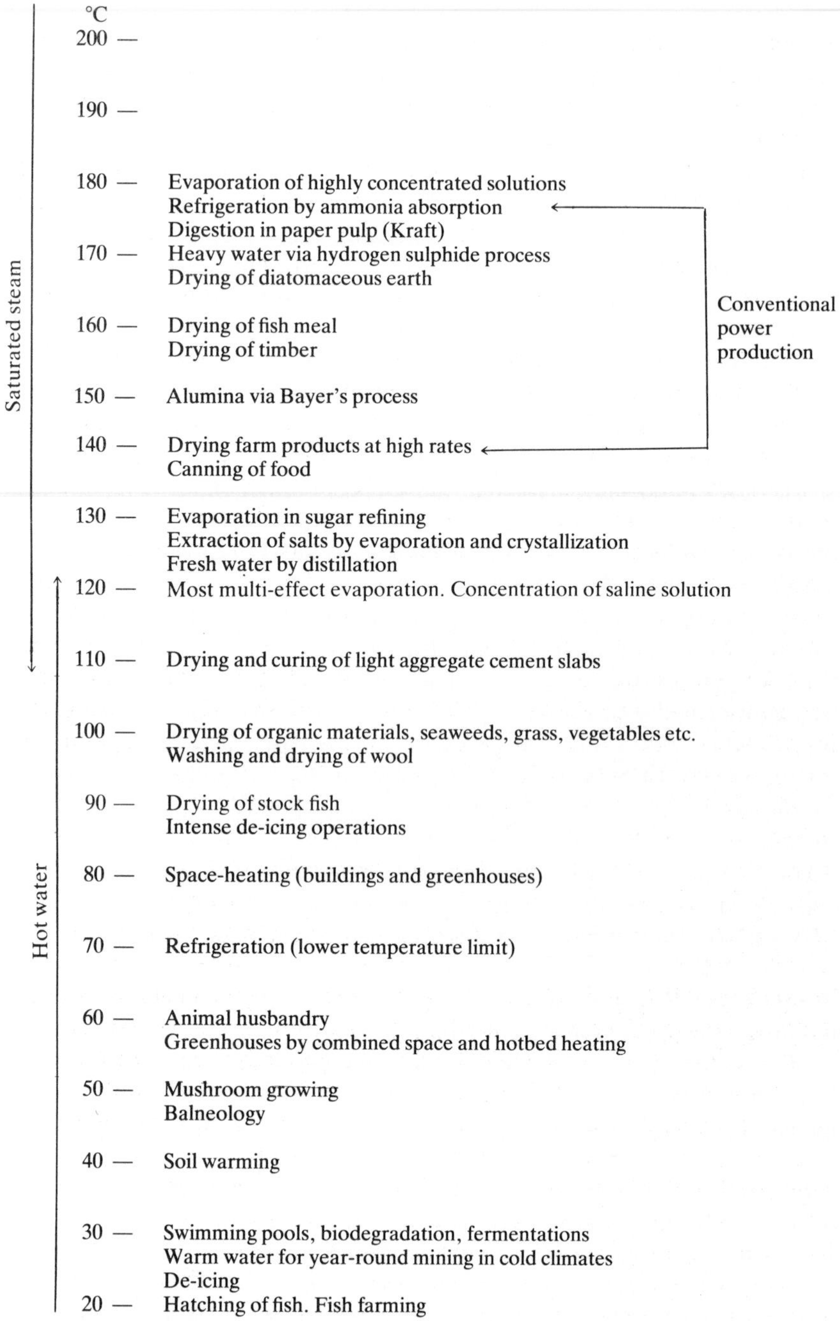

°C

200 —

190 —

180 — Evaporation of highly concentrated solutions
      Refrigeration by ammonia absorption
      Digestion in paper pulp (Kraft)
170 — Heavy water via hydrogen sulphide process
      Drying of diatomaceous earth

160 — Drying of fish meal
      Drying of timber

150 — Alumina via Bayer's process

140 — Drying farm products at high rates
      Canning of food

Saturated steam

Conventional power production

130 — Evaporation in sugar refining
      Extraction of salts by evaporation and crystallization
      Fresh water by distillation
120 — Most multi-effect evaporation. Concentration of saline solution

110 — Drying and curing of light aggregate cement slabs

100 — Drying of organic materials, seaweeds, grass, vegetables etc.
      Washing and drying of wool

90 — Drying of stock fish
     Intense de-icing operations

80 — Space-heating (buildings and greenhouses)

70 — Refrigeration (lower temperature limit)

Hot water

60 — Animal husbandry
     Greenhouses by combined space and hotbed heating

50 — Mushroom growing
     Balneology

40 — Soil warming

30 — Swimming pools, biodegradation, fermentations
     Warm water for year-round mining in cold climates
     De-icing
20 — Hatching of fish. Fish farming

Figure 2.1 Approximate temperature requirements of geothermal fluids for various applications. (From Lindal [13])

temperatures rather than at a single value. But the diagram clearly shows that heat of almost any grade obtainable from geothermal fluids can be put to good use. Clearly 'fire heat' – i.e. the very high temperatures needed for the pottery and glass industries, smelting, etc. – could not fall within the geothermal 'net' (at any rate not until or unless we become capable of utilizing hot lavas or magma in the conceivable remote future) but the scope of using natural steam and hot water is clearly very great. It should also not be forgotten that something like two-thirds of the world's consumption of energy is for non-electric purposes. In short, geothermal energy is far too versatile an asset to be used for power generation only.

Fig. 2.2 shows a more comprehensive chart of temperatures and uses to which low-grade geothermal energy can be put.

Reistad [15] emphasizes some of the points made here and presents a most revealing breakdown of the different uses of energy in the USA (which may be taken as fairly typical of most highly industrialized countries) under various headings – electricity generation, residential, commercial, industrial and transportation. The electricity is then re-allocated to the other four sectors according to the requirements of each. Each sector's energy needs are further broken down into the various applications such as space-heating, cooking, refrigeration, process steam, etc. and a final allocation is then made to one or other of two groups – 'potential geothermal use' and 'non-geothermal use'. The first group excludes all uses requiring temperatures exceeding 250°C: it also excludes transportation and such applications as cooking, which have intermittent demands. Reistad's final conclusion is that about 41% of the USA's energy requirements could be provided by geothermal energy *were it available*. If the upper temperature limit were dropped from 250°C to 200°C the proportion of energy that could theoretically be supplied geothermally would scarcely be affected – a drop merely from 41% to 40% – because very few processes use temperatures between 200°C and 250°C. If the temperature limitation were further dropped to 150°C and 100°C, the percentage of national energy requirements falling within the 'potential geothermal use' group would become about 30% and 20% respectively. This interesting analysis clearly shows that the problem of geothermal application is one of *availability* rather than *applicability,* as the uses to which geothermal energy could be put, if it were available, would be enormous. Reistad's exclusion of transportation (which could partly be provided by electrified railways and – perhaps in future – by battery cars) and of cooking on the grounds of intermittency of demand shows that his conclusions tend to err towards the conservative side. There is growing evidence of a much keener awareness of the wideness of applicability of earth heat – of its immense versatility. The present imbalance between applicability and availability will, it is hoped, be rapidly moderated.

In 1974 the Geo-Heat Centre was established at the Oregon Institute of Technology at Klamath Falls, Oregon, USA. This centre performs valuable work in research and development of direct applications of geothermal heat, especially of low grade, for space-heating, agriculture, horticulture, aquaculture and industry. Geographically it is well placed for the pursuit of its

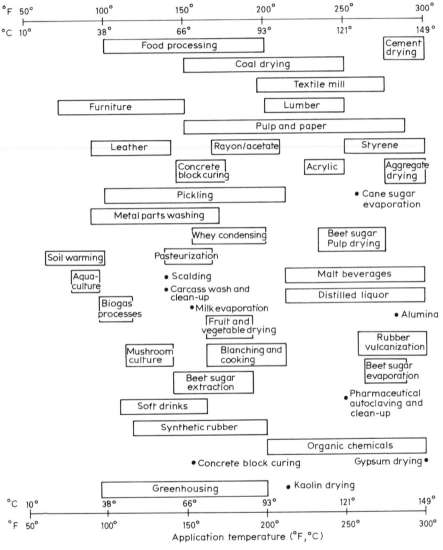

Figure 2.2 Application temperature range for some industrial processes and agricultural applications. (Reproduced from pp. 4–26, Geothermal Resources Council Special Report No. 7, 1979 [14])

aims, for the township of Klamath Falls is sited over a low-temperature field (of not more than 115°C). The area is already perforated with more than 500 wells that are put to a variety of practical uses. An informative quarterly bulletin is published by the centre.

One of the most promising economic potentialities of earth heat is the establishment of dual or multi-purpose plants. Different processes often require different grades of heat. Thus a power plant, for example, taking fairly high-pressure steam could exhaust (either wholly from a back-pressure turbine, or partially from a pass-out turbine) low-pressure steam that could be used to supply a district heating system, a farm, a distillation plant, a heat-intensive industry, or at least a swimming pool. Conversely, an industry requiring high pressure steam could exhaust low-pressure steam to supply condensing turbines. Or again, with a 'wet' geothermal field the steam could be used for one purpose and the hot water for another. The possible combinations of processes are very great. Sometimes more than two end products could be obtainable – e.g. if minerals could be extracted from the thermally spent fluids after they had fulfilled two or more pre-liminary functions. In this way the most efficient practicable use could be made of earth heat (and perhaps of chemical by-products) in a manner similar to that used in 'total energy' or 'co-generation' plants that are rapidly gaining favour in more conventional contexts.

Dual or multi-purpose plants are not without their own peculiar prob-lems, the chief of which is the matching of variable demand patterns of the two or more end products; but such problems are not necessarily insuper-able, particularly if storage facilities are available, either for hot fluids or for the end products.

3 The structure of the Earth

Before attempting to study or to understand geothermal phenomena, the reader would be well advised first to learn something of the main facts and theories concerning the physical structure of the earth. Although we are rapidly becoming familiar with the conditions of outer space we have, until fairly recently, been surprisingly ignorant of what is going on only a mile or two beneath our feet. That the nether regions of the earth are hot is a fact that has long been known to every schoolboy. The fact, no doubt, was originally deduced from the visual evidence of volcanoes and other surface thermal manifestations that are to be found in several parts of the world. It was given non-scientific, but highly emotional support by the widespread belief, entertained by many of the world's religions, in the existence of a Hell.

3.1 Concentric layers of the earth

The only parts of the earth with which we are directly familiar are the crust and the atmosphere. But these together account for less than half of 1% of the total mass of the planet. The remaining 99½% or more lies concealed beneath the crust, and our knowledge of its nature is indirect, largely deduced from the study of earthquake waves and of lavas and from measurements of the outward flow of heat from the interior towards the surface. Nevertheless, this indirect knowledge has enabled us to build up a fairly clear and consistent picture of the structure of the earth, which is now believed to consist of five distinct concentric spheres, assembled in 'onion fashion'. These, passing from the outside towards the centre of the earth, are the *atmosphere* (itself a complexity of sublayers), the *crust* (including the land masses, the seas and the polar icecaps), the *mantle,* the *liquid core* and finally the *innermost core* which is probably solid. Both temperatures and densities rise rapidly as the centre of the earth is approached (Fig. 3.1). Thus we have a picture of a very hot planet, efficiently lagged by a thin crust of low

*For fuller information on the structure of the earth, and on the bearing of this subject upon geothermics, see [16–20].

20

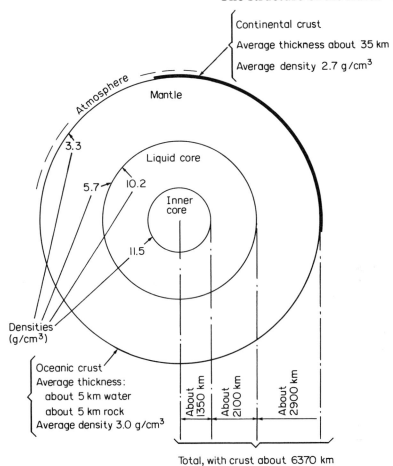

Continental crust
Average thickness about 35 km
Average density 2.7 g/cm³

Atmosphere

Mantle

3.3

Liquid core

5.7   10.2

Inner core

11.5

Densities (g/cm³)

Oceanic crust
Average thickness:
  about 5 km water
  about 5 km rock
Average density 3.0 g/cm³

About 1350 km

About 2100 km

About 2900 km

Total, with crust about 6370 km

Figure 3.1 Concentric layers of the earth. (After Bullard [20])

thermal conductivity. Thanks to this insulating crust, the conditions on the land surfaces and in the seas are sufficiently temperate to support life. Although the vast quantities of heat stored within the entire planet are of course 'geothermal' it is only with the crust and the upper layers of the mantle that we need concern ourselves for practical purposes.

3.2 The crust

Let us take a closer look at the crust (Fig. 3.2), which bears to the rest of the planet a similar relationship to that of an eggshell to an egg, except that its thickness is far from uniform. This illustration shows a section (not to scale) through the earth's crust at an 'inactive' coastline such as the eastern seaboard of North America. (The meaning of the word 'inactive' will later

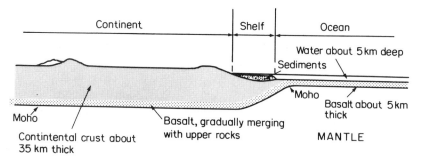

Figure 3.2 Schematic section through the earth's crust at an inactive coastline (e.g. that of eastern North America). (After Bullard, Fig. 1 [17])
Note: Under the continent the proportion of basalt probably increases with depth, but it does not necessarily form a separate layer. A great thickness of sediments usually underlies the continental shelf.

be explained). The crust forming the bed of the ocean is very thin (about 5 km). The continental crust is about seven times as thick – even more in mountainous districts – and is fringed with a sedimental 'shelf' of varying width. It is in this shelf that off-shore oil and gas may sometimes be found. Relatively shallow waters, up to about 200 m, cover the shelf, but the depth of water in mid-ocean averages about 5 km and reaches about 8 km in certain 'trenches'. The boundary that separates the mantle from the crust is known as the Mohorovičić discontinuity (after its Yugoslavian discoverer), or 'Moho' for short. At this boundary there is a sudden change in the velocity of earthquake waves, which indicates a corresponding change in material composition and physical state. The temperature at the Moho is believed to be generally of the order of 600°C, while at the centre of the earth it may perhaps reach about 6000°C.

3.3 The seismic zone

This simplified model here described is not, however, everywhere in a state of equilibrium. High stresses can be built up in the crust and upper mantle, and when these stresses are relieved by material movements the released forces give rise to earthquakes. Seismologists are now able to detect many thousands of earthquakes every year, varying in intensity from the lightest of tremors, imperceptible to the human senses and detectable only by the most delicate devices, to violent shocks capable of devastating cities – such as those that destroyed Lisbon in 1755, San Francisco in 1906, Tokyo in 1923 and Tangshan in 1976. These particularly severe earthquakes release enormous quantities of energy, incomparably greater than that emitted by the largest of man-made nuclear bombs. It is now possible to map with precision the intensities and locations of earthquakes and the depths at which they originate: it is found that with very few exceptions they are confined to ribbon-like zones that spread over the face of the globe more or less in the manner shown in Fig. 3.3. These zones are collectively called the 'seismic belt', much of which lies beneath the oceans.

3.4 Plate tectonics

The explanation of this seismic zone pattern baffled investigators for a long time, but since the late 1950s a body of very impressive evidence has been assembled by Sir Edward Bullard of Cambridge University, and others. It now seems almost certain that the crust of the earth consists of several discrete 'plates' – six major and a few smaller ones – that are in a state of continual relative motion at average rates of a few centimetres *per annum*. These relative motions imply that in some places, where the plates are moving apart, the crust must be splitting; while at others, where plates are colliding, the crust must be crumbling. Fig. 3.4 shows a rough sketch of the two processes. That part of the seismic belt that runs down the middle of the Atlantic Ocean is typical of crust-splitting: the gap formed between the two separating plates is continuously being replenished by the upward flow of hot magma from the mantle to form a steadily spreading ocean floor. The Andes, on the other hand, are typical of colliding plates, where the crumpling has thrown up great mountain ranges. This crumpling effect is modified by the fact that one of the colliding plates – the denser, basaltic plate – yields to the other and slides beneath it at a downward angle, finally merging into the mantle whence it originated many millions of years ago. A collision, or subduction, zone, such as that shown to the right of Fig. 3.4 is 'active', as distinct from the 'inactive' coastline of Fig. 3.2.

According to the theory of Continental Drift (or Plate Tectonics as it is sometimes known) the seismic belt follows the lines along which the plates are either separating or colliding. For the motion of the plates is believed to occur in a series of jerks, large or small, giving rise to major and minor earthquakes respectively. These jerks occur whenever the built-up stresses exceed the frictional and tensile resistances to motion. The motive power behind the plate motion is not understood with certainty but could well arise from convection currents in the mantle which, though acting as a solid when transmitting seismic waves, behaves like an extremely viscous liquid under the influence of forces maintained over very long periods. In fact the material of the mantle 'creeps' like other materials under the combined influence of high temperatures and mechanical stresses. In some places, contiguous plates are neither colliding nor separating, but sliding past one another in a shearing motion. Such as the San Andréas fault in California which, being kinked or 'stepped' instead of having a plane boundary, can give rise to such dangerous situations as that which caused the destruction of San Francisco in 1906. The kink tends to resist the sliding motion and thus

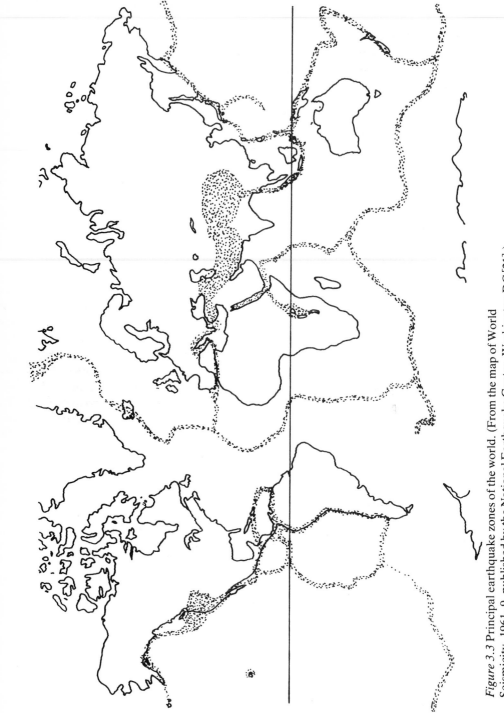

Figure 3.3 Principal earthquake zones of the world. (From the map of World Seismicity, 1961–9, published by the National Earthquake Centre, Washington DC [21].)

causes the build-up of very high stresses until it finally gives way to the accompaniment of a very severe seismic shock. The San Andréas fault has a shearing movement averaging about 2 inches p.a., combined with a small measure of rotation that causes a slight rifting in the Gulf of California.

A feature common to crust-splitting and plate-colliding zones is the presence of volcanoes – active (Plate 1), dormant or extinct. The mid-Atlantic seismic belt coincides with a long range of submarine volcanoes, while the presence of visible volcanoes round the periphery of the Pacific Ocean has caused that zone to be known as the 'Belt of Fire'. There are at present more than 460 active volcanoes above sea-level, of which 384 are situated in and around the Pacific. In a few places the mid-Atlantic Rift rises above sea-level, appearing as volcanic islands such as Iceland, the Azores, St Helena, Tristan de Cunha and others, as shown to the left of Fig. 3.4. Crust-splitting zones sometimes appear also in the continental land masses – e.g. the Great African Rift Valley, along which there is a line of volcanic cones, including Mt Kenya and Mt Kilimanjaro. Common features of plate-collision zones are the presence of a deep ocean trench on one side of a mountain range and of volcanoes at some distance behind the range on the other side. For example, the Andean volcanoes of Chimborazo and Cotopaxi lie to the east of the main sierra, while the Peru/Chile ocean trench lies at a comparatively short distance to the west of the South American continent. As a corollary to the presence of volcanoes there must be, or have been, magmatic penetrations of hot material from the mantle into the crust as a result of the disturbances at the edges of the plates. Such penetrations

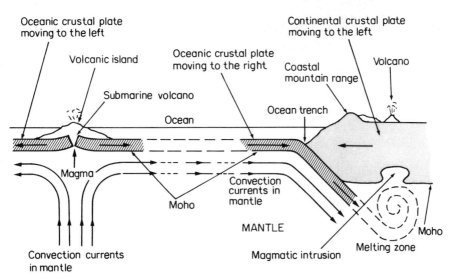

Figure 3.4 Crustal plate movements (not to scale).

may be complete, when lava issues from an active volcano, or they may be partial, when they take the form of magmatic intrusions, or pockets of mantle material bulging into the crustal rocks to within perhaps a few kilometres of the surface. Partial intrusions are sometimes volcanoes in embryo, though they may perhaps cool off before they can break the surface.

The explanation of the observed pattern of crustal structure at collision zones may perhaps be somewhat as illustrated to the right of Fig. 3.4. The descending plate, forcing its way into the reluctant mantle, generates intense frictional heat and causes a melting zone to form at some distance inland. The hot, relatively fluid mantle material would be fairly buoyant and could penetrate the base of the continental mass where the high temperature may have rendered the crustal rocks more or less plastic. In the course of time some of these intrusions could work their way to the surface and erupt as volcanoes.

The theory of Continental Drift dates from 1915 when Alfred Wegener (mentioned in Chapter 1) drew attention to the fact that the east coast of South America would fit snugly against the west coast of Africa (Fig. 3.5) like pieces of a jigsaw puzzle. He also pointed out that geological and botanical similarities were to be found along the two coasts, and accordingly suggested that the two continents had once been joined together and had later split and drifted apart. He even carried the idea to an extreme by proposing that at some time in the remote past there had existed only a

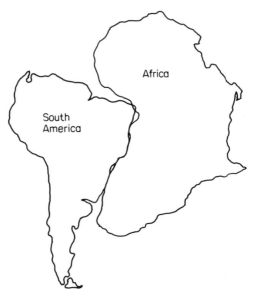

Figure 3.5 Wegener's observation of continental 'fit'.

single, very large continent, which he named Pangea, embodying all the land masses of the world. The whole surface of the earth, according to his theory, was either ocean of part of Pangea: there were no islands. Then at some date, of the order of 100 million years ago, Wegener alleged that unexplained internal convulsions within the core(s) of the earth must have caused this super-continent to break up, with its disintegrated fragments drifting apart until they formed the present configuration of land masses that we know today.

Wegener's theories were at first received very sceptically, but nevertheless our modern conception of Continental Drift and Plate Tectonics stem from his original rather vague hypotheses. It has now been discovered from very accurate geodetic measurements that the Americas are in fact moving away from Europe and Africa at a slow, but detectable average rate of the order of 10 cm per annum. Similar relative motions have been detected between other crustal masses. Moreover, if the continental *shelf* be taken instead of the coastline, the apparent 'fit', not only between South America and Africa but also between North America and Europe, is too good to be explained away by mere chance (Fig. 3.6). Similar good fits are to be found elsewhere; for example, between the west coast of the Arabian peninsula and the African Red Sea coast.

Wegener's postulation of Continental Drift posed some awkward questions at first; for the rates of movement, although they seem slow in absolute terms, are relatively fast on a geological time scale. The separation of the Americas from Europe and Africa to their relative positions at an average rate of 10 cm a year could have been achieved in about 50 million years, while the movement of the Tonga Isles from the Eastern Pacific Rift would have taken about twice as long. Geologists tell us that the age of the earth is of the order of 4500 million years. Why, after surviving for more than 4000 million years, should Pangea suddenly have started to disintegrate? What cataclysmal event could have accounted for it? Will the dispersing land masses ultimately coalesce again into some new Pangea within a few hundreds of million years from now, and will the whole process be repeated? Had disintegration and re-coalescence in fact occurred several times before the last Pangea started to break up?

It is now generally believed that the crustal plates have been in a state of relative motion ever since an early stage of the earth's history, and that these plates are continuously being created at splitting zones and simultaneously destroyed at collision zones. The plates, as it were, 'float' upon the mantle, and in the course of the ages they may re-form and change their directions, like ice floes in an eddying pool, under the influence of magmatic forces as yet not properly understood. There was probably never a Pangea.

Modern theories of plate tectonics, as broadly described in this chapter, are not without critics; but the evidence in support of these theories is very

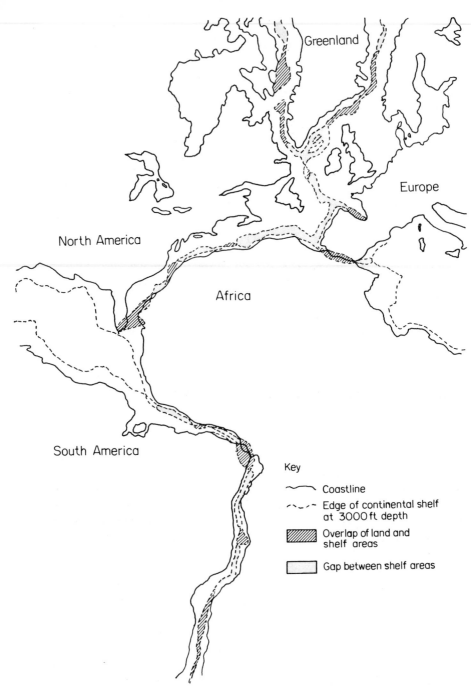

Figure 3.6 Probable arrangement of continents before the formation of the Atlantic Ocean. (After Bullard [19])

Key

~~~ Coastline

~-~-~ Edge of continental shelf at 3000 ft depth

▨ Overlap of land and shelf areas

▢ Gap between shelf areas

impressive. Apart from the improbably good fit of the continents as shown in Fig. 3.6, it is highly significant that whereas earthquakes at splitting zones generally originate at shallow depths, those that occur on one side of collision zones are often very deep-seated (at depths of up to about 650 km), as would be expected where a downward moving plate is forcing its way into a resisting mantle (Fig. 3.4). The existence of deep ocean trenches along one side of what are believed to be collision zones, as shown in Fig. 3.4, is also significant; and what better explanation can be offered of the presence of coal seams in temperate and even Arctic regions (e.g. Spitzbergen) that have clearly been formed by the decay of *tropical* vegetation, than that such parts of the crust have wandered in the course of millions of years far from the torrid latitudes where the primaeval forests once grew? Finally there is further evidence provided by what Bullard aptly descibes as 'a kind of geological tape-recorder' to be found in the directional magnetization of the magnetic elements in the ocean floor. It is known that the orientation of the earth's magnetic field is not only constantly shifting but that it has reversed its polarity at various times during the history of the planet. As the newly formed ocean floor cools at the sides of the mid-ocean rift, the magnetic elements of the solidifying rock become magnetized in the direction of the earth's magnetic field. Records of this directional magnetization, obtained quite simply by towing a magnetometer behind a ship, show symmetrical bands of similarly magnetized ocean bedrock on either side of, and more or less parallel with, the rift; thus suggesting a fairly steady outward movement of the sea floor away from the rift [22].

The most critical of detectives, judges and juries could hardly fail to be impressed by this imposing mass of evidence or to reject 'coincidence' as a satisfying explanation. While much remains to be explained, it seems reasonable to accept that the crust and outer layers of the mantle have behaved, and continue to behave, more or less in the manner here described. The earth appears to be a gigantic heat engine, whose precise method of working is not yet fully understood but whose internal energy motivates the crustal creep and the changing patterns of magmatic convection within the mantle.

The importance of the theory of Plate Tectonics to the geothermicist lies in the fact that, as far as is known, hyperthermal fields (see Chapter 5) are to be found only within the seismic belt; that is, within zones of crustal weakness where volcanic activity is, or has been, intense. It is therefore within the seismic belt that exploration for high temperature fields at moderate depth should be confined. It is true that lower temperature geothermal fields have been found outside the seismic belt – e.g. in Hungary, in the Caucasus, in the Arzacq Basin of France, in Switzerland, Australia and doubtless elsewhere.

The existence of these zones of relatively low-grade heat could perhaps be attributed to any or all of the following conditions:

(i) There could be parts of the crust that bear abnormally high proportions of radioactive rocks (see Section 3.5).

(ii) The crust could well be relatively thin in places, so that the Moho is at shallower depth than in more normal 'non-thermal' areas. (This emphasizes that Fig. 3.2 is somewhat stylized: there is no reason to suppose that the Moho is uniformly horizontal beneath the continental masses).

(iii) The temperature of the mantle just beneath the Moho is not necessarily everywhere the same: there could be local magmatic disturbances that give rise to hot spots in certain places.

The Pannonian Basin in Central Europe provides an excellent example of the second and third of these conditions, for the crust in that region is only about 25 km thick in places, or about two-thirds of the normal crustal thickness in the continental masses, and a rising 'plume' of convected magma brings very hot mantle material into contact with the underside of the thinned crust. These phenomena are due to various orogenic influences associated with the formation of the Alps [23].

The fact that the known high-temperature fields are all found in or very close to the seismic belt does not imply that such fields are to be found anywhere and everywhere within that belt. They occur only at certain places within the belt where the geological and hydrogeological conditions favour the formation of high-temperature fields: hence their comparative rarity.

3.5 Temperature gradients and heat flow

Apart from the visual evidence of a hot interior to the earth, as provided by volcanoes and other natural thermal phenomena, we have the less spectacular, but even more convincing evidence provided by the fact that as we penetrate the earth's crust by drilling or by mining operations we invariably experience a rise in temperature. Observed thermal gradients vary widely from place to place, being as low as about 10°C per kilometre in some places; but a fair average in the non-thermal regions of the earth would be of the order of 25 to 30°C per km. In a few places, specially favoured by nature, very much higher gradients may be encountered – e.g. gradients of 200 to 800°C/km have been observed at Larderello. The mere existence of a thermal gradient implies a continuous outward flow of heat from below ground to the surface.

Thermal gradient and heat flow must not be confused with one another. Both are found to vary from point to point in the earth's crust. The outward *conductive* heat flow at any point in the solid crust will, of course, be proportional to the product of the thermal gradient and the thermal conductivity of the rock at that point. Thus a high thermal gradient could be associated with low heat flow if the rock conductivity is poor. But in permeable zones containing fluids, the outward heat flow is no longer purely conductive but is aided by convection, so that this simple relationship is no longer valid. Differences in thermal gradient in impermeable zones are accounted for partly by differences in thermal conductivity and partly by the presence of local hot spots at depth.

Accurate measurements of heat flow have been made in many parts of the world – even in the ocean bed – and it is found to vary quantitatively from place to place, being generally lowest in the ocean bed far from submarine rifts, and highest in parts of the seismic belt. The worldwide average value is about $1.4\,\mu cal/cm^2\,s$ or 0.06 Watts (thermal)/m^2. In certain places, however, the local escape of heat can very greatly exceed this average, largely because of hot fluid emissions from below ground. For instance, the extent of the Wairakei field in New Zealand is about 18 km^2, and the natural outward heat flow from the whole area was estimated at 1860 MBtu/h before the field was first exploited. This figure is equivalent to about 30 W/m^2, or approximately 500 times the world average. The escape of heat through active volcanoes, though locally very impressive, forms a negligible fraction of the total outflow of heat from the earth – this total amounting to something like $2\frac{3}{4} \times 10^{16}$ cal/h, or the thermal equivalent of about 30000 million kilowatts. But it is of little use to lament this apparently wasteful loss of energy, because the natural outward heat flow from below ground is far too diffuse and of far too low a grade to be of any value to us.

It might be thought that this vast natural outward heat flow originates entirely from the huge quantities of stored heat within the mantle and cores of the earth beneath the crust. It is true that part of it does in fact arise from this source and is conveyed outwards mainly by the thermal conductivity of the rocks, aided in places by fluid convection in permeable zones. Nevertheless, in non-seismic areas this usually accounts only for part of the total heat flow: nearly all of the balance is believed to be derived from radioactivity in the crustal rocks. It is the presence of this radioactivity that largely confused Lord Kelvin when he attempted to account for the age of the earth in terms of original stored heat and of other factors such as shrinkage energy. Most of the earth's radioactive elements are believed to occur in the crust – a fact that may cause some surprise since the radioactive elements are the heavier ones, which could reasonably be expected to gravitate towards the central core of the earth. But as, for the most part, these elements have entered into chemical combination with lighter elements and have been dissolved in relatively light-weight rocks, they have tended to concentrate in the outer layers of the planet.

There are three other internal sources of heat which, though of relatively minor importance, could collectively account for appreciable quantities of energy. These are:

(i) The heat released by exothermic chemical reactions within the crust.

(ii) The friction generated in faults, where the sliding action of huge masses of rock exerting enormous gravitational and tectonic pressures against one another is caused by the readjustment of tectonic stresses.

(iii) The latent heat released by the crystallization of solidification of molten rocks on cooling.

4. The heat energy resources of the Earth

4.1 Stored heat

The total heat content of the earth is not easily assessed; for although the structure of the planet is believed to be more or less as described in Chapter 3 it is not possible to be dogmatic about the physical properties of the extremely dense matter under immense pressures deep within the mantle and cores, since the conditions prevailing there cannot be reproduced in the laboratory. Such properties as specific heat and thermal conductivity at great depths can only be guessed at. Thermal gradients, and therefore the temperature distribution over the radius of the earth are consequently subject to great uncertainty. Nevertheless it seems probable that the total heat content of the earth, reckoned above an assumed mean surface temperature of 15°C, is of the order of 3×10^{27} kgcal.

Beyond noting that this figure is vast, the reader is advised to forget it, as being of no further interest; for all but a tiny fraction of it must forever remain inaccessible to man. It would be better to lower our sights and consider the heat content of the crust alone which, also reckoned above 15°C, is of the order of 1.3×10^{24} kgcal. This is perhaps less than 1/2000th part of the total heat content of the earth, but it is still an immense figure. However, most of this too is not, and never will be, available for man's use; for apart from technological limitations there must certainly be safety constraints upon the amount of stored heat that could be extracted from the crust without risking cataclysmal consequences of a seismic or other adverse nature. Nevertheless, it will here be shown that even a very small fraction of the total stored crustal heat could satisfy a large part of our energy needs for a very long time. The *usable* quantity of crustal heat is clearly not limited by storage considerations but by the constraints of technology and safety.

4.2 Specific crustal heat

Perhaps the most convenient way of expressing the crustal heat resources is by considering a vertical column of the continental crust having one square kilometre of horizontal projected area. Assuming the thickness of the crust, and therefore the height of the column, to be 35 km (see Fig. 3.2), the volume of the column will be 35 km³. If we take, as reasonable assumptions, a mean crustal density of 2.7 (see Fig. 3.1) and a mean specific heat of 0.2 cal/g, then by cooling the column we could recover the following approximate quantities of heat:

$$
\left.
\begin{array}{l}
1.89 \times 10^{13} \text{ kgcal, or} \\
2500 \text{ MWyt, or} \\
79\,000 \text{ TJ, or} \\
2.7 \text{ Mtce}
\end{array}
\right\}
\begin{array}{l}
\text{per km}^2 \text{ of land surface} \\
\text{per °C or average crustal} \\
\quad \text{cooling}
\end{array}
$$

These figures may be regarded as the specific crustal heat. They apply only to the crust beneath the land surfaces, and they allow for stored heat alone. They disregard all chemical energy in the form of fuels stored in the crust, all energy emitted by radioactive crustal rocks and all heat conducted from the hot mantle beneath. Moreover they claim to be no more than approximate averages: local differences in crustal thickness, in rock densities and in specific heat will ensure that the specific crustal heat will differ quite widely from place to place. The fact that some parts of the crust are hotter than others owing to local thermal anomalies does not *per se* affect the value of the specific crustal heat, which is expressed in terms of temperature *drop,* and not of temperature.

'Average crustal cooling' does not of course imply evenly cooling the whole thickness of the crust through the same, small temperature drop, but the cooling of certain layers through quite substantial drops while leaving the remaining layers at their original temperatures. Being 'specific', the figures in the form expressed above give no indication of how much heat could be recovered *in toto,* as this would have to depend both on the *possible* and the *permissible.*

4.3 The possible: problem of accessibility

Crustal heat is now commercially extractable only at those comparatively rare 'hydrothermal' places (see Chapter 5) where permeable formations containing hot fluids exist within economic range of the drill, so that some of the fluids may be brought to the surface for useful applications of one kind or another. In future, however, it may become possible to gain access to far larger quantitites of crustal heat by using technologies that have not yet been perfected but which are being pursued by vigorous research. The goals that are being sought are as follows:

(i) *Geopressurized fields* (see Section 5.16) which exist at very great depths in a few places – notably in the southern USA and in Hungary. The thermal content of these fields (apart from the far greater chemical energy of the associated methane and the modest hydraulic energy arising from high fluid pressures) is considerable; but their great depth means that drilling costs are very high.

(ii) *Magma chambers* (see Chapter 18). Suspended at various depths in certain parts of the crust are pockets of magma of temperatures ranging from about 600 to 1300°C. Their individual heat contents are immense, but their frequency of occurrence is not very great, and the problems of dealing with the high temperatures involved have not yet been mastered.

(iii) *Hot dry rocks at moderate depths* (see Chapter 18). There are many such rocks from which heat can already be extracted by methods that have shown considerable technical success, but which have yet to be commercially established.

(iv) *Hot dry rocks at great depths.* There are far more hot dry rocks at depths that are at present beyond the economic range of the drill. After the necessary technology for (iii) has been perfected, these will become accessible only when new, commercially acceptable, means of very deep penetration have been mastered.

There is little doubt that some, if not all, of these resources will become accessible in time. If so, the concept of specific crustal heat will acquire much greater significance; for more and more square kilometres of land areas will become amenable to cooling by artificial means. At present only very few parts of the world can be cooled at all by the extraction of heat, and they are not easily quantifiable; but if the fourth of the resources listed above should ever become commercially accessible, we could 'mine' heat in any part of the world where energy is needed. It would then theoretically become possible to apply the specific crustal heat to most of the land surfaces of the earth. These total about 148 million km^2, but much of them would be of no interest or of too difficult access. It is possible, however, that we shall learn how to extract heat from beneath the continental shelves, as with petroleum. If, say, 100 million km^2 – about two-thirds of the total land areas – could be exploited for heat extraction, then the recoverable heat would be approximately:

$$
\left.
\begin{array}{l}
1.89 \times 10^{21} \text{ kgcal, or} \\
2.5 \times 10^{11} \text{ MWyt, or} \\
7.9 \times 10^{12} \text{ TJ, or} \\
2.7 \times 10^{8} \text{ Mtce}
\end{array}
\right\}
\begin{array}{l}
\text{per °C of average} \\
\text{crustal cooling}
\end{array}
$$

These figures are still without meaning until we know the *permissible* degree of cooling. But if 1°C of average cooling were possible – and that represents only about one-third of 1% of the total local crustal heat above 15°C – the recoverable heat would amount to about 27 300 times the total primary energy consumed by the world in 1980. Limitations of grade, and the peculiar needs of transportation (which at present accounts for about 10% of the world's total primary energy consumption) would necessitate continued reliance upon fuels to some extent – at least for a fairly long time. Nevertheless, it is clear that the heat of the crust could make enormous contributions to the world's energy needs as soon as the necessary technology is acquired. There are good reasons for believing that it will be acquired to a large extent at least.

4.4 The permissible: constraints upon crustal cooling

There must be ultimate constraints upon the permissible degree of crustal cooling imposed by the risks of seismic shocks, thermal ground shrinkage or pollution of one kind or another. All these factors must be very carefully monitored and studied before it will be possible to make any firm pronouncements as to how much average cooling could be tolerated. In the absence of some such pronouncement the world's total geothermal potential must remain a matter for speculation.

It is of interest to examine the possible climatic effects of excessive withdrawal of crustal heat. Almost all of the energy produced and consumed by man must ultimately re-appear as heat surrendered to the atmosphere and seas. In 1980 9338 Mtce, or 6.54×10^{16} kgcal, were thus released into the earth's surface environment. In the same year the outward heat flow from below ground, at a steady rate of about 30 000 million kWt (see Section 3.5) added $3 \times 10^{10} \times 8760 \times 860$, or 2.26×10^{17} kgcal to that environment. Also in the same year the earth received solar energy at a mean rate of about 175 000 TW, of which 52 000 TW were reflected off the outer atmosphere into space and the balance of 123 000 TW, or $1.23 \times 10^{14} \times 8760 \times 860 = 9.27 \times 10^{20}$ kgcal entered the atmosphere. Thus the quantity of solar energy entering the atmosphere was about 4000 times as much as that which flows outwards from below ground, and about 14 000 times as much as was being consumed by man in 1980. Clearly man's rate of energy consumption could be increased by a huge factor before any anxiety need be felt about affecting the climate. Most of the solar energy received into the atmosphere is ultimately radiated back into space, especially at night; but some of the balance is absorbed by photosynthesis to form organic matter in the creation of animal and plant life, part of which is ultimately transformed into fossil fuels. Before it is lost into space, the solar energy entering the atmosphere is first shuffled about through the weather cycle, by the lifting of water vapour and the creation of wind and waves; but the energy thus absorbed is later again degraded into heat as the water vapour is condensed and falls back into the seas and as the wind and wave energy is dissipated in frictional heat of very low grade.

The measurement of the extent of average crustal cooling is not easy; but it can be shown that at Wairakei, New Zealand, heat has already been extracted since the field was first exploited in the 1950s to an extent of about 2500 MWyt per km² of field area. From the figures mentioned in Section 4.2, this would suggest that the crust has already been locally cooled there through about 1° C. There are, however, three modifying considerations:

(a) There are good reasons for believing that the crust at Wairakei is substantially less than 35 km thick owing to the suspected presence of a magmatic intrusion beneath the field.
(b) It is not positively known to what extent the hot fluids have been extracted only from that part of the crust that lies vertically beneath the superficial confines of the field, or how much has been drawn in from further afield by way of deep flow paths.
(c) As already stated in Section 3.5, the natural heat flow at Wairakei was about 1860 MBtu/h before exploitation began. This is equivalent to 545 MWt, or about 25 to 30% of the rate at which heat is now being artificially extracted. The figures of specific crustal heat in Section 4.2 have been calculated for normal areas where the natural heat flow is so small that it can be ignored; but this could not apply to Wairakei, for although the natural heat flow from below will probably have been modified by exploitation, it is still likely to form a considerable fraction of the amount being artificially withdrawn.

Consideration (a) would imply that the average local crustal cooling at Wairakei has been more that 1°C, whereas considerations (b) and (c) would imply the reverse. Although too much should not be read into the Wairakei figures, there would at least seem to be some *prima facie* evidence to suggest that the average crustal cooling at Wairakei has been about 1°C already (1980). Yet the power plant there continues to operate well, and there is every confidence that the field is still good for many more years of useful service. Only one adverse effect of exploitation has been observed so far at Wairakei: there has been considerable ground subsidence. But this has been attributed not to thermal shrinkage but to the removal of some 1400 million tonnes of water from the underlying aquifer, thus depriving the over-burden of physical support. Had the greater part of this water been reinjected into the ground, as is now commonly practised elsewhere, the subsidence would undoubtedly have been far less.

4.5 Attempted estimation of usable geothermal energy with existing technology

The notion of specific crustal heat, as propounded above, is perhaps rather academic since what is *possible* will change with time and what is *permissible* cannot be clearly defined. It does, however, serve to show that a very small mean temperature drop in the crust could provide a very great quantity of energy; and the evidence of Wairakei is somewhat encouraging as a crude indication of what might be permissible. What is of more immediate interest is to have an approximate idea of the amount of geothermal energy that could now be won *with existing technology* and assuming favourable economic conditions. Any attempt to estimate this must be highly speculative, as huge tracts of the world still remain geothermally unexplored, and unfortunately it is very difficult to reach realistic conclusions from what has been published.

4.5.1 *World geothermal resources.* The World Energy Conference (WEC) organization has made such at attempt and has presented its conclusions in a publication *World Energy Resources: 1985–2020* [24], in which the following geothermal potentials are suggested for the generation of electric power:

Resource base, taking into consideration the continental land masses to a depth of 3 km and reckoned above 15°C	4.1×10^{25} joules
Only 2% of this is assumed to be of adequate temperature for electricity generation	8.2×10^{23} joules
Allowing for overall recovery and conversion efficiencies (of 2.2% combined, by implication), the energy convertible into electricity would be	1.8×10^{22} joules
Of this amount it is assumed that 20% is convertible by existing technology	3.6×10^{21} joules

This last figure is equivalent to 1.14×10^8 MWye, which is about 120 times the present annual world production of electricity.

The basis of these estimates is not very clear. The resource base of 4.1×10^{25} J tallies with the heat content of the upper 3 km of the crust if calculated in accordance with the mean density and specific heat suggested in Section 4.2 and a mean temperature gradient of 27.235°C/km. This gradient is of the same order as the probable mean for non-thermal areas; but if this were intended as the basis of calculation the temperature would nowhere exceed 96.7°C – (3×27.235)+15 – which is too low for large-scale power generation. The resource base would seem to have been estimated on the basis of a lower mean gradient for most of the continental masses and a small proportion of 'hot spots' to make up the balance: in fact the resource base has been broken down by the authors into different temperature bands, showing 12.13% of the heat at over 100°C and 2.86% at over 150°C. As the authors refer to 'deposits' and to 'current technology' it must be presumed that they only had hydrothermal regions in mind as exploitable hot spots, and that they excluded hot dry rock from their reckoning. Though the estimate may well be of the right order, it is suggestive of rather insecurely based guesswork, especially in the 2.2% and 20% assumed in the third and fourth terms respectively; but it is probably as much as can reasonably be expected in view of the incomplete data available.

The authors of the WEC survey also quote a figure of 2.9×10^{24} J as the estimated recoverable *thermal* energy theoretically suitable for direct applications at lower temperatures. This is about 10000 times the present annual world consumption of primary energy without regard to grade. No suggestions are made by the authors as to what proportions of this energy are likely to be usable, in view of the inability to transmit heat over more than fairly short distances. It is not at all clear how the gross figure of 2.9×10^{24} J has been deduced, for it amounts to about 7% of the estimated resource base and implies the cooling of the whole of the crust beneath the land masses through an average drop of about 0.25° C.

The geographical distribution of the resource base has been given by the WEC as:

	%
North America	20.99
Central America	0.66
South America	13.91
Western Europe	3.90
Eastern Europe	17.09
Asia	20.75
Africa	13.67
Pacific Islands	9.03
	100.00

4.5.2 *National geothermal resources.* There have been many national attempts to estimate the geothermal resources of individual countries, a few of which are shown in Table 4.1. To have true meaning, such estimates should be expressed in energy units. In some cases this has been done: in others, power potentials have been presented together with associated periods of years – which amounts to the same thing. Sometimes, however, only power potentials or heat flows have been quoted without any specified times: such figures are of limited significance. To aid comparisons where possible, estimates that have been expressed in different units have been converted in the table into MWye for power resources and into joules for heat resources. (10^{18} joules are 5.2% less than 1 quad, or 10^{15} Btu.) The WEC world figures quoted in the last subsection have been brought forward into Table 4.1 for convenience.

The table achieves little more than to show how difficult it is to place much confidence in geothermal resource estimates. This is apparent from the wide range of the five Japanese power estimates, calculated by different methods; by the improbably high proportions of the world geothermal power estimates formed by the two Japanese MWye figures; by the surprisingly low

Table 4.1 National and world estimates of geothermal resources.

Country	Electric power				Beneficial heat			Ref.
	MWe	Years	MWye	% of world potential	As quoted	Joules	% of world potential	
Hungary					10^{16} kgcal economically recoverable	39.7×10^{18}	0.0014	[25]
Iceland	3 200	50	160 000	0.014				[26]
Indonesia	8 000 to 10 000	NS						[27]
Italy	2 000	50	100 000	0.088	Reserve -12×10^6 MWyt	378×10^{18}	0.013	[28]
Japan†	20 000 32 000	5000 NS	100×10^6	87.7	2×10^{17} kgcal	837×10^{18}	0.029	[29] and [30]
	8 650 40 000 40 000	NS 1000 NS	40×10^6	35.1				
Mexico	4 000	ca.38	152 000	0.133				[31]
New Zealand	2 000	NS						[32]
USA*	95 000 to 150 000	30	2.85×10^6 to 4.5×10^6	2.85–3.95		230–350×10^{18}	0.008–0.012	[33]
USSR					250 to 300×10^6 Gcal p.a.	1.046×10^{18} p.a.		[34]
World			114×10^6	100		2.9×10^{24}	100	[24]

NS Not specified.
*Excluding national parks.
†Five estimates calculated by different methods.

proportions of power potential implied by the USA, Iceland and Mexican figures; and by the still lower proportions of the beneficial heat figures of all the quoted countries in relation to the world estimate. The unfortunate, but inescapable, conclusion is that too little is yet known about accessible geothermal resources for them to be assessed with any degree of confidence.

4.6 Renewability of geothermal energy

Is geothermal energy a *renewable* source? From the gigantic figures quoted in Section 4.5.1 it could, in one sense, be regarded as so abundant as to be virtually an infinite source of energy; but on a local scale a field can become exhausted through prolonged exploitation. The Tuscan field in Italy can maintain output only by drilling more and more bores over an ever widening area, and there are signs that the boundaries of the field have been reached and that the stored energy is beginning to run down. In Wairakei, New Zealand, pressures and temperatures have been slowly declining for some time. If a field is run down to an uneconomically low temperature for the purpose for which it is being exploited, it might in theory recover in the course of a very long time; but this would involve periods so long as to be of no commercial interest, for thermal conduction through rock is an extremely slow process, while other forms of heat infeed (see Chapter 5) are small and unpredictable. If a field were tapped only to the extent of the rate of heat infeed at depth, it would be truly renewable – at any rate for as long as the infeed is maintained – but for economic reasons it will always certainly pay to extract heat more rapidly than the rate of heat infeed. White [35] has suggested that the rate of artificial heat draw-off may exceed the rate of natural heat infeed by a factor of about 5. Similar considerations would apply to the extraction of heat from hot dry rocks.

In local 'hot spots' it is not possible to know whether the heat infeed is constant, rising or falling; for these are more or less long-term transient phenomena. But over the greater part of the land masses it would not be unreasonable to assume that a condition of approximate thermal equilibrium prevails. Over an immense time scale it is probable that the earth is cooling: ultimately, in fact, it *must* cool owing to gradual loss of stored heat and exponential decay of radioactivity, unless it is first swallowed up by the solar expansion foreseen by the astronomers. But over periods of human interest – mere millennia – it is probable that the earth will cool at an almost imperceptible rate. This is not as obvious as it may seem; for it is contended by some that the earth may still be heating up very slowly owing to an excess of radioactivity over the surface heat loss [36]. If this were so, there would have to be negative temperature gradients in the lower part of the crust; and as yet no evidence of this has been found. However, any net rate of cooling or of heating is likely to be so slow that the average infeed of heat into the crust is almost equal to the average outward heat flow, totalling about

3×10^{10} kWt for the whole planet, or about 60 kWt per sq. km of surface (average).

Now let it be assumed that a typical column of crust in a non-thermal area has been exploited to an extent that has caused its mean temperature to have fallen through $1°$ C. If artificial heat extraction were then to cease, the cooled crustal column would have to rely on this inflow of 60 kWt/km² for its ultimate recovery to the temperature that prevailed before exploitation began. It is true that if a pocket of crust at some particular depth had been cooled through a large temperature drop (affecting the *average* temperature of the column by no more than $1°$ C) than the heat flow from below would increase owing to the higher temperature gradient created between the cooled pocket and the Moho; but above the pocket the direction of heat flow would be reversed into a downward direction. There would also be lateral changes of heat flow from adjacent uncooled zones of crust; but although the *pattern* of heat flow would be different, long-term recovery would be effected by the mean rate of heat flow that had existed before exploitation began.

The cooling of a typical crustal column of 1 km² horizontal section through $1°$ C mean temperature drop should, according to Section 4.2, release about 2500 MWyt of heat. Recovery of this heat from natural outward heat flow should therefore take.

$$\frac{2500 \times 1000}{60}, \text{ or about } 42\,000 \text{ years}$$

This, for all practical purposes can be regarded as infinite. In short, heat extracted from 'typical' areas of the crust should be regarded as non-renewable. While this applies to average crustal conditions, the situation would be different in places of abnormally high heat flow such as are usually to be found in hyperthermal areas. For instance, as has been noted in Section 3.5, the natural heat flow at Wairakei before exploitation was about 500 times the world average of 60 kWt/km². Heat recovery at Wairakei could therefore be theoretically attained in 42 000/500, or about 84 years per $°$ C of average crustal cooling, and it has been shown in Section 4.4 that the extent of mean crustal cooling has been about $1°$ C since exploitation began. This is probably an over-simplification; for the field is not being exploited over its full extent, nor has local heat flow ceased since exploitation began. From the commercial viewpoint even 84 years is rather long for the resource to be regarded as renewable.

Of course, if smaller quantities of heat are extracted, recovery times would be proportionately reduced. But in typical non-thermal areas, even the extraction of only 1 MWyt/km² would require 17 years for recovery. In fact in such areas any rate of heat extraction higher than *ca.* 60 kWt/km² – a very low level of exploitation indeed – could not rely on full renewability. All large-scale exploitation should be regarded as non-renewable.

4.7 Conclusion

This book is mainly about *hydrothermal* phenomena, though these are not the only sources of geothermal energy. Published estimates of hydrothermal resources are inconclusive and inconsistent; but it is clear that they could provide a moderate, but useful contribution to the world's energy needs. However, when the technology of heat extraction from hot dry rock, magmatic pockets and geopressurized fields has been mastered and established commercially, far larger energy resources will become available. Geothermal energy cannot be regarded as a renewable resource if developed on any but a very small scale.

5 The nature and occurrence of geothermal fields

5.1 The comparative rarity of geothermal fields

From what has been said in Section 3.4 it will be understood that geothermal fields are rarities. Those of higher temperature are to be found only in a few parts of the seismic belt, which itself covers but a very small fraction of the earth's surface. Fields of lower temperature, although not confined to the seismic belt, are also to be found only in certain favoured places. By far the greater part of the earth's surface is non-thermal.

5.2 Classification of 'areas'

It is convenient to classify all the areas of the earth's surface into three broad groups, so:

(i) *Non-thermal areas,* having temperature gradients ranging from about 10 to 40°C per kilometre of depth.

Thermal areas. These are of two kinds:

(ii) *Semithermal areas,* having temperature gradients of up to about 70°C or 80°C per kilometre of depth.

(iii) *Hyperthermal areas,* having temperature gradients many times as great as those found in non-thermal areas.

(The absence of specific temperature gradients in the case of hyperthermal fields is deliberate, for there are places – e.g. Lanzarote in the Canary Islands – where gradients can more conveniently be expressed in terms of degrees per centimetre rather than per kilometre [37]. In fact on Montaña del Fuego in Lanzarote it is possible to burn one's fingers by scraping away a few inches of the ground surface with one's hands. It is for this reason that the rather vague expression 'many times as great' has been used).

It could be argued that a more logical basis for classifying the areas of the world would be in terms of *heat flow* rather than temperature gradient; but whereas heat flow must be deduced after making allowance for the thermal conductivity of rocks and for fluid escape, temperature gradient is an easily measured and simple function that can usefully serve as a preliminary basis for this very broad and rough classification which, it is emphasized, has no sanction of classic authority whatsoever; it is purely a suggestion of the author's. Others might well prefer some different means of classifying areas. It should also be understood that the gradients used in the above definitions are only approximate figures, and not clearly defined demarcations.

5.3 Classification of 'thermal fields'

It should be noted that up to this point no mention has been made of the word 'field', but only 'area'. Every point on the earth's surface lies within some *area,* but by no means does every point lie within a *field.*It is now necessary to make a distinction between a 'thermal area' and a 'thermal field'. A thermal *area*, if associated only with rock of low or zero permeability below the surface, is, with the techniques at present available to us, incapable of commercial exploitation. A good example of a thermal area which can scarcely claim to be a thermal field is to be found at Pathé in Mexico. Here there is a temperature gradient of about 550°C/km, but despite the drilling of 17 bores, some of which are of diameters up to 12 inches, the amount of steam won is so feeble that only about 150 kW of power can be generated (despite a 3500 kW installation). This is due to inadequate subterranean permeability.

A thermal *field* may be defined as a thermal area (either semi- or hyper-) where the presence of permeable rock formations below ground allows the containment of a working fluid, without which the area could not be exploited. The working fluid – water and/or steam, normally associated with certain gases – serves as a medium for the conveyance of deepseated heat to the surface. Since geothermal fields require both water and heat for their existence, they are often referred to as 'hydrothermal systems'.

Like 'areas', geothermal fields may also be conveniently classified into three types, as follows:

(i) *Semithermal fields*, capable of producing hot water at temperatures up to about 100°C from depths of 1 or 2 km.

Hyperthermal fields. These are of two kinds:

(ii) *Wet fields*, producing pressurized water at temperatures exceeding 100°C, so that when the fluid is brought to the surface and its pressure reduced, a fraction is flashed into steam while the greater part remains as boiling water.

(iii) *Dry fields*, producing dry saturated, or slightly superheated steam at pressures above atmospheric.

Hyperthermal wet and dry fields are sometime referred to as 'water-dominated' and 'steam-dominated' fields respectively.

5.4 Low-grade aquifers

There is another phenomenon that has come into prominence in recent years, namely the *low-grade aquifer*. These are sometimes to be found in *non-thermal* areas, and are capable of producing useful hot water of low grade (up to about 70° C or so) simply by virtue of the temperature gradient, low though that be, if the aquifer is at the right depth. The depth of the aquifer is of great importance. It must not be too shallow, or the water will only be tepid: nor must it be too deep, or the cost of drilling into it will be excessive despite the higher temperatures encountered at greater depth. If a mean annual surface temperature of 15°C be assumed, then in a non-thermal area having a temperature gradient of, say 30° C/km, temperatures of 60 and 75° C could be encountered at depths of 1.5 to 2 km respectively. If aquifers of good permeability can be located at such depths, then water at these temperatures could be commercially useful, especially if boosted at times of peak demand by means of fuel or by the use of heat-pumps. It would seldom be worth drilling more deeply than about 2 km to reach low-grade aquifers unless the temperature gradient were appreciably above normal – say about 35° C/km – or unless there are grounds for believing that an aquifer of very great vertical thickness is to be found not far below that depth.

An example of a low-grade aquifer is to be found in the Paris Basin (France), where it is being extensively put to use for heating buildings and greenhouses. Others are known to exist elsewhere in France, in some other European countries and doubtless in other parts of the world also.

Low-grade aquifers cannot properly be regarded as geothermal 'fields', for a field implies the presence of an anomaly, whereas low-grade aquifers occur in normal non-thermal areas wherever the geological formation happens fortuitously to be convenient.

The commercial value of these aquifers is very dependent on the density of the local heat market. They would seldom be worth exploiting except in or fairly close to densely populated areas.

5.5 Distribution and potential values of areas, fields and low-grade aquifers

There seems to be some kind of perverse law of nature whereby the world distribution of areas, fields and low-grade aquifers is more or less in inverse ratio to their actual or potential economic worth. Hyperthermal fields, both dry and wet, are of great commercial value but also of great rarity. Non-thermal areas lacking low-grade aquifers are at present completely worthless as sources of energy, but form by far the greatest part of the land areas of the world. Even among hyperthermal fields, the dry variety is cheaper and simpler to harness but is rarer than the wet variety. In between the extremes of valuable hyperthermal fields and (at present) worthless non-thermal areas there are exploitable zones, semithermal hot water fields and low-grade aquifers, of which the former are the more valuable but the

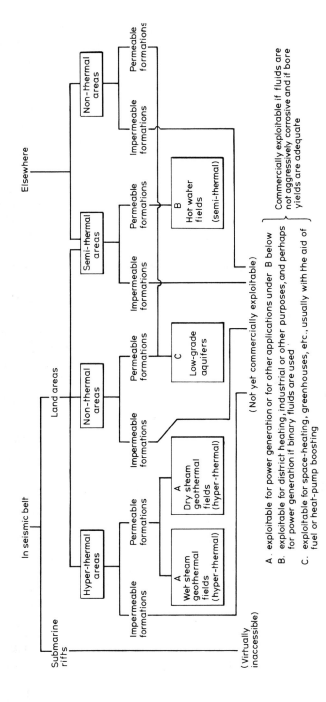

Figure 5.1 Classification of 'areas', 'geothermal fields' and low-grade aquifers.

rarer; and the hyper- and semithermal areas lacking permeable formations. Of the latter, which are not yet capable of exploitation but which soon may well be, the semithermal areas are the commoner but would be of less value than the hyperthermal areas after the art of exploitation has been mastered. Thus nature seems to have contrived that actual or conceivable commercial worth is associated with rarity, while abundance applies only to resources of doubtful or no economic merit.

Although low-grade aquifers have hitherto been developed only on a very limited scale, it is possible that they are far more widespread than was formerly believed. Tikhonov and Dvorov [38] have claimed that 50–60% of the territory of the USSR overlies economically exploitable thermal waters. If this is so, it is possible that some of these waters may owe their heat to the presence of semithermal fields caused by one or more of the three conditions mentioned near the end of Section 3.4; but it seems more probable that most of them would be due to the presence of low-grade aquifers in non-thermal areas.

Fig. 5.1 shows the occurrence and groupings of fields and areas. Considering the fact that the seismic belt forms such a small fraction of the total surface of the earth, and that much of it lies beneath the oceans, it is not surprising that hyperthermal fields – wet or dry – are comparatively rare phenomena. It will be noted that in this figure all 'areas' not associated with permeable formations are marked as 'not yet commercially exploitable'; but this does not imply that they never will be exploitable. More will be said about this in Chapter 18, but meanwhile it is well to remember that even in non-thermal areas there is a great deal of heat some few kilometres below the surface. These hot rocks at great depth, which will almost certainly be impermeable, will not necessarily always remain inaccessible. It is probable that in non-thermal areas there is low-grade heat at about 100° C only 3 or 4 km below ground, and high-grade heat at about 250° C at 9 or 10 km depth. Even now we are technically capable of reaching these depths, though it would not be economic to do so for the sake of heat of these grades; but the problems of exploitation at such great depths involve more than mere penetration.

The presence of a hyperthermal field is usually, but not always, accompanied by surface manifestations of thermal activity, such as geysers and fumaroles; but the converse is not necessarily true. Surface manifestations may simply be heat leakages through occasional cracks in a generally impervious formation. Likewise, excellent fields have been detected – e.g. at Monte Amiata in Italy – in places totally devoid of any visible surface thermal phenomena whatsoever.

5.6 Speculation concerning the nature of geothermal fields

The structure of geothermal fields, both hyper- and semithermal, and the mechanism of the upward transfer of deep-seated heat to higher levels are still subject to a certain amount of speculation. As underground fields are invisible, it is necessary to rely more upon our powers of deduction that upon direct observation. Nevertheless, from the evidence collected over the years in the course of exploration and exploitation it is possible to construct a 'model' consistent with the observed behaviour of the field and capable of describing the underground processes which could cause that behaviour. The probable validity of such a model must stand or fall by the accuracy with which predictions drawn from it are fulfilled. The most obvious and practical predictions of importance are the locations of promising drilling sites for gaining access to hot exploitable geothermal fluids. Hence if drill sites chosen according to the model shows a high success ratio, the implication may reasonably be drawn that the hypothetical model fairly closely resembles the actual physical features of the field. If the success ratio is low, then clearly the model must be faulty. The validity of the model can be further checked in the course of time against predictions of down-hole changes of pressure and temperature, and other symptoms of field behaviour. So much exploration and research has now been carried out in the last two or three decades that the fog of uncertainty is gradually lifting. Working hypotheses have been advanced that have generally fitted fairly well with the observed facts; and the present consensus of opinion is that the nature of thermal fields is likely to be somewhat as broadly described in the next two sections of this chapter, though new theories and models may well be advanced in future to cause some of the previously held theories to be modified.

5.7 Model of a hyperthermal field

A hyperthermal field requires five basic constituents:

  (i) A source of heat.
 (ii) A layer of bedrock.
(iii) An 'aquifer', or permeable zone of fractured and fissured rock capable of entraining a large quantity of water and/or steam.
(iv) A source of water replenishment to make good the fluid losses induced by nature or by artifice from the aquifer.
 (v) A 'caprock' to prevent the wholesale loss of heat and vapour from the field into the atmosphere.

Such a combination of five widely different features cannot be expected to occur very often; hence the comparative rarity of hyperthermal fields. The relative arrangement of these features is probably somewhat as illustrated in Fig. 5.2.

In seeking to account for the first requirement – a source of heat – it surely cannot be without significance that hyperthermal fields occur only within the

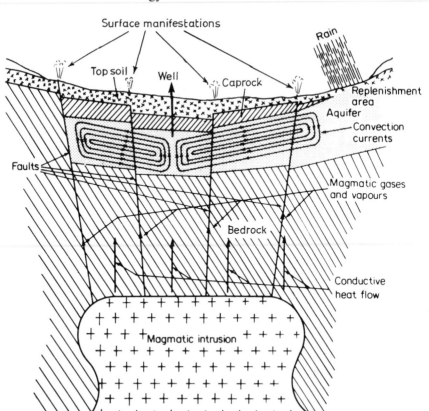

Figure 5.2 Stylized model of a hyperthermal field.

seismic belt. If the theory of Plate Tectonics outlined in Chapter 3 is broadly correct, then hyperthermal fields will occur only at those places of crustal weakness where adjacent plates are in relative motion and at which the intrusion of magma into the crust is liable to occur (Fig. 3.4). It is therefore generally believed that the main source of heat beneath a hyperthermal field is a *magmatic intrusion*. Such an intrusion, bulging into the body of the crust, has the effect of reducing the thickness of the thermally insulating crustal cover. Moreover, magmatic intrusions will tend to occur where the temperature of the magma is higher than the general temperature level of the mantle just beneath the 'Moho' in undisturbed places, because of the high frictional forces beneath subduction zones and because of the lateral tearing action that tends to ease the upward passage of deep-seated magma beneath rifting zones. Hence the high outward heat flow in hyperthermal fields may chiefly be due to these two causes – local hot spots in the mantle and local thinning of the crust. Other possible contributory sources of heat are those mentioned at the end of Chapter 3, high concentrations of local radioactivity in the crustal rocks, and the direct ingress into the aquifer of intensely hot magmatic gases forcing their way through faults in the bedrock.

The second requirement of a hyperthermal field – a layer of bedrock – needs no special explanation, as it will usually form part of the mainly basaltic crustal rocks. As the bedrock generally occurs at great depth – probably from 2 to 5 km at its upper boundary – the prevailing lithostatic pressure will usually be so high as to render it more or less impermeable by squeezing out of existence any fissures, especially horizontal ones, that may have been temporarily formed at moments of high mechanical stresses. Nevertheless, despite this re-closing tendency, it is possible that vertical or near-vertical faults of infinitesimal width could perhaps survive at these great depths under the influence of extremely high-pressure magmatic gases and vapours flowing upwards from far greater depths.

The third requirement of a hyperthermal field is the aquifer, or permeable zone of rock. The word 'aquifer' of course means 'water-bearer', but the term is also used to describe *steam*-bearing permeable zones that sometimes occur in hyperthermal fields. The permeability of the aquifer is probably accounted for by the incidence of frequent and intense thermal and mechanical stresses that have been induced at times of volcanic activity, resulting in cracks and fissures in rock of relatively low mechanical strength.

Before discussing the fourth requirement of a hyperthermal field – a source of water replenishment – it is well to mention that there are three types of water with which the hydrogeologist and 'field model-maker' are concerned:

(i) *Meteoric waters,* which is merely a 'highbrow' term for rainwater or the waters released from melting snows and ice at the earth's surface.
(ii) *Magmatic waters,* which originate from the released vapours from the water of crystallization when magmatic material fully solidifies on cooling.
(iii) *Connate waters,* as implied by the adjective, are waters 'born with' some rock formation – e.g. sea-water trapped in marine rock formations, or water or crystallization released from certain rocks when their stability has been disturbed by chemical or physical changes.

The water within a hyperthermal field, and its associated vapour, could be any of these types. First, the permeable aquifer may outcrop at some distance from the field, as shown in Fig. 5.2, in which case meteoric waters will continually be flowing into the aquifer to make good any *natural* losses occurring at points of surface activities (e.g. hot springs, geysers, etc.) together with any *artificial* fluid losses arising from the draw-off through boreholes when a field is exploited: meteoric waters can also sometimes penetrate downwards through faults at the site of a field, where there are no uprising hot fluids. Secondly, it is possible that magmatic water vapour under very great pressure and at very high temperature may rise through the bedrock from the magmatic intrusion beneath and enter the fissures and voids of the aquifer. Finally, connate waters could occur within a field, but as they would lack the means of replenishment they play only a minor part in most thermal fields.

The last requirement of a hyperthermal field – an impermeable caprock to form the 'lid of the kettle' that prevents the wholesale escape of heat and vapour into the atmosphere – has caused much speculation in past years. The presence of such a convenient 'lid' in all known hyperthermal fields seemed to be too fortuitous. In a few cases, of course, layers of impervious rock may have been deposited quite naturally in the course of geological evolution. But in others it seems probable that in the remote past permeability was not confined to the aquifer alone, but extended right up to the surface, in which case there would have been a continuous escape of geothermal fluids and of heat to the surface for a very long time by way of fissures in the upper layers of the rock formation. If this were so, then the very escape of fluids could have caused a self-sealing process that could ultimately block the fissures and render the upper layers of the rock impermeable, thus forming a caprock that would prevent further escape of heat and fluids. Such self-sealing action could be caused by two chemical processes; first, the deposition of minerals (mainly silica) from solution, and secondly the hydrothermal alteration of the upper rock causing kaolinization. This theory was advanced in 1964 by Facca and Tonani [39]. Evidence of the suggested processes can clearly be seen in certain fields, and the theory enables the far-fetched assumption of coincidence to be dispensed with. Moreover, the formation of caprock by self-sealing suggests that hyperthermal fields may be commoner than hitherto suspected; for the process would tend to repress surface manifestations, so concealing evidence of an underlying field.

Thus we have a plausible general model of a hyperthermal field in its simplest form, as illustrated in Fig. 5.2. Some years ago it was commonly believed that magmatic vapours were probably responsible for supplying both the source of heat and a large fraction of the water or vapour held in the aquifer; but isotopic measurements have now largely disproved this theory. Water of magmatic origin differs from water of meteoric origin in its deuterium content. The ratio of hydrogen to deuterium (H/D) in rain water is approximately 6800:1, whereas in magmatic steam (or 'juvenile' steam as it is sometimes called) the ratio is about 6400:1; hence it is a simple matter to determine the relative proportions in a mixture of meteoric and magmatic waters. It has now been clearly demonstrated that the magmatic water content of all known hyperthermal fields is very small – perhaps as much as

5% but never more than 10%. The bulk of the water appears to be of meteoric origin, having penetrated from the surface through permeable formations to great depths where it has been heated, mainly by conduction through the bedrock from below, though perhaps sometimes assisted by the upward infiltration of small quantities of magmatic vapours.

A field such as that illustrated in Fig. 5.2 would almost certainly be 'wet', yielding steam/water mixtures up bores sunk through the caprock into the pressurized aquifer beneath. Convection currents would be expected to flow within the aquifer; and in the process of rising to the under side of the caprock, hotter water from below would tend to boil and release steam. But with a configuration such as that shown in Fig. 5.2 the steam could not accumulate to any great extent: it would tend to slide along the under side of the caprock towards cooler, outer zones where it would be condensed and rejoin the convection water stream. Although only two convection loops are shown in Fig. 5.2, there could well be several such loops or circulation cells.

Sometimes, if the aquifer is slightly dome-shaped, hot water may collect in the lower part while steam collects in the upper part. Where this happens, as in Tuscany, the field would be 'dry' (although by drilling very deeply into the aquifer boiling water might be encountered). Dry hyperthermal fields are less common than wet.

The caprock will often be faulted in a few places, and it is by way of such faults that small quantities of thermal fluids will sometimes escape upwards and appear as surface manifestations.

Certain structure forms are sometimes to be found in parts of the crust that have been subjected to strong tectonic forces: these are shown in

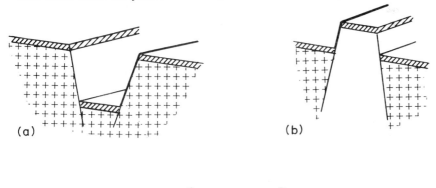

Figure 5.3 Some important earth structure forms. (a) Graben; (b) Horst; (c) Caldera.

simplified sketches in Fig. 5.3. A *graben* is a longitudinal down-faulted block: a *horst* is an uplifted block (the converse of a graben): a *caldera* is a more or less circular depression formed by the collapse of a volcano as a result of the withdrawal of the magma from beneath. (The withdrawn magma may have erupted elsewhere or it may simply have migrated internally along lateral 'dykes'. The shape of a graben is conducive to the formation of a hyperthermal field, for the trough can be partially filled with volcanic debris at a time of great tectonic activity and later overlaid with sedimentary deposits, thus providing the aquifer and caprock respectively. In the absence of sedimentary deposits the upper layers of debris could attain impermeability in time by the self-sealing process described above. The floor of the graben, being nearer to the source of disturbance that caused the down-faulting, would be liable to receive an excess of heat from beneath. A caldera has been subjected to such intense mechanical and thermal stresses that complex faulting will have caused the formation of extensive permeability which, in time, may become covered with caprock as in the case of a graben. The fact that magma had erupted from a caldera in relatively recent geological times will almost certainly mean that there is still a large store of residual heat beneath the fractured zone.

5.8 Model of a semithermal field

Semithermal fields are probably far commoner than hyperthermal fields, but as they are less prone to advertise their presence by the display of surface manifestations they are more likely to escape detection until they have been revealed by thorough exploration. With the growth of systematic geothermal prospecting in many countries, more and more of these fields will probably be discovered and their exploitation will become an activity of growing importance.

A semithermal field could sometimes take much the same form as that of a hyperthermal field as illustrated in Fig. 5.2, but with the heat flow from depth insufficiently intense to induce the relatively high temperatures associated with hyperthermal fields. However, as the hot water entrained in the aquifer of a semithermal field seldom reaches a temperature of 100°C, no caprock is needed to retain the heat and pressure within the aquifer; nor does it normally occur. The thermal gradient and depth of the permeable aquifer of a semithermal field should be sufficient to maintain convective circulation, but the temperature in the upper part of the reservoir is unlikely to exceed 100°C, partly because there is probably no caprock to hold down a pressure build-up above atmospheric, and partly because there may be mixing with cool ground waters. Re-charge with meteoric ground waters can of course occur at any point in the field except where there may be fortuitous formations of impermeable layers of rock above the aquifer. The problems of accounting for the source of heat in semithermal fields that do not occur near the boundaries of tectonic plates is not simply answered. It could be due to one or more of the three causes mentioned near the end of Section 3.4. After all, the local thinning of the crust virtually amounts to a 'magmatic intrusion'.

5.9 Actual hyperthermal fields

It is emphasized that Fig. 5.2 is stylized. Individual fields may well be of a more complicated structure. Nevertheless, a close examination of the geology of known convective hyperthermal fields enables the five basic features (see Section 5.7) to be identified. Two of the simplest fields will now be described.

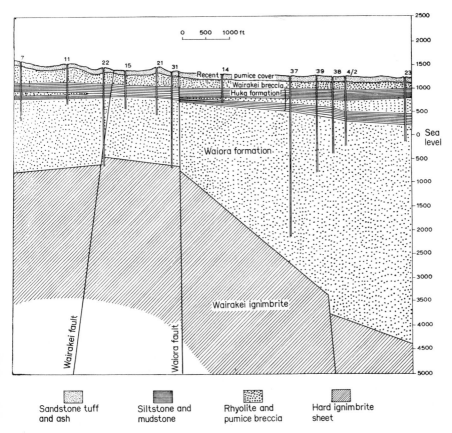

Figure 5.4 Geological section of the Wairakei thermal field, New Zealand.
Note: Some of the bores, sunk to various depths, which lie approximately along the plane of the vertical section are shown (with their local identification numbers). (Reproduced from Fig. 1 of [40])

5.9.1 *Wairakei.*   Fig. 5.4 shows a simple geological section through this field, which occurs in the Taupo volcanic zone in the North Island, New Zealand. In this case there are believed to be two layers of bedrock. The upper layer, which has been partially but not fully penetrated by drilling, is an almost impermeable formation of ignimbrite, extensively faulted as shown in the figure. Beneath the ignimbrite, but not shown in the figure, and at a depth of about 4000 m, a sedimentary basement of greywacke is believed to exist. The aquifer is a clearly defined layer of permeable pumice breccia known as the Waiora formation, of thickness ranging from about 350 to 1400 m. The caprock is no less clearly defined, and consists of the impermeable 'Huka formation', composed of siltstones, mudstones, pumiceous sandstones and diatomites of thicknesses varying from about 70 to 170 m. The caprock is covered with 80 to 160 m or so of pumice and breccia. Water replenishment of the field is believed to originate from two sources. Mostly it is probably from rainwater precipitated upon outcropping parts of the Waiora formation beyond the confines of the caprock, but to some extent also it is believed to come from below, through the faults in the ignimbrite. Some of this second source of water may be of meteoric origin, having penetrated from far away into the boundary between the greywacke basement and the ignimbrite, while some of it may be of magmatic origin having forced its passage through faults in the greywacke. The source of heat is probably for the most part conduction through the bedrock from a deep-

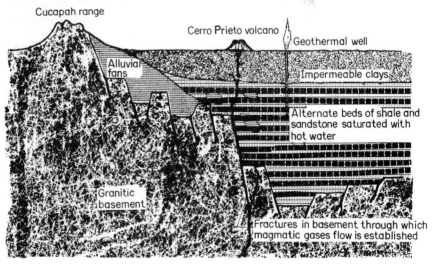

Figure 5.5 Geological section of the Cerro Prieto thermal field, Baja California, Mexico. (Reproduced by kind permission of the Comision Federal de Electricidad, Mexico, from their illustrated brochure describing the Cerro Prieto venture, 1971 [41])

seated magmatic intrusion, though a fraction may be conveyed through the bedrock with magmatic vapours.

5.9.2 *Cerro Prieto*.   This field, situated a few kilometres to the south of Méxicali – a township near the western end of the US/Mexican frontier – is related to the San Andréas fault system which extends from the Gulf of California to Cape Mendocino between San Francisco and the Oregon State boundary. A cross-section of the field is illustrated in Fig. 5.5. Here there is a granitic basement, faulted in many places, lying at a depth of about 2500 m beneath most of the field but rising to the west to outcrop in the Cucapah mountain range. A layer of impervious plastic clays, about 700 m thick, serves as a caprock, while between this layer and the granitic basement lie alternate beds of shale and sandstone which are sufficiently permeable to form an aquifer. A re-charge area for replenishing the aquifer with meteoric water is provided by the alluvial fans that form the eastern slopes of the Cucapah range, while the presence of the extinct Cerro Prieto volcano shows that magma has at some time in the past penetrated to the surface, and is suggestive of the present existence of a magmatic intrusion at depth, to provide the source of heat. It is also possible that conductive heat through the granitic basement may be supplemented to a small extent by very hot magmatic vapours rising through the faults in that basement from below.

5.9.3 *Other fields*.   These two fields closely follow the conventional pattern of a convective hyperthermal field as illustrated in Fig. 5.2. Space does not permit the description of other fields in this chapter, but Facca [42] examines the structures of three other fields – the Geysers (California), Otake (Japan) and Larderello (Italy). The first and third of these three fields yield dry superheated steam, while Otake (like Wairakei and Cerro Prieto) is a wet field. Although the structure of these three fields lacks the extreme simplicity of Wairakei and Cerro Prieto, an underlying resemblance to the archetype of Fig. 5.2 can be detected in every case. Structural models of several other hyperthermal fields have also been described in the proceedings of the UN Geothermal Symposium, San Francisco, 1975 [3].

A description of the unusually complex field of Krafla, Iceland, where the magmatic intrusion has risen to quite a shallow depth beneath the ground surface is to be found in [43].

5.10 Blowing of wet hyperthermal bores

It is of interest to consider what would happen if a very long vertical tube, having walls of thermally insulating material and blocked off at the base, were filled with water and steadily heated at the bottom of the tube with high-grade heat. The presence of convection currents in such a tube may virtually be ruled out if the heat is uniformly applied across the whole area of

the base. The temperature of the lowest layer of the water would steadily
rise, and although a small amount of heat would be conductively transferred
to higher layers, the lowest layer would ultimately boil and give off steam.
This steam, being buoyant, would rise into a slightly higher and cooler layer
and would condense, thereby raising the temperatures of this second layer
until it, in turn, would boil. However, the hydrostatic pressure in this second
layer would be slightly lower than that in the bottom layer, so it would start
to boil at a very slightly lower temperature than that at which the bottom
layer boiled. Steam from this second layer would then rise into a still higher

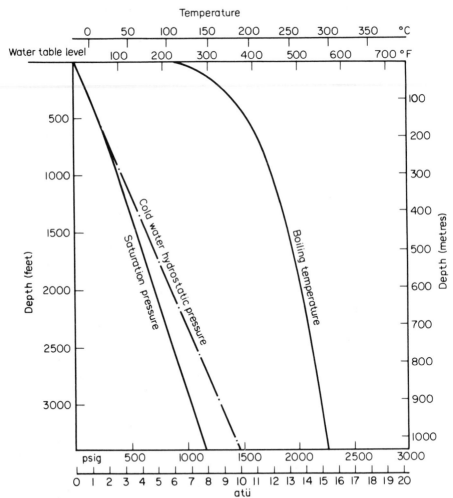

Figure 5.6 Variation of boiling temperature, saturation pressure and cold water
hydrostatic pressure with depth below water table.

layer, condense, and cause boiling at a rather higher level and at a rather lower temperature. The process would continue until the entire column of water is boiling; the temperature at each level in the tube being at the boiling point corresponding with the hydrostatic pressure at that level. The resulting temperature/depth relationship would then be as shown in Fig. 5.6 (which allows for the falling density of water with rising temperature and depth), and the continued addition of further heat from below would cause the emission of steam from the top of the tube at atmospheric boiling point.

A water-filled aquifer receiving a steady supply of heat from below would tend to imitate this curve to some extent; but the relationship between depth and temperature could be disturbed by the cooling influence of ground waters near the surface and by other complex factors. Near the outer boundaries of the convection loops (Fig. 5.2) the temperatures of the descending convection currents would be lower than those of the boiling point/depth curve, as the waters would not be descending unless they had cooled appreciably. Elsewhere the relationship could approximately hold. In the middle of the field, where a 'plume' of heated water is rising more or less vertically at the point where two convection loops are contiguous, the falling hydrostatic pressure would cause the water to boil and the temperature to drop as the fluid rises; and the steam bubbles so formed would give buoyancy to the fluid and would thus assist the convection process. If the steam at the top of this plume can only dissipate itself through paths of high flow resistance, the pressures and temperatures in the plume zone may exceed the values shown in Fig. 5.6.

When bores were first sunk at Wairakei, it was observed that the enthalpy of fluids collected by shallow bores of only a few hundred feet depth was about 575 Btu/lb (320 cal/g) whereas that of deep bores of 2000–3000 ft depth was about 460 Btu/lb (255 cal/g) – about 20% lower. This rather surprising observation was explained by the fact that the shallow bores were almost certainly collecting, with the hot water, a certain amount of steam that was boiling off from the hotter waters rising from below.

Actual underground temperatures in a wet field will seldom follow exactly the boiling-point/depth relationship, not only for the reasons given above but also because of changing patterns of ground-water flow and of convection currents, local differences of flow resistance, the presence of large masses of relatively impermeable rocks possessing 'thermal inertia' to temperature changes and other conceivable complexities. Measurements in static (non-flowing) bores sometimes even reveal temperature inversions.

If the temperatures in a non-discharging bore at all depths lie to the left of the curve of Fig. 5.6, the bore would be incapable of spontaneously 'blowing' because the vapour pressure would everywhere be less than the overlying hydrostatic pressure. To start such a bore discharging, it would be necessary to reduce the hydraulic pressure – either by using compressed air to blow out some of the 'top' water or by injecting some effervescent – so that the vapour pressure at depth could then take control. Once a bore starts to 'blow', it should, if the permeability of the aquifer at the base of the bore is adequate, continue to blow indefinitely owing to the reduced density of the bore fluid as it flashes off steam when rising up the bore, unless the permeability at the base of the bore is very low. However, if a bore is sunk to a point where the temperature lies to the right of the curve of Fig. 5.6, the mere opening of the valve at the top of the bore should set off spontaneous and sustaining blowing. A proposal has been advanced [44] for stimulating the upward flow of bore water by injecting into the base of the bore a highly volatile 'secondary fluid' such as freon or isobutane, and using that fluid to actuate a binary fluid generation cycle (see Section 10.2.6).

Sometimes, if a bore diameter is rather too great and the aquifer permeability inadequate, the volume of flashing steam as the fluid rises up the bore will be insufficient to sustain buoyancy, and the well will suddenly quench itself.

5.11 Blowing of dry hyperthermal bores

In a steam-dominated field a large volume of steam is trapped beneath the caprock under pressure owing to its inability to escape – by lateral migration into cooler surface waters, by leakage through surface manifestations or by artificial draw-off through bores – as rapidly as it is being generated at depth. The underlying water, deep down in the field, must boil at a temperature corresponding not to the hydrostatic pressure alone but to the sum of the hydrostatic pressure plus the steam pressure above its surface. A bore sunk into the steam zone will, owing to the prevailing pressure, automatically 'blow' without stimulation.

5.12 The phenomenon of superheat in dry hyperthermal fields

If, as is generally believed, a dry hyperthermal field consists of a steam zone overlying a deep-seated water zone in the aquifer, it might have been expected that the steam drawn off from the upper zone would be dry saturated, since the two phases are in contact with one another. Both at the Geysers and at Larderello, however, the steam at the wellheads is moderately superheated – by about 12°C at the Geysers and about 70°C at Larderello. Moreover, the enthalpy of the steam at Larderello is about 703 cal/g (1265 Btu/lb), which is greater than the maximum attainable enthalpy of dry saturated steam (669.8 cal/g or 1205.6 Btu/lb). At the Geysers the enthalpy of the steam is just about the same as that maximum. At one time it was believed that the superheated condition was due to the admixture of very hot magmatic vapours rising from great depth with dry saturated steam at higher levels, but it is known that the proportion of magmatic steam would be insufficient to account for the observed degree of superheat. Another theory entertained was that meteoric waters, after filtering down through a permeable zone, encountered very hot impermeable bedrock and immediately flashed into superheated steam. However, as it is now almost certain that there is in fact a water phase always present in the lower part of a steam-dominated field, this theory is no longer tenable.

The probable explanation of the superheat is two-fold. First, it is known that steam fields, before exploitation, produced dry saturated steam. The process of exploitation has lowered the water/steam boundary level, because the inflow of recharge waters is insufficient to make good the mass of total drawn-off steam. It is believed that the water level has dropped several hundreds of metres since the Larderello field was first exploited [45]. In doing so it has left 'high and dry' large quantities of rock that have retained much of their original heat at a higher temperature than that at which the steam is boiling off at the water surface. The passage of dry saturated steam from the boiling water surface to the bores, through this hot rock, would cause superheating. Secondly, the mere throttling of dry saturated steam when flowing through a permeable or porous resistance to flow must also cause superheating; for throttling is an isenthalpic process which – as can be seen from a Mollier chart – both superheats and (paradoxically) cools it slightly owing to the Joule–Thomson effect.

5.13 Fluid losses from a field

Except where reinjection is practised (see Chapter 16) or where down-hole heat-exchangers are used (see Chapter 11) the exploitation of a hydrothermal field will involve the continuous removal of fluid from below ground. Even when reinjection is practised, less fluid will be replaced than is withdrawn. As a result, there will be a depression of the water table in the aquifer which would ultimately drain a field of all its fluid content if it were not for a compensating inflow of 'new' fluid through the normal replenishment channels and from beyond the confines of the field. With a dry field, the lowering of the water table will be very pronounced, as already mentioned in Section 5.12. This is because the possible paths of inflow into a dry field are very restricted. With a wet field there is a greater degree of communication with outside sources of water, and the depression of the water table will be less marked. It is, of course, the very depression itself that causes the inflow of outside waters by creating an impelling hydrostatic head; and the greater the depression the greater will be the inflow. Hunt [46] has described how the fluid loss from the Wairakei field has been monitored by gravimetry, which has revealed that during the early years of exploitation only about 20% of the water withdrawn was being replaced by natural inflow, whereas after a few years there was little, or no, net fluid loss. it would thus seem that exploitation ultimately leads to a state of stability in which a balance is achieved between fluid withdrawal and inflow. A flow balance, however, does not imply a *heat* balance. More will be said about this in Chapter 17, Section 5.

5.14 Very deep, wet, hyperthermal fields

Reverting to Fig. 5.6, it will be realized that at some very great hydrostatic pressure the boiling point must rise to the critical temperature for water – 374° C or 706° F. This temperature would only be reached at a water depth of about 3500 m, say 11 500 ft, and it is doubtful whether any formation could retain permeability, and thus act as an aquifer, at such depths at which the lithostatic pressure would be extremely high. However, if such a deep aquifer could in fact be found, the curious condition would arise of a very deep layer of boiling water floating on the top of a 'cushion' of very high pressure superheated steam, in the same manner as a globule of water dropped onto a very hot metal plate will float above the plate in the 'spheroidal state', carried on a very hot thin layer of superheated steam.

5.15 Blowing semithermal bores

Except where artesian or 'geopressurized' conditions prevail (see Section 5.16) a semithermal field will not 'blow' spontaneously without aid, as the vapour pressure is insufficient to lift the hot water to the surface. In some cases it may be necessary to raise the water by the continuous action of a submersible pump. In other cases, once a flow has been established by initial pumping from great depth, it may be sustained by thermo-syphon effect by virtue of the higher specific gravity of the less hot overlying water by comparison with the hotter water in the rising well. It all depends upon the depth below the ground surface at which the water table occurs, and upon the depth of the aquifer below that.

5.16 Geopressurized fields

Although not falling within the broad classification of geothermal fields given in Section 5.3 and Fig. 5.1, there is also a type of 'freak' field known as the 'geopressurized' field. Such fields are found along the north side of the Gulf of Mexico, mostly in the States of Louisiana and Texas [47], and also in Hungary [48]. An interesting feature of these fields is that they may occur in *non-thermal areas,* where the temperature gradients range from about 27 to 30° C/km only. It is only on account of their great depth (up to about 6000 m, or about 20 000 ft, in places) that temperatures ranging from rather less than 93° C (or about 200° F) to rather more than 150° C (or about 300° F) are encountered. These fields are filled with pressurized hot water of 'connate' origin at pressures ranging from about 40% to about 90% *in excess* of the hydrostatic pressure corresponding to the depth. The explanation of these very high pressures is believed to be that gradual subsidence *via* growth faults has led to the ultimate isolation of trapped pockets of water contained in alternating layers of sandstone and shale. The trapped fluid supports a substantial share of the weight of the overburden and has prevented the full compaction of the formation. The pore pressures in these trapped pockets have values intermediate between the hydrostatic and lithostatic pressures. Thus one particular well in the Gulf of Mexico area, 4900 m (about 16 000 ft) deep and having a reservoir pressure of 871 atmospheres (12 800 psi) at the bottom, can give rise to a net *wellhead* pressure of about 439 atmospheres, or 6450 psi at no flow – an excess pressure of about 95% of the hydrostatic value. Geopressurized fields can produce energy in three forms – *thermal* by virtue of the fluid temperature, *hydraulic* by virtue of the excess water pressure over that of the vapour, and *chemical* by virtue of the calorific value of the methane gas that is normally associated with the water. It is the methane that is of the greatest commercial value. In terms of total energy content the geopressurized resource is believed to be immense; but the economics of its exploitation have yet to be justified. Active research continues.

5.17 Reservoir capacity

When plans are being drawn up for exploiting a hydrothermal field, it is very useful to have at least a rough idea of the total amount of exploitable energy available, so that the rate of heat extraction may be related to commercial and economic considerations. Over-exploitation could result in the run-down of a field in a shorter time than the expected economic life of the exploitation plant and equipment; and that would be tantamount to a lamentable waste of capital expenditure and, in consequence, excessive production costs.

Estimates of the thermal capacity of a field can be made only very roughly, as they must depend upon the accuracy of the 'model' built up by the earth scientists. Whilst the methods commonly used for the making of such estimates are likely to err on the conservative side, it would be most unwise to tempt Providence by exploiting a field at anything approaching the maximum theoretical possible rate. For example, let it be supposed that the capacity of a field (in terms of power production) has been estimated at 10 000 MW-years. If a 25-year commercial life could be assigned to the power plant it could be argued that as much as 400 MW of plant could safely be installed (assuming that an adequate power market is available), so that the resources of the field would be fully used up by the time that the plant would have been written off. Quite apart from the philosophical aspects of whether it is wise to exploit a country's resources quickly on the 'short-life-and-a-gay-one' principle or to eke them out slowly on the 'thought-for-the-morrow' principle, it would be extremely unwise to install the full 400 MW at the outset. In the first place the estimate, though believed to be cautious, might in fact be over-optimistic. Secondly, the run-down of a field through excessive exploitation would be a gradual process involving, perhaps after a few years, a slow decline in steam temperatures and pressures; so that even though plenty of energy might still remain in the ground, it would not be possible to obtain full output from the installed plant. It would be far wiser to install a smaller plant at first so that if, by any chance, signs of declining steam conditions should be observed before the end of that plant's commercial life, then a second-generation plant could later be installed specially designed to operate at reduced pressure. In general, it is wise to develop a field rather cautiously, step by step, and to observe the behaviour of the field very closely during the years of exploitation. As more and more knowledge of the field is accumulated it will become steadily easier to plan the future rate of exploitation.

Banwell [49] has devised a well-argued method of estimating the thermal capacity of a wet field, based on the assumption of a permeable aquifer of known area and depth impregnated with water, everywhere at boiling point according to the temperature/depth curve of Fig. 5.6. He assumes reasonable values for porosities and specific heat of the rock and for energy conversion efficiencies and, taking into account the stored heat both in the water and in the rock, he deduces the family of curves shown in Fig. 5.7, which shows the energy stored in a field in terms of its horizontal extent in area, and of its depth. It will be noted that two horizontal scales are provided – one for depth and the other for enthalpy at the lower limits of the field. The

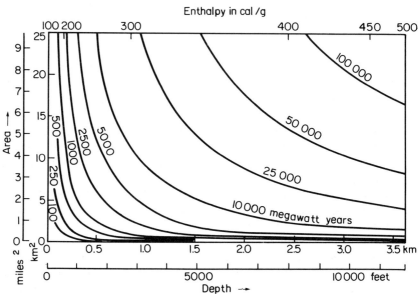

Figure 5.7 Approximate estimation of the power potential of a hyperthermal field. (After Fig. 4 [39]).

area and depth of the field have to be deduced from geophysical and other exploration data. The enthalpy, which follows from the depth and assumed temperature/depth relationship, should be confirmed by geochemical methods where possible. Clearly Banwell's premises cannot be precise, and the depth is likely to be uncertain; so his curves should be regarded as no more than a rough approximation. As they allow for stored heat alone, and make no allowance for continued heat infeed from depth, the curves are likely to err on the side of caution. The rate of heat extraction from a field during exploitation may, of course, greatly exceed the rate of heat infeed at depth; but the latter could be quite considerable, and could even rise temporarily as the field temperature falls after long exploitation. The pressures and temperatures in a wet field will decline in the course of exploitation owing at first to the fall in water level and later to the ingress of cooler waters from outside. Although these inflowing waters will gather stored heat from the hot rocks that have been drained of their original fluid content, the original temperatures could not be maintained indefinitely. Very probably the boundaries of a field will be ill-defined, and *warm* waters will intervene to some extent between the hot waters of the original field proper and the cold waters from further away. Moreover, very hot waters may well flow in from afar at great depths to help replenish the depleted aquifer, so that the *heat* flow could be sustained to some extent in this way. After more than 25 years of exploitation it was established at Wairakei that the heat flowing into the field was something like 70% of the heat being extracted: an approximate *flow* balance had already been achieved by then. When that field was first exploited pressures fell rather rapidly, but later they approached fairly stable values. Obviously a day must come, if exploitation continues at a higher rate than the heat inflow, when the permissible rate of extraction must suffer. That point will not come suddenly: there will be forewarning in the form of falling temperatures as waters from outside gradually extract the stored heat in the aquifer rocks and cool them. Early in 1968 the draw-off at Wairakei was deliberately reduced to about one-third of the normal rate for about three months, in order to see whether the field would show signs of 'recovery'. There was an immediate, but slight, rise in field pressures.

Estimating the thermal capacity of a dry field is less easy than for a wet one, for it is necessary to know the level of the water/steam boundary that is believed to exist at depth. If a water level can clearly be detected, and if the temperature at its surface and the probable depth of water beneath that surface are known, then it is relatively simple to calculate the approximate amount of heat contained in the lower, water-saturated part of the field, including heat stored in the rocks if reasonable assumptions are made as to their permeability and specific heat. In the upper steam-filled part of the field, the greater enthalpy of the steam is more than offset by its lower density in comparison with the same volume of water at the same temperature, so that the heat content of the steam could range from about 1.8% at 6.8 atü (100 psig) to about 4.4% at 23.8 atü (350 psig) of that of the same volume of water at the same temperature. If the dimensions of the field are known, however, the heat contained in the steam and rock above the water boundary can also be calculated for different pressures and for different assumptions for the specific heat and permeability of the rock. Hence an estimate of the total heat stored within a steam-dominated field may be calculated; but the calculations cannot be expressed in the form of a family of curves as in Fig. 5.7 because everything will depend upon the depth at which the steam/water boundary occurs and upon the geometry of the aquifer.

It might be expected that pressures would decline with exploitation more rapidly in dry than in wet fields, as the ingress of fresh fluid into the former is more restricted; but as the fluid enthalpy in a dry field is much higher than in a wet, the rate of fluid extraction will be far less for a given scale of utilization. At the Geysers field in California signs of very slow pressure decline became evident after about 15 years of steadily increasing exploitation.

5.18 Bore yield decline

The inevitable slow decline in the heat potential of a wisely exploited field can usually be concealed for many years (as at Larderello) by sinking new bores or deepening existing ones so as to tap new zones of the aquifer and thus maintain a fairly constant output from the field. This, however, is a process that cannot be prolonged indefinitely because the potential of a field is not infinite. But although the life of a whole field may be extended for a very long time by avoiding over-exploitation, the life of individual bores is

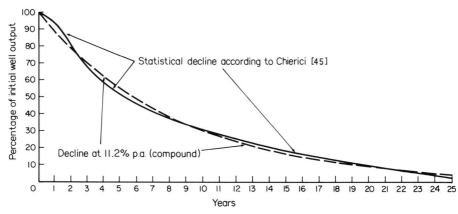

Figure 5.8 Bore aging at Larderello, Italy.

definitely limited. With the passage of time the heat yield from an individual bore will usually decline fairly rapidly: in other words a bore will 'age'. This aging process may in part be due to chemical deposition, particularly silica, within the fissures and pores of the aquifer. In dry fields an additional cause of aging is probably the increasing length of the path to be traversed by the steam from its point of generation (the steam/water interface) to the point of entry into the bore, as the water table falls in the course of exploitation. Chierici [45] has published a statistical relationship between the output and the age of the Larderello bores (Fig. 5.8). His original figure was plotted to a logarithmic horizontal axis, but, when plotted to linear axes, as in Fig. 5.8, the curious pattern of decline can perhaps be seen more clearly. The curve suggests an average aging rate of only about 7 or 8% for the first 1½ years, followed by a more sustained decline of about 12% p.a. over the next 20 years or so. A simplified approximation, which agrees with Chierici's curve at three points (after about 2, 9 and 20 years) would be to assume a steady compound aging rate of 11.2% p.a. (also shown in Fig. 5.8). Similar aging patterns have been observed for wells in other fields, both wet and dry, though the decline rates are individualistic to each field. Economically, a bore will have a finite life – usually about 10 years – after which its heat yield will become insufficient to justify the continued tying up of the capital invested in wellhead equipment and branch pipelines. After this life has been reached, it will pay either to deepen the bore or to scrap it and sink a new one elsewhere; in the latter case removing the wellhead equipment and whatever pipework can be salvaged to the new site.

6 **Exploration**

6.1 General

Geothermal exploration is a subject so wide as to justify many separate books, each fully devoted in its own right to a single branch of the several exploratory sciences involved. To attempt to cover the whole subject in a single short chapter could well be regarded by any exploration specialist almost as an impertinence; but the reader is reminded that the purpose of this book is to introduce him in a rather general way to the geothermal sciences – not to instruct him fully in all of them. No more can therefore be attempted here than to describe very superficially some of the exploration techniques and their interrelationships, and to refer the reader to the bibliography from which more detailed information may be acquired. These exploration techniques will here be treated separately under each of the several disciplines involved; but it is emphasized that all these disciplines are inter-dependent. Each specialist must rely to a large extent upon his technical colleagues, versed in other disciplines, to supplement his own observations and to support or cast doubts upon his own deductions. In short, geothermal exploration is essentially a matter of *teamwork,* and the team members may aptly be likened to a squad of detectives and forensic experts in search of evidence.

6.2 The aims of exploration

The aims of the geothermal explorer, of whatever discipline, are as follows:

(i) To locate a geothermal field or low-grade aquifer.
(ii) To decide whether a field, if found, is semithermal or hyperthermal.
(iii) To decide whether a hyperthermal field, if located, is steam- or water-dominated.
(iv) To define as closely as possible the location, area, depth and probable range of temperatures of any located field or low-grade aquifer.
(v) From all this, to estimate the order of magnitude of the heat or power potential of any located geothermal resource, and the grade of heat obtainable therefrom.

In the not far distant future the geothermal explorer is likely to become interested not merely in locating and evaluating geothermal fields and low-grade aquifers (see Section 5.3 and 5.4) but also hyper- and semithermal *areas* (see Section 5.2), the commercial exploitation of which is becoming increasingly probable. Even now, the discovery of thermal areas in the course of the search for thermal fields should be carefully noted for possible future use.

6.3 Likely locations of geothermal fields [50]

Of all the possible sources of exploitable geothermal energy, high-temperature hydrothermal fields are commercially the most rewarding. Such fields are invariably located in young orogenic zones, particularly where there has been recent volcanism within these zones. 'Recent' is a relative term, and volcanism may be a process of long duration. The dating of volcanic phenomena that have given rise to the present existence of hyperthermal fields must therefore be an imprecise exercise. A magmatic intrusion older than a few million years would either have broken the surface long ago or it would have cooled by now past the point of geothermal interest; but an intrusion that *started* in the late Cenozoic era (less than 16 million years ago) could well have initiated a convective hydrothermal system within the last million years or less. The Wairakei field is believed to be about half a million years old [51], while in Iceland volcanic activity is a continuing process capable of producing immediate effects – e.g. the disturbances of 1975 which adversely affected the steam production at Krafla.

It is in areas of known or suspected 'recent' volcanism, within the seismic belt, that exploration for hyperthermal fields should therefore be concentrated, though the experience at Krafla suggests that locations within an area of *present* volcanic activity should be avoided. The occurrence of surface thermal manifestations may be regarded as a reasonably promising sign – though by no means a *sine qua non* – of the probable presence of a hyperthermal field not far away. Nevertheless such manifestations do not necessarily imply the existence of a field in the immediate vicinity, as hot fluids can escape to the surface from below ground along inclined faults or fissures from a source fairly distant from the visible surface phenomena. Useful fields can, however, occur in areas totally devoid of any surface evidence whatsoever, and these must be sought in the light of the known geologic environment.

Semithermal fields, not being restricted to the seismic belt, can best be sought in areas exhibiting abnormally high temperature gradients.

Low-grade aquifers can be inferred only by studying the known geology, as they offer absolutely no surface evidence of their existence. Areas worthy of investigation are those where water-bearing strata such as sandstones and limestones are depressed to form deep basins, where thermal gradients are rather higher than normal, and where there is evidence of faulting which could allow hot waters to rise more closely to the surface and thereby reduce drilling costs [52].

Geopressurized fields are normally discovered accidentally in the course of oil exploration: it would be well to place continued reliance upon such accidental discovery, because these fields occur at such very great depths that systematic exploration would be too speculative and expensive.

6.4 Analysis of records

Before undertaking any systematic exploration programme it is important that all available local physical and chemical data first be collected and carefully digested. Even in the most remote parts of the world there will usually be *some* relevant field data available, much of which will have been amassed for purposes other than geothermal exploration. All recorded data such as topography, meteorology, geology, hydrogeology, observations of hot springs, geysers and fumaroles, geochemistry and geophysical measurements should be carefully collected and reviewed. From all these data it may be possible to select promising regions, or even more narrowly defined areas, that are likely to repay closer investigation. It is important that data of interest should not be confined to areas that are obviously or apparently 'thermal'; for the more information that is available concerning non-thermal areas surrounding a thermal area, the more effectively can thermal anomalies be characterized.

6.5 The function of the geologist [50, 53]

The rôle of the geologist in geothermal exploration is particularly difficult to describe, and is made more arduous by his use of a jargon and vocabulary that is more or less unintelligible to the layman! McNitt [50] emphasizes that in geothermal exploration 'the geologist must be a generalist, sufficiently familiar with a great variety of specialized exploration, drilling and well-testing techniques to ensure their proper coordination in the overall exploration effort'. He goes on to say that 'geology is more of a subjective discipline in which conclusions are based on a minimum of information that is often internally inconsistent', and that the geologist's tasks include collaboration with the geophysicist and geochemist 'in choosing the specific exploration techniques which he believes will yield the best results' and to 'resolve the inevitable conflicting interpretations resulting from these techniques'. All this makes it sound rather as though the geothermal geologist should be a qualified witch doctor! This is not to suggest that the author disputes what Dr McNitt has said; but the quoted remarks do serve to show how difficult it is to explain in simple words the rôle of the geothermal geologist.

Reduced to its simplest terms, it could be said that the task of the geothermal geologist is to deduce, as accurately as possible and as funds and circumstances permit, a three-dimensional 'model' of the geological structure of an allegedly thermal region to as great a depth as may be practicable, and to deduce promising drilling sites therefrom. By means of surface geological mapping; by the study of the tilt of outcrops; by the results of any cored soundings which may have been obtained by previous investigators (perhaps in pursuit of minerals or of ground water) or by his own exploratory drillings which he may consider to be justified; by the observation of faults and of surface thermal manifestations; by his deduction of the presence of 'caprock' formations; by all these means he should try to build up his hypothetical model, to suggest what zones may be permeable and contain hot fluids, and to pinpoint promising drilling sites. He may further suggest from where such zones are recharged with water and whence the source of heat originates. Since all but a tiny fraction of his model will be concealed from the eye, he must rely more upon deduction than upon direct observation; and it is on this that his skill mainly rests. Supporting evidence from his exploration colleagues of other disciplines will be of great help to him in building up his model. The geologist furthermore can sometimes deduce rough evidence of subterranean temperatures simply by detecting the presence of hot spring deposits that betray the temperatures of the waters that brought them to the surface. Thus calcium carbonate in the form of travertine suggests an aquifer temperature of not more than 120–140°C, while opaline silica (sinter) implies a probable temperature below ground of at least 180°C.

6.6 The function of the hydrogeologist

The task of the hydrogeologist is to collaborate closely with the geologist and to deduce the probable paths along which water will flow underground, through and between the various geological strata and boundaries of the geologist's model, and generally to assist the geologist to build up that model. The hydrogeologist (who may sometimes be the geologist himself, using a different facet of his knowledge, rather than another individual) should offer a plausible explanation as to how the thermal fluids reach the permeable zones of the alleged field, how they escape therefrom to the points where natural surface thermal phenomena are observed, how they are contained so as to prevent them from escaping elsewhere, and how they may be expected to behave when provided with artificial escape paths in the form of drilled boreholes. He must study gradients, porosities and permeabilities of the constituent geological formations; and he will have to rely to a large extent upon the opinions of his exploration colleagues to confirm or to challenge his theories. He, in collaboration with the geochemist, should attempt to differentiate between magmatic and meteoric waters; and he may need to use isotopes to put to the test his theories of underground flow paths [54, 55].

6.7 The function of the geophysicist [56–58]

The science of geophysics is basically that of seeking and interpreting anomalies of all kinds. In, say, a very large flat prairie or homogeneous formation it would be expected that the intensity of gravity, the rate of heat flow from below ground, the electrical resistivity of the surface soil and subsoils, the magnetic qualities of the subterranean rocks and various other physical properties would be identical at all points in the prairie, except perhaps near the edges. On the other hand, in mountainous regions containing various mineral deposits and different geological formations, it is only to be expected that these physical properties will vary from place to place. The task of the geophysicist in geothermal exploration is to measure as accurately as possible several physical properties in many places, to detect anomalies, to plot them where possible in the form of isotherms, isogals, isoresistivity lines or other equipotential 'contours' so that the anomalies may clearly be recognized; and then to interpret them as plausible evidence of underground formations. The geophysicist, in his rôle of detective, has little more than 'finger-prints' to aid him in solving a mystery. But, as with his other exploration colleagues, he has *their* evidence and *their* interpretations to which he may refer for corroborative, or perhaps conflicting, evidence. It is only to be expected that a geothermal field – which in a general way may be described as a large volume of steam and/or hot water retained in permeable rocks – is likely to give rise to various anomalies by comparison with surrounding norms. Hence it is not surprising that geophysics can serve as a powerful tool in the detection of thermal fields.

Many techniques are available to the geophysicist, varying widely in reliability and cost. Fortunately, some of the most useful techniques are the cheapest; and it is on these that he should first concentrate before calling upon his resources of more expensive and sophisticated methods to such an extent as he may consider to be justified.

One of the cheapest, simplest and most useful weapons in the armoury of the geophysicist is the thermometer, which may take many forms – the geothermograph, the Amerada gauge, thermo-couples, thermistors, platinum resistance thermometers or even mercury maximum thermometers; each has its own particular applications. By embedding thermometers of suitable type at various depths and at many points over a wide area and by observing their readings, the geophysicist can deduce temperature gradients and (with the help of the geologist) heat flow rates and can thus detect circumstantial evidence of more deeply seated local hot spots [59–61]. If he is a good diagnostician he will be sceptical of *obvious* evidence derived from such temperature measurements, because they may perhaps be disguised by cross-flows of cool ground waters that will tend to displace the apparent position of hot spots away from their true positions. Deductions such as these form part of the skills of the geothermal geophysicist who, in this particular instance, would be aided by the hydrogeologist. Temperature measurements in close proximity to surface thermal manifestations should, as far as possible, be avoided as suspect. The geochemist, as will later be shown, can also supplement direct temperature measurements with *deduced* temperatures arrived at from chemical evidence; so that he too may aid the geophysicist in his temperature investigations.

The measurement of natural outward heat flow from a thermal field should by no means be confined to ground conduction measurements; for the amount of heat escaping to the atmosphere by conduction through the ground surface from a hot field beneath will usually form only a fraction of the total natural outward heat flow, the balance occurring at hot springs, geysers, etc. The amount of heat escaping by these latter courses will usually be disguised by dilution from local cool surface waters, but the degree of dilution can often be estimated fairly closely by chemical methods. To estimate the *total* natural heat flow from a thermal area it is necessary to take into account *all* heat and mass losses in and around the margins of the area, including possible seepages into rivers and lakes and ground water movements across the area. This can be a difficult and complicated task, but estimates with a reasonable degree of accuracy can usually be obtained by exercising great care and judgement and, after allowing for the degree of dilution, an approximate estimate may be made of the enthalpy of the escaping fluids by dividing the deduced heat loss by the estimated mass loss [56].

Total natural heat loss over an area is regarded as a measurement of some importance in the assessment of a geothermal field, but it would be fallacious to assume that the rate of natural heat loss is necessarily the same as the rate of heat *inflow* from below the field, for to do so could be to presuppose a state of equilibrium. It is always possible that a field is still in the process of forming and that the quantitiy of stored heat is accumulating; or alternatively that it has passed its zenith and is beginning to decline by surrendering more heat than it is receiving. The difference between the heat inflow and the heat outflow is of course a measure of the changing quantity of *stored* heat in the field. It may take many years of careful observation to arrive at a reasonably reliable estimate of heat inflow.

In addition to thermometric studies, *electrical resistivity* measurements can also be performed fairly cheaply and quickly and can yield valuable information [62, 63]. They are achieved by injecting current into the ground through suitably spaced electrodes, and measuring the voltages between these electrodes. The techniques of resistivity methods can be adjusted so as to separate the horizontal from the vertical resistivity components within the area of survey. The interpretations of resistivity observations is a highly skilled art, for the geophysicist may be beset by false clues. Not only can differences of resistance be caused by the physical differences in the rocks between the electrodes, but also by the presence of steam or of electrically conductive thermal waters held within the permeable rock zones. The chemical constituents of the thermal fluids will further complicate matters. The geophysicist, in conjunction with the geologist and geochemist, must learn to differentiate between the various causes of resistivity anomalies. In general, a hot water field will tend to produce a zone of *low* resistivity, mainly on account of the dissolved salts, whereas a steam field will tend to show up as a *high* resistivity zone. However, these broad tendencies, as already stated, can be masked by misleading factors that confuse the interpretation of the observed results. The presence of clay formations, for example, can give rise to low resistivity and can thus masquerade as hot water fields that may well be non-existent. Other geological factors can similarly ensnare the unwary.

Further techniques in the weaponry of the geophysicist include:

gravity measurements [62];
seismic measurements, reflective and refractive [63];
micro-wave techniques;
electro-magnetic techniques [64];
radio frequency interference;
ground noise measurements [65];
micro-seismicity [66, 67];
audio-magneto-tellurics;
aerial scanning of infra-red radiation [68].

All these methods, some of which are costly to perform, may be regarded as reserve tools to be used only when the cheaper techniques of thermometry and resistivity produce apparently conflicting evidence or irreconcilable interpretations; but sometimes they can be justified for other reasons. It has been claimed, for example, that the presence of both vertical and horizontal fissures can be predicted by seismic measurements and could thus indicate promising drilling sites for penetrating highly permeable zones that might be capable of yielding large qualities of thermal fluids. One or the more interesting electro-magnetic techniques is based on the Curie point phenomenon – the fact that at 578°C rocks lose any magnetic properties they may possess at lower temperatures. Where the Curie point can be detected at relatively shallow depths, proof of a high thermal gradient is provided [69]. Each technique has its potential value for different circumstances, and the judgement of the geophysicist is required as to which method is best suited to resolve any special problems encountered.

Scanning by artificial earth satellites is a technique of growing importance, as very long range photography can reveal anomalies that are difficult to detect at closer quarters [70, 71].

6.8 The function of the geochemist [72]

Geochemistry affords one of the cheaper tools available to the geothermal explorer. It can even sometimes prevent the unnecessary wastage of exploration effort and expense at the very outset by demonstrating that some particular area is likely to prove to be geothermally worthless.

The geochemist's first task is to analyse the chemical constituents of natural thermal discharges, where such exist. The results of his analysis can 'serve as an important guide for decision making on subsurface exploration by drilling' [72]. After exploration bores have been sunk, the chemical analysis of deep geothermal fluids 'provides information on flow patterns of water and assists in selecting improved drilling sites' [72].

The chemistry of thermal fluids emitted from hot springs and fumaroles is determined by the interaction of underground fluids and rocks with which they come into contact. If the fluids are hot, chemical equilibrium will be achieved quite rapidly, and certain chemical features of fluids discharged at the surface are indicative of the temperature at which this equilibrium has been attained. Thus, the geochemist may be regarded as a highly sophisticated 'thermometrician' and can so aid the geophysicist and his other exploration colleagues in deducing the temperatures prevailing at depth in a thermal field. However, like his fellow investigators, the geochemist must always be on his guard to recognize falsified evidence resulting from dilution from cool surface waters and from chemical reactions between rising thermal fluids and rocks lying not far beneath the surface.

The three principal indicators of deep reservoir temperatures, to be sought in hot spring chemistry, are *silica, magnesium* and *sodium/potassium ratios*. Silica concentrations are more reliable for hot springs of high discharge than for those of low discharge or minor seepages. Magnesium is of limited value as a temperature indicator, but its total absence could be suggestive of economically useful reservoir temperatures (at least 200° C), as magnesium is retained in clay materials which are stable at high temperatures. The atomic ratio Na/K is a fairly reliable indicator of deep temperature – the lower the ratio the higher the temperature. The presence of free hydrogen in fumaroles is another indicator of subterranean temperatures exceeding 200° C [72].

The geochemist also has other useful analytical tools to help him deduce the quality and temperature of deep thermal fluids, both gaseous and aqueous, which are of the utmost interest to the specialist but which need not here be pursued. He is also able to detect the presence of valuable mineral constituents in thermal fluids that could be of interest to industrial geothermal developers, and he is able to form approximate estimates of the relative proportions of magmatic to meteoric waters present in the reservoir, thus aiding the conception of the field model. He must exercise the utmost care in his sampling methods.

Even after exploitation, the geochemist's art can be a valuable tool for detecting changes of temperature, water levels and water quality in a reservoir, and so in aiding an understanding of a field's behaviour. For example, a change in the chemical quality of bore water after a long period of exploitation could be indicative of the inflow of external waters into an aquifer, and even of their source.

6.9 Exploratory drilling

The final aim of all preliminary exploratory work is to select promising sites for exploratory drilling. Until such drilling has been undertaken, all the evidence collected by the team of earth scientists should be regarded more or less as circumstantial. No detective can carry his investigations to their logical conclusion until he has obtained *proof* of his deductions. The object of exploratory drilling is to seek this proof.

From the combined evidence provided by all his team members, the project manager must sooner or later select promising sites for exploratory drilling which, he hopes, will provide him with the necessary supporting evidence to prove that his model of the field is broadly valid. Of course it is always just possible, if circumstances should be particularly unfortunate, that in the course of the preliminary exploration the considered collective opinion of the team will be that no commercially exploitable field exists at all, in which case the relatively high cost of drilling operations can be saved and the expenses of the abortive project may thus be limited to a reasonable sum. However, assuming that the manager and his team are of the opinion that a useful field does exist, then, after making his best possible choice of site in the light of the evidence available, the manager will sink a bore to a depth at which he hopes to penetrate to a zone of permeable rock impregnated with hot thermal fluids. Expense and time may be saved if the first few exploratory holes are 'slim – i.e. of small diameter: larger bores may follow after confidence in the field has been established. It may be necessary to sink the first hole to a greater depth than originally intended, to allow for inaccuracies of the model of the field, if the evidence collected during the drilling process – e.g. cores, down-hole temperatures – is sufficiently encouraging. Quite possibly the first chosen site may yield a negative result; but if so, this should not necessarily be regarded as unduly discouraging. For every hole drilled, even non-productive ones, will provide additional evidence that will enable the project manager to modify his model so as to approach, step by step, more closely to the truth. Until a hole has been sunk, the explorer has only the evidence of his eyes and his instruments on which to construct his model. However, as soon as he has penetrated deeply with the drill – even if the bore should be totally unproductive – he can gain the direct evidence of cores, of deep temperature measurements and of chemical analysis at depth that will greatly reinforce his overall knowledge of the field, so that his next choice of drilling site should have a greater probability of success. His very first bore, regardless of its productivity, will have added a dimension to the explorer's knowledge.

Before many exploratory bores have been sunk, unless he has been particularly unfortunate, the project manager should succeed in striking a productive bore, and from then onwards the distinction between exploratory drilling and production drilling becomes steadily less clearly defined. After two or three good producing bores have been won, the main work of exploration has virtually been completed. A plausible model of the field will have been conceived and the pattern of future productive drilling will have been broadly determined, though both the model and choice of drilling sites may well have to be modified to some extent as the evidence of each completed bore is added to the sum total knowledge of the field.

The work of exploration may, of course, still continue into areas beyond the presumed boundaries of the field until those boundaries have become fairly clearly defined. Thereafter, exploration, as such, more or less ceases and 'field management' takes its place (Section 17.5).

6.10 The value of preliminary work

By far the most expensive part of an exploratory programme is the drilling, and it is of the utmost importance to economize on this activity without prejudice to the overall efficacy of the programme. In the early days of geothermal development, before much was known about the structure and behaviour of geothermal fields, reliance was placed to a large extent upon wild-cat drilling at sites chosen by hunch rather than by reason. If hot springs or geysers abounded in a region, then holes were drilled at any convenient points in the vicinity. Sometimes this 'hit-or-miss' approach paid off handsomely; but more often it led to a low drilling success ratio, even though the presence of a useful field might be revealed thereby. Larderello, the Geysers and Wairakei were all developed to a large extent in the early days by means of wild-cat drilling. Of course as the number of successful bores rose, the choice of further drilling sites became less speculative; but there is little doubt that with the knowledge, experience and sophisticated technology now available the costs of developing these fields could have been substantially reduced.

The modern trend is to spend more time and effort in gathering as much data as possible by the methods briefly described in this chapter before a single bore is drilled. Meidav [69] has expressed the opinion that as pre-drilling geoscientific studies have amounted in the past to only about one-eighth of the drilling costs, great economies could be effected by intensifying these preliminary studies before initiating any drilling work. The additional expense of these studies would be small by comparison with the large savings that could be expected as a result of a greatly improved drilling success ratio.

6.11 The extent of worldwide geothermal exploration by late 1980

Considering that little geothermal exploration had been undertaken anywhere until about 1950, it is impressive to contemplate the enormous extent of such exploration that was initiated in the next 30 years. By the end of 1980 thermal fields or low-grade aquifers were already being *exploited* in the following countries (which exclude several others where balneology was the only application of geothermal energy). In nearly every case exploitation had been preceded by exploration:

Australia (H,I)
Austria (H,F)
Chile (P)
China, The People's Republic (P, H, I, F, M)
Czechoslovakia (H,F)
El Salvador (P)
France (H,F)   *Key*
Hungary (H,I,F)   P   power generation
Iceland (P,H,I,F)   H   space-heating/domestic
Indonesia (P)   hot water
Italy (P,H,I,F)   I   industry
Japan (P,H,I,F,M)   F   farming
Mexico (P)   M   mineral extraction
New Zealand (P,H,I,F)
Philippines (P)
Romania (H,F)
Taiwan (M)
Turkey (P,F)
USA (P,H,I,M,F)
USSR (P,H,F)

In all these twenty countries geothermal exploration is still being actively pursued. Whereas in the USA geothermal exploitation had, until recently, been confined to the western mainland states it is now extending further afield, and exploration is being carried out in the eastern states, Hawaii and Alaska. In the USSR, where exploitation was first achieved in Kamchatka, there have now been considerable developments in the Caucasus and in other parts of the Union, while exploration is being undertaken in many parts of the country.

By 1980 geothermal exploration programmes had either been proposed or, in most cases, set in motion in the following 44 countries in addition to those twenty listed above:

Algeria	French Antilles	Netherlands
Argentina	Germany, Federal	New Britain (Rabaul)
Belgium	Republic	Panama
Brazil	Greece	Peru
Bolivia	Guatemala	Poland
Bulgaria	Honduras	Portugal (Azores)
Canada	India	Santa Lucia
Colombia	Iran	Somalia
Costa Rica	Israel	Spain
Denmark	Jordan	Switzerland
Djibouti	Kenya	Tanzania
Ecuador	Korea (North)	Thailand
Egypt	Korea (South)	Tunisia
Eire	Montserrat	United Kingdom
Ethiopia	Nicaragua	Yugoslavia

Thus no less than 64 countries are now becoming involved, to a greater or lesser extent, in geothermal development.

This list is not necessarily complete, as certain countries have not publicized their activities; but even so it is imposing, and indicative of the tremendous leap forward that has occurred within about 30 years in international concern for geothermal energy. Some of the listed countries hope to find, or have already found, high-grade heat suitable for power generation: others expect to win only low-grade heat for other applications. It should be noted that the list contains such countries as Belgium, Denmark, West Germany, the Netherlands and the United Kingdom, all of which until quite recently had been regarded as geothermally worthless. This is partly due to the highly successful economic development of non-thermal areas in France, and partly to the hopes that are now being felt of our ability to exploit hot dry rock in semithermal areas within the fairly short-term future.

7 *Drilling*

7.1 General

Drilling for steam and hot water, like exploration, is an important branch of technology in its own right, to which no more than very superficial treatment can be given in a single chapter. For more detailed descriptions, the reader is referred to [73–87]. In hyperthermal fields drilling depths generally range from about 500 m to 2000 m, although a few bores may lie outside these limits: depths of 600 to 1500 m would cover the majority. In semithermal fields and in low-grade aquifers depths averaging about 1800 m are typical, but in some places – e.g. Klamath Falls, Oregon, USA – the aquifer is close to the surface, and well depths may range from about 30 to 300 m. In geopressurized fields it may be necessary to penetrate as deeply as 6000 m or so.

For very shallow wells penetrating into hot water fields the old-fashioned cable tool rig may be used [88]. This breaks up the rock by percussion, caused by alternately raising and dropping a heavy cylindrical hammer bit. The crushed rock is then mixed with a water slurry and baled out. But for nearly all geothermal wells it is now necessary to employ rotary drilling, to which process this chapter is mainly devoted.

7.2 Cellars

Before rotary drilling operations are begun it is necessary to construct a concrete cellar (Figs 7.1 and 9.2(a)) about 10 ft × 8 ft × 10 ft deep internally, to support the weight of the drilling rig and later to accommodate the wellhead valving and expansion spool. A concrete stairway normally provides access to this cellar, which is also served with means of drainage (by gravity wherever possible). Consolidation grouting should be injected into the surrounding ground, where this consists of pumice or other weak formations, over a diameter of about 60 ft and to depths ranging from about 50 ft at the perimeter of the grouted area to about 100 ft beneath the cellar. This grouting not only gives support to the cellar but it can also serve to

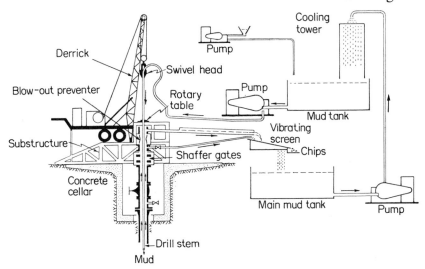

Figure 7.1 Typical rotary drilling rig and mud circulation arrangements. (After Fig. 2 [73])

deflect away from the wellhead any steam that may accidentally ascend to the surface along the *outside* of the bore and its casings: such escapes of steam, though fortunately rare, have been known to occur, and the consolidation grouting provides a degree of safety by ensuring that the ensuing blow-out is diverted sufficiently far away from the cellar as not to endanger access thereto. Consolidation grouting may add 5–10% to the cost of a bore, but this may be regarded as a worthwhile insurance [83].

7.3 Optimum bore diameter

It is important that the diameter of a bore be neither too small nor too large. Too small a bore will restrict fluid production by offering a high frictional resistance to its upward flow. Too large a bore, on the other hand, will cost much money and time in sinking, thereby staking an excessive financial risk in view of the possibility (never entirely absent) that the bore may cease to be productive after a relatively short period of service, or even that it may be non-productive from the outset. Moreover, in a wet field, a large bore might strike a formation – permeable, but of high resistance to flow – incapable of producing a sustained output of geothermal fluid against the large mass of water that would tend to fill the bore, so that the well would become quenched. Theoretically there is an optimum bore diameter which is a function of the flow resistance within the bore itself, the flow resistance within the permeable formation from which the bore is fed with fluid, the cost of the bore, the probable success ratio in winning productive wells and the value attached to the geothermal fluid(s) when won. This optimum diameter more or less defies calculation and clearly cannot be determined even empirically at an early stage of development: it must be chosen in the light of the results from the first few production bores. At first it is advisable to drill bores of rather moderate diameter, and to increase the size if and when the results appear to justify doing so. Although the best choice of bore diameter must to some extent depend upon the 'quality', or wetness, of the fluid, the figures quoted in Table 7.1 may be taken as fairly typical of optimum bore sizes. The terms used in the 'classification' column will be explained in Section 7.5.

Table 7.1 Guide to optimum bore sizes (from [89]).

Steam flow (tons/h)	Open hole (inches)	Casing size o.d. (inches)	Casing classification
10–25	17	$13\frac{3}{8}$	Surface
	$12\frac{1}{4}$	$9\frac{5}{8}$	Intermediate
	$8\frac{5}{8}$	7	Production
	$6\frac{1}{4}$	$4\frac{1}{2}$	Slotted liner
25–50	18	16	Surface
	$14\frac{3}{4}$	$11\frac{3}{4}$	Intermediate
	$10\frac{5}{8}$	$8\frac{5}{8}$	Production
	$7\frac{5}{8}$	$6\frac{5}{8}$	Slotten liner (o.d. of coupling skimmed by 1/16 inch)
50–80	22	18	Surface
	17	$13\frac{3}{8}$	Intermediate
	$12\frac{1}{4}$	$9\frac{5}{8}$	Production
	$8\frac{5}{8}$	7	Slotted liner

N.B. Inches are the usual standard. 1 inch = 2.54 cm.

7.4 Rotary drilling

The methods and equipment normally used for geothermal drilling do not basically differ very greatly from those used for winning petroleum or natural gas; but the harder rock formations and higher temperatures usually encountered in geothermal drilling, together with the possibility of striking highly corrosive fluids, call for certain improved techniques, materials and equipment. Penetration is normally effected by the chipping, spalling and abrasive action of special serrated tri-cone 'bits' (Plate 6) of very tough steel, the depth of the serrations being dependent upon the hardness of the rocks encountered. The bits are rotated, together with a hollow 'drill stem' to which they are attached, by means of a mechanical drive at the surface – usually a diesel engine.

Plate 6 Rotary drilling bits, as used at Wairakei, New Zealand [90]. *Left.* Tri-cone bit, with small heap of chippings. *Right.* Coring bit, with piece of core. (By courtesy of the New Zealand Ministry of Works and Development)

A drilling rig with all its appurtenances is quite an elaborate installation. First there will be a tall lattice steel tower, or derrick, containing a pulley system for positioning and withdrawing drill stems and casing so as to give access to them for examination and for joining sections together before they are lowered into the ground. Then there will be power units for rotating the drill, operating the derrick and driving the auxiliary pumps, air compressor, etc. There will be a rack for carrying a stock of casing pipes and drill stems ready for joining together and insertion into the ground; and there will be a circulating system for pumping, cooling, screening, settling and storing the cooling mud (Fig. 7.1 and Plate 7).

The purpose of the cooling mud [78, 79] is to cool and lubricate the bit and drill stem, to wash away rock cuttings from the hole, to prevent the bore walls from caving in and to cool the surrounding ground. The mud is forced downwards through the hollow drill stem and returns upwards through the annular space outside it. Various cooling muds are used, all having a higher specific gravity than water. For temperatures up to about 150° C bentonite or other clay-based muds are satisfactory, but at high temperatures such muds tend to 'gel' and the filtrate to increase in quantity. Chrome lignite/chrome ligno-sulphonate (CL/CLS) muds are then used, which are alkaline and contain a small amount of bentonite and traces of sodium hydroxide and of a de-foamer. Muds discharged from up a bore are screened to remove the rock chippings and are passed through a cooling tower (aided by forced draught if need be) before being pumped down the hole in re-circulation. Cooling through about 10 to 15° C is usually possible with a cooling tower using

Plate 7 Drilling rig in the Wairakei field, New Zealand. Note the mud cooling tower (right background) and the rack (foreground) for casings and drill stems. Several wellhead silencers can be seen in the background. (By courtesy of the New Zealand Ministry of Works and Development)

natural draught, though this depends upon the local climate: with the help of a fan the cooling range can be increased to about 20 or 25° C under typical conditions.

As an alternative to mud, compressed air is sometimes used [80]. Air drilling is cheaper and quicker than mud drilling and avoids possible damage to the production zone from circulating mud which may enter the zone when drilling. Nevertheless, this method cannot be used in formations that are very wet or which tend to slough. A common practice is to use mud until after the production casing (see Section 7.5) has been placed in position, and to use air drilling when penetrating the production zone.

In the course of drilling, fissures and voids are liable to be encountered at various depths. If these are indicative of a useful producing 'horizon' at which hot thermal fluids can be won, they may be left undisturbed and a slotted casing inserted, but at shallower depths they will usually be un-productive, or even counter-productive, either by allowing cool ground waters to enter a bore or hot bore fluids from lower down to escape into higher formations. In such cases they should be sealed off by injecting cement, and later protected by means of an inserted casing. The presence of such fissures and voids will be evidenced by a loss of mud circulation.

7.5 Casings

One of the most important factors in drilling is the provision of adequate steel casings to the correct depths. There will normally be as many as four concentric casings in a single well, all made of high quality steel (J55 to API standards usually, but acid-proof casings should be used where thermal waters are highly corrosive). These casings are rigidly fixed by means of cement to the surrounding rock (except in the case of perforated linings) and are joined to one another by means of screwed couplings. The depth to which each casing is sunk will depend upon the nature of the geological strata through which the bore penetrates. Fig. 7.2 shows a typical arrangement of casings as adopted at Wairakei, New Zealand. The largest, or *surface* casing extends to about 60 to 80 ft depth. The upper end terminates at the wellhead cellar (Fig. 7.1) and the casing extends through the upper stratum of unconsolidated pumice, thereby providing a firm locating support to the system of inner concentric casings that are later inserted. The second, or *anchor* casing extends downwards to 300 or 400 ft depth through the zone of soft breccias into the upper layers of the Huka mudstone caprock which effectively anchors the lower end. The third, or *production* casing extends downwards through the firm breccias as deeply as may be necessary (say, 1000 to 2000 ft typically) to seal off non-productive or counter-productive fissures and to reach the productive zone to the aquifer. The lowest part of the bore may sometimes be left unsupported, but more generally it is provided with a slotted liner (Plate 8) extending to the bottom of the bore and standing about ½ inch clear of the bore walls. The purpose of the slotted liner is to act as a strainer to hold back the larger pieces of rock which may here and there break off from the bore walls. The length of the innermost slotted liner may be anything, according to the levels at which productive horizons are struck, but only a few enter the top of the ignimbrite bedrock beneath the aquifer, as nothing is to be gained by deepening them further. Fig. 7.2 and the above description apply particularly to Wairakei: other fields will have different casing arrangements appropriate to the local geology and to the depths of the productive zones. For very deep holes it may be necessary to increase the number of concentric casings to more than the conventional three, as illustrated in Fig. 7.2, excluding the slotted liner.

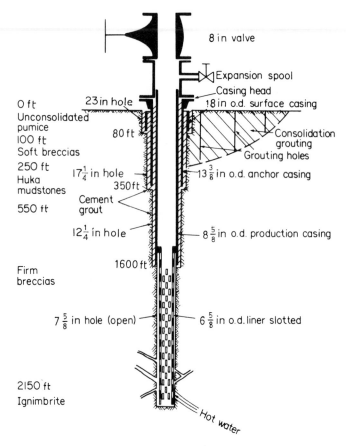

Figure 7.2 Casing layout and geological formations at Wairakei, New Zealand. (After Fig. 1 [73])

In the early days of drilling at Wairakei, slotted liners were not used. When a newly drilled bore was first 'blown' it was not unusual for large quantities of rock and debris to be discharged up the bore and scattered over the countryside. After a while, the quantity of this discharge usually fell off to negligible amounts. The theory was held that the thermal fluids entering the base of the bore at high velocity broke down the bore walls, ejected the debris and formed cavernous chambers around the base of the bore. As these chambers grew in size, the yielding area increased, the fluid velocity at the chamber walls fell, and the disruption declined. Later it was found that bore walls continued to collapse spasmodically long after a bore had been

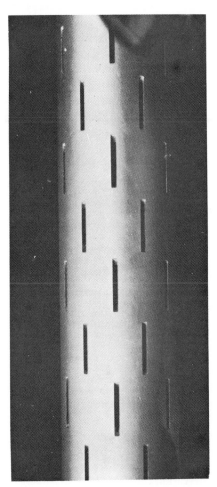

Plate 8 6⁵/₈-in slotted casing, as used at Wairakei, New Zealand, 28 slots per foot of liner [91]. (By courtesy of the New Zealand Ministry of Works and Development)

deemed to have been cleared: so to prevent the discharge of debris into the wellhead equipment, slotted liners were then fixed as a standard precautionary practice. This practice is now adopted in most other fields.

In shallow, low-temperature wells such as are used for certain direct applications of earth heat (see Chapter 11) the lowest, or intake, section of casing (corresponding with the slotted casing of Fig. 7.2) may be separated from the bore wall by a fairly wide annular space filled with gravel, which acts as a sand filter.

7.6 Cementing [82]

It is of the utmost importance that all casings, except the perforated liners, be firmly cemented into position. This is effected by injecting cement slurry down the casing by means of a plug which forces the slurry to flow around the bottom, or 'shoe' of the casing and to rise up the surrounding annular space until it reaches the surface. Ordinary Portland cement is suitable for temperatures up to about 150° C but special cements containing various proportions of silica flour, perlite, fly-ash and other ingredients must be used for higher temperatures. Faulty cementing can give rise to subsequent troubles such as collapsed casings, and it is very important that the greatest care be exercised when cementing. Casings are held concentrically during the cementing process by means of spacers, so as to give uniform annular sections.

7.7. Blow-out prevention [83]

In case of a blow-out during drilling, two safety devices are provided. First are the Shaffer gates (Fig. 7.1) which can be closed almost instantaneously either manually or by means of compressed air. The gates have semi-circular rubber rams that can fit closely round the drill-stem, drill collar or casing, as may be appropriate, so as to seal off the surrounding annular space. Above the Shaffer gates is a hydraulically actuated rubber blow-out preventer which is sufficiently flexible to close tightly round a drill string or casing, as may be required.

7.8 Coring

By using special annular bits (Plate 6), core samples may be extracted from a bore so that a continuous geological record can be made available of all the penetrated strata. Coring is rather a slow process which should be performed only when the geologist is in need of the valuable information that it can contribute towards the knowledge of a field.

7.9 Drilling times, penetration rates and bit life

The rate of drilling is of course influenced by such factors as rock hardness, ease of access to the site, the power and efficiency of the rig, the quantities of casing required, the bore diameter, the lengths to be cored and the depths to be penetrated into the productive aquifer. The periods shown in Table 7.2 are not unreasonable expectations under favourable conditions [74].

Table 7.2 Typical drilling times for different depths.

Depth (m)	Actual drilling (days)	Finishing and testing (days)
500	15–30 ⎫	
1000	25–45 ⎪	10
1500	35–55 ⎬	
2000	50–70 ⎭	

Actual penetration rates are of course dependent upon the weight carried by the bit and on the speed of rotation, but if either is excessive too much heat will be generated and the circulating mud or air will be unable to cool the bit effectively and a blow-out may occur. Under Wairakei conditions [73] typical bit loadings range from 150 to more than 1300 lb per inch of bit diameter, with increasing depth. Rotary speeds are of the order of 90 to 100 rev/min and penetration rates range from 23 to 40 ft/h (the lower limit applying to the larger diameter surface casing holes, and the upper limit to the deep small diameter holes in the production zone). In igneous rocks at fairly great depth penetration rates may be as low as 30 ft/day [92].

As might be expected, the life of a drilling bit falls off as the rate of penetration rises, but there is certainly no simple inverse relationship between these two variables: the economic bit life is affected by several factors such as depth, temperature, tripping rate and other drilling cost elements. For instance, the halving of the penetration rate might perhaps increase the bit life as much as ten-fold; but it could well be economic to accept a shorter bit life because of the cost savings that would arise from a higher penetration rate [93]. Tri-cone bits are generally serviceable at temperatures of well over 250°C, their life being only slightly affected by temperatures below that level; but at higher temperatures deterioration becomes more rapid, especially over 300°C.

7.10 Corrosion and erosion

Some thermal fluids are highly corrosive and require the use of special resistant steel casings and wellhead equipment materials. If the fluid is too corrosive, a bore – or even perhaps a whole area – may have to be abandoned. Erosion of production casings and wellhead equipment can be caused by the ejection of sand and larger rock particles spalled from the bore walls, and this can lead to fractures and blow-outs. Slotted liners will normally reduce erosion to acceptable limits, though special tough quality steels may in rare instances have to be used.

7.11 Maintenance or enhancement of output

Sometimes the deposition of silica or of calcitic salts on the bore walls in a wet field can result in a marked decline in fluid yields. Chemical deposition commonly occurs just above the slotted casing in the lower part of the production casing, where flashing of the uprising hot water causes the concentration, and consequent precipitation, of the solubles, as illustrated in Fig. 7.3. Usually, and fortunately, this is a slow process that can be remedied, at intervals of a year or two, by reaming out the deposit so as to restore the full internal diameter to the bore. The volume of deposit accumulated in the particular hole illustrated in Fig. 7.3 was 108 ft³ after discharging for 19 months. In a few very persistent cases the build-up of

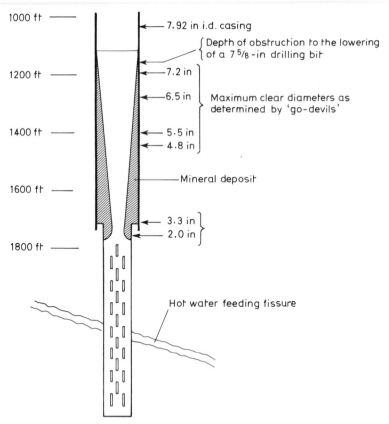

Figure 7.3 Typical mineral deposit build-up in a bore casing (not to scale). (After Fig. 3 [84]) The measurements shown in this figure are from an actual case at Wairakei.

deposits has been so rapid that frequent reaming has proved to be uneconomic, and the bore has had to be abandoned. However, at Cerro Prieto in Mexico it has recently been found possible to ream out a scaled-up bore without quenching it, merely diverting the discharge to the bypass and making use of a water-cooled Shaffer gate: this has made it economic to accept more frequent reaming operations.

Sometimes, where the bore fluid yield has been disappointing, it has been possible to increase the flow by perforating the production casing by means of a bullet type gun [85] consisting of a special barrel, having four chambers per foot, each housing a hardened ¾-in steel bullet (19 mm) and an explosive charge. The barrel is lowered to a depth where the geologists believe there may be a sealed-off producing horizon, the charges are detonated and the bullets pierce the casing and its surrounding grout and penetrate into the rock formation. Appreciable yield increases have sometimes been achieved in this way. Reaming will not of course influence the decline in bore yield due to chemical deposition within the fissures and pores of the aquifer, which can partly account for the decline illustrated in Fig. 5.8 and referred to in Section 5.18, but bullet perforation *could* help to offset this phenomenon. So too, perhaps, could hydro-fracturing (see Section 18.3.2).

7.12 Directional drilling [73]

Standard techniques have been developed in the oil drilling industry, and are adaptable to geothermal drilling, for producing a *deviated* bore that does not descend vertically. The ability to aim a bore in any chosen direction can be extremely useful for either of two specific purposes:

(i) It can provide means of sealing off a 'rogue' bore that has broken out of control (see Section 7.13).
(ii) It offers great possibilities of economizing in surface pipework and in land acquisition costs, while at the same time giving access to parts of an aquifer that underlie difficult or inaccessible terrain.

Regarding point (ii) it may happen that parts of a field underlie a built-up area, valuable cultivated land, a beauty spot, expensive land, or a terrain that is unstable, very steep or even precipitous, where the cost of levelling and stabilizing a suitable wellhead area would be excessive. In such cases it should be possible to tap the underlying aquifer by drilling deviated holes in the manner shown in Fig 7.4. There could of course be legal problems which would have to be investigated, regarding the rights of the owner of the surface land overlying the tapped area, particularly if subsidence risks are foreseen; but there could be many occasions where deviated drilling would bring to a relatively small collection area the potentialities of a widely spread aquifer. An indirect advantage of doing this would be that the quantity of surface collection pipework could be greatly reduced, and it might even be

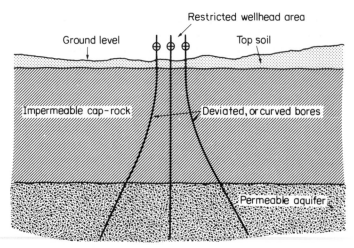

Figure 7.4 Suggested method of using deviated, or curved bores, to tap an extended producing zone from a restricted wellhead area.

possible to place the collection area close to the exploitation plant. The differential costs of the longer deviated holes, by comparison with shorter vertical holes, would have to be weighed against the saved costs of reduced surface pipework (see Section 14.10). The method has been put to great advantage in France (see Section 11.2.5).

7.13 Mishaps [86]

Many kinds of mishap can occur during and after drilling, such as steam blow-outs, drill-pipe sticking, parting of a drill-pipe joint or the breaking off of a drill collar, casing collapse, parted casing couplings, casing rupture due to excessive erosion or thermal shock. To pursue all these potential troubles and to describe the appropriate remedial procedures would require too much detailed description for a book of this type, and the reader is referred to other more sophisticated papers and articles in the references [74, 83, 86, 87]. In general, most mishaps can be prevented by the exercise of great care in the selection of bits, ensurance of adequate cooling during drilling, proper cementing of casings, consolidation grouting round cellars and the avoidance of thermal shock by heating or cooling bores very slowly when starting up or servicing.

At Wairakei a spectacular blow-out occurred in 1960 when a casing ruptured at about 600 ft below the surface and huge quantities of steam escaped to the surface and formed a miniature volcano close to the wellhead. This was followed by the collapse of a hillside, and the whole area soon resembled an inferno. By the taking of prompt action the equipment was

rescued and no-one was seriously hurt. The trouble was ultimately overcome by the very skilful drilling of a deviated, or curved, hole some 200 ft
away from the original bore [74, 90]. This new hole was so accurately aimed
that it struck a point very close to the original bore at a depth of about 1500 ft
just below its production casing (see Fig. 7.5). Cement and gravel were then
injected down the new bore so as to seal off the 'rogue' bore at its base. The
surface eruption immediately started to decline and it soon became possible
to fill the whole of the defective bore with cement, thus completely blocking
off the ruptured casing. The deviated hole was then cleaned out, cleared of
its cement plug at depth, and deepened, thus becoming a new and useful
production bore. This incident clearly illustrates the value of directional
drilling for the first purpose mentioned in Section 7.12.

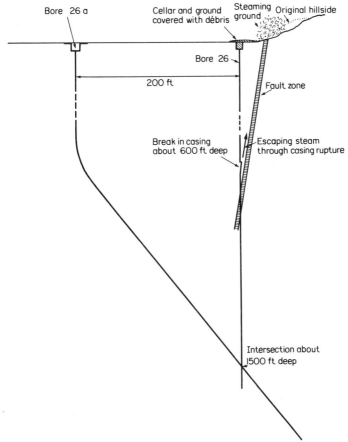

Fig. 7.5 'Rogue' bore (No. 26) at Wairakei, with deviated rescue bore (No. 26a) (not
to scale). Simplified version of more detailed sketch appearing in official Wairakei
brochure published by the New Zealand Government [90].

Another dramatic mishap occurred in Mexico in 1967 at the Cerro Prieto field when a wellhead pipe fractured below the shut-off valve. Vast quantities of steam and boiling water were escaping into the air, and it was dangerous to approach the well too closely. By the skilful cutting away of the jagged upper end of the fractured pipe by means of a blow-lamp; the careful lowering, by means of a derrick, of a new open valve with attached pipe stem over the jet of escaping fluid until the pipe stem butted against the stub end of the faulty well; the welding together of the new pipe stem to the old stub end; the welding of a safety collar round the new welded joint; and by the final closing of the new valve; the well was ultimately brought under control.

Other serious mishaps have occurred on rare occasions elsewhere; but the experience gained from each has been put to good use, so that such misadventures are now exceedingly rare.

7.14 Well spacing

No hard-and-fast rule can be laid down concerning well spacing. Clearly, the more closely the wells are spaced relatively to one another the less money will have to be spent upon collection pipework, but if they are too closely sited they may interfere with one another's performance. The mutual interference of wells is largely fortuitous, depending upon the random fissure patterns within the aquifer. Matsuo [74] suggests an empirical well spacing of 100 m for bores of 500 m depth to 300 m for bores of 2000 m depth. Budd [94] has established by means of a reservoir simulation model a relationship between well spacing and the rate of declining yield for the Geysers field, California, as illustrated in Fig. 7.6. The choice of optimum spacing then becomes a straight question of economics in terms of the prices of land, of drilling and of surface pipework. Budd qualifies the curves by stating, as would be expected, that they would be influenced by bore depth. A typical

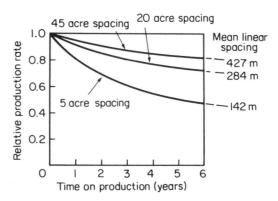

Figure 7.6 Effect of well density on production rate in terms of time (Geysers field). After Fig. 5 [94])

bore depth in the Geysers field is about 6000 ft (1800 m) according to p. 3 of Budd's paper, and if this depth be applied to Matsuo's very rough relationship between depth and spacing it would seem that the 20-acre spacing curve would be about appropriate. This curve shows an average annual decline over six years of about $5\frac{1}{3}\%$, which compares with about 11% p.a. for Larderello (Fig. 5.8); but as the Larderello bores are on average about half as deep as the Geysers bores, the decline would be expected to be more rapid in the former case, though the magnitude of the difference is perhaps rather surprising. It is only to be expected, however, that each field will have different permeability characteristics, and therefore interference characteristics. At Wairakei, the average bore spacing is about 100 m and the average bore depth about 600 m, which does not greatly differ from Matsuo's suggestions.

7.15 Maximum economic depths for conventional drilling

The question of drilling costs will be dealt with in Chapter 14, but it is here necessary to state that the cost/depth relationship of conventional drilling is such as to impose a maximum depth beyond which it is just not worth drilling deeper to seek geothermal fluids, because the incremental cost of each additional metre penetrated rises too rapidly. It is stressed that this maximum depth, whatever it may be, is imposed only by *economic* constraints – not technological; for even now we are capable of penetrating about 15 km or so into the crust by conventional means. It is not of course possible to define a maximum economic depth for geothermal drilling with any precision, because so much depends upon the quality of the fluid tapped, the geology, and on local factors such as accessibility to the drilling sites; but it is seldom worth penetrating more than 2.5 to 3 km for the purpose of exploiting hyperthemal fields. When drilling for oil or gas, far greater depths are economic because of the greater quantities of energy that can be extracted through each bore. With geopressurized fields, which occur at very great depths – of the order of 6 km or so – the economics of exploitation have not yet been properly established; but it would certainly never pay to drill into these fields for the sake of the heat energy alone: it is only because of the additional methane content that such very deep penetration could ever be justified. (The hydraulic energy associated with geopressurized wells is very small by comparison with the thermal and gas energy.)

One of the secondary factors that contribute to the imposition of an 'economic depth' for a geothermal bore in dry steam fields is the higher steam friction induced by long bores, thus reducing the yield from a well of any chosen diameter. Budd [94] has demonstrated this clearly in Fig. 7.7 which shows that at a wellhead pressure of 165 psia doubling the depth of an 8.5-in bore in the Geysers field could reduce the steam output by about 20%. He has even deduced a conceptual 'no depth' output from the same sized

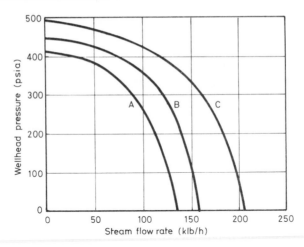

Figure 7.7 Steam flow rate versus wellhead pressure for two well depths for a typical steam well at the Geysers field, California (Fig. 2 [94]). A, 8½-in hole, 10000 ft deep; B, 8½-in hole, 5000 ft deep; C, theoretical performance with zero well depth.

bore, which suggests that at 5000 ft depth steam friction would rob a bore of about 25% of its intrinsic potential. Similar considerations would apply also to wet fields, but to a lesser extent, because the friction effect would to some extent be offset by improved enthalpy.

When the art of 'heat mining' in dry formations has been mastered it may be both necessary and economically justifiable to penetrate to greater depths than in hydrothermal fields; but it is still too early for positive pronouncements, as the economics of this activity are still in the stage of investigation. Heat mining in non-thermal areas of normal temperature gradient must remain a dream until some quite new method(s) of very deep penetration have been developed, with totally different cost/depth relationship from that of conventional drilling.

7.16 Turbo-drilling

With conventional drilling the mechanical power drive is placed at the ground surface and the driving torque is transmitted to the bit through the long drill stem. In very deep holes this driving torque may cause two or three complete rotations of the stem, so that the bit lags behind the drive by several hundred degrees of angle. Rotating drills stems are apt to whip, and thus to damage themselves, the casings and unlined bore walls. Various attempts have therefore been made to fix the driving unit at the base of the hole, close to the bit, so that the stem – though still twisted by the torque – need not rotate. Electric motors have been used for this purpose, but probably the most successful drill without a rotating stem is the Bristol–

Siddeley turbo-drill, designed by Sir Frank Whittle. This incorporates a compact multi-stage turbine motivated by the drilling mud, an epicyclic speed reduction gearbox, and a device for automatically adjusting the pressure on the bit according to the cutting resistance so as to ensure a fairly constant speed of rotation and to reduce mechanical shocks. The drill claims improved penetration rates and easy adaptability to directional drilling. Although it has not yet seen a great deal of service there is no reason to doubt its reliability, and it is a finely engineered device. The cost pattern of using it at various depths has not yet been well established, but it seems likely to be comparable with that of conventional drilling. The turbo-drill does not of course enable the drill stem to be dispensed with as the device has to be supported and the driving torque resisted; but the fact that the stem does not rotate eliminates the tendency to whip and avoids the consequential damage.

7.17 Improvements in conventional drilling techniques

The economic limit to depth referred to in Section 7.15 has been reinforced to a certain extent by the technical constraint of *temperature;* for at great depths – even in zones of fairly moderate gradient – the rock becomes too hot for the durability of various components and materials used in the rotary drilling process, such as bits, blow-out preventers, logging tools, muds and packers. This constraint is not exhibited by a sudden cut-off point beyond which further penetration becomes impossible: rather it is a steady trend that raises the costs incurred and the troubles encountered as the rock temperature increases – at first slowly up to about 260–280°C, and above that more markedly. The temperature constraint may therefore be regarded as one of the several influences that determine the maximum economic depth for the exploitation of geothermal fields.

Although this economic depth must have restricted geothermal development to some extent, its influence has probably been less than might have been thought; for the likelihood of finding the high permeability required for the retention of thermal fluids will decrease with depth owing to the rising lithostatic pressure that will tend to crush out of existence any voids or fissures that might once have been formed. Nevertheless, there are doubtless deeper aquifers than those that have hitherto been exploitable, in which the reduced permeability is more than offset by the higher enthalpies of the fluids they contain. Any way of cheapening drilling costs would thus not only produce direct savings *per se* but could also bring within the range of accessibility hotter fields at greater depths.

In recent years intensive research has been concentrated upon every aspect of conventional drilling with the aims of reducing costs and of increasing the economic depth for geothermal exploitation. These efforts have attained quite a large measure of success, and further improvements

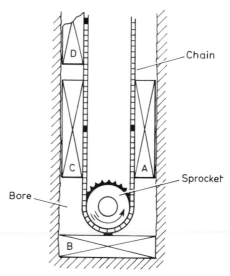

Figure 7.8 Diagrammatic sketch of the down-hole changeable bit drill developed by the Sandia National Laboratories, Albuquerque, New Mexico, USA. This sketch is purely diagrammatic, and makes no attempt to show how the cutters can flex as they turn into the service position. Nor is the manner of cutter attachment nor of cutter drive shown.

continue to be made. As *time* is one of the most influential factors in determining drilling costs, the use of more expensive materials may often prove to be a worthwhile economy if the average rate of penetration can thereby be sufficiently speeded up.

There have been notable improvements in recent years in the cutting qualities and durability of drilling bits at higher temperatures and in hostile environments. The use of tungsten carbide and artificial split-resistant diamonds, with improved bonding into corrosion-resistant matrices, has given improved abrasiveness and extended bit life: high-temperature bearing seals and lubricants for tri-cone bits have been developed: optimum rake angles and weight-bearing intensities have been determined by the use of computers: high-temperature drilling fluids have been developed having corrosion-resistant properties for the protection of bits and drill stems in hostile environments: special steels have been developed for drill stems, less vulnerable to stress corrosion and fatigue: the sensitivity and reliability of logging instruments have improved, so that subterranean conditions can be quickly and accurately monitored. All these improvements have steadily reduced drilling costs in terms of real money and have steadily extended the range of economically accessible depths.

The development of non-conventional methods of very deep penetration, depending upon altogether different principles from those used in modern drilling practice will be discussed in Chapter 18; and it is largely upon them that the success of future heat-mining operations will depend. But there are two novel methods of drilling that have recently appeared which may

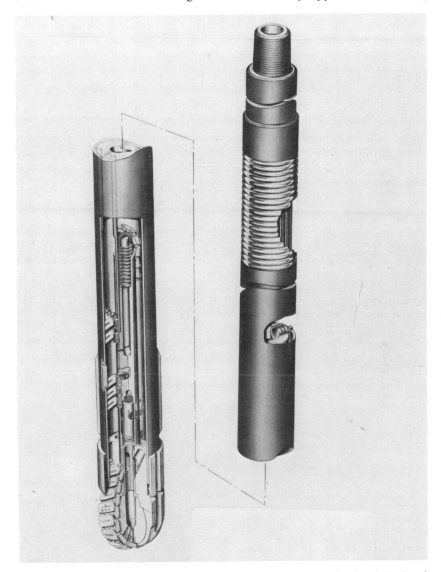

Figure 7.9 Continuous chain drill bit (prototype II) developed by the Sandia National Laboratories, Albuquerque, New Mexico, USA. (By courtesy of Sandia Laboratories)

legitimately be regarded as 'conventional' in principle even though of unconventional and original design. Both are aimed to reduce the time wasted by the present need to withdraw and replace the whole of a drill stem whenever a worn bit has to be replaced. It is this wastage of time, more than any other single factor, that accounts for the rapid rise in drilling costs with increasing depth. The two novel methods concerned are the 'changeable bit drill' and the 'continuous chain bit', both developed by Sandia National Laboratories of Albuquerque, New Mexico, USA [95]. The down-hole changeable bit drill (Fig. 7.8) contains a series of new cutter heads so that as one wears out a new one can be moved into position without retracting the drill stem from the bore. The continuous chain bit (Fig. 7.9) relies on a similar principle: the chain consists of a series of diamond or tungsten carbide drag cutting elements. When one set of cutters becomes dull, the chain is rotated so as to bring a new set of cutters into position. Several sets of cutters can now be accommodated within a single head, so that the withdrawal time can be reduced by a large factor.

8 Bore characteristics and their measurement

8.1 General

To succeed in bringing to the surface hot water and/or steam from below ground is, of course, only the first step in geothermal exploitation. Before we can know to what practical purpose and on what scale these fluids can be applied we must first study their available quantities and their qualities, both physical and chemical. Several important measurements must be taken as soon as possible after a new bore has been 'blown' (or, in the case of certain semithermal fields and low-grade aquifers, pumped or thermo-syphoned) in order to ascertain the physical properties and energy potential of the discharged fluids. It will also be necessary to determine the *chemical* characteristics of the fluids, but these will be dealt with in Chapter 15. The most important variable physical features of a productive bore are as follows; and as some of them will be interdependent, much of the gathered information will take the form of a series of curves rather than fixed figures:

  (i) The wellhead pressure.
 (ii) The wellhead temperature.
(iii) The yield, or mass flow, of the steam (if any).
(iv) The yield, or mass flow, of the hot water (if any).
 (v) The fluid enthalpy.
(vi) The fluid 'quality' – i.e. the dryness or wetness factor.

Although loose references will later be made to 'typical' bores, it is important to recognize at the outset that well characteristics will vary widely from field to field and from bore to bore. Strictly speaking there is no such thing as a 'typical' bore any more than there is a 'typical' man or woman, but a certain broad range of features can be discerned from the study of many fields, so that orders of magnitude at least may roughly be characterized.

8.2 Wet hyperthermal fields

With bores yielding water/steam mixtures, saturated conditions must of course prevail, so the variables (i)/(ii), and (v)/(vi) must be interdependent. The relationships between the variables can be obtained accurately from the steam tables, or approximately from Figs 8.1 and 8.2. The fluid yields, variables (iii) and (iv), will differ widely from bore to bore and from field to field, and must be determined by test measurements: they will vary with the wellhead pressure. These yields will be dependent upon the temperature at depth and upon the flow resistances through the fissures of the aquifer and up the bores; and since these factors will differ in every case, the flow characteristics of every individual bore will be unique to that bore. Fig. 8.3 shows some typical bore characteristics at various exploited wet fields; but the word 'typical' requires further qualification. First, for each field a random number of bores has been taken according to the available published data, and average yields have been deduced therefrom. The characteristics of any one of the selected bores may well differ appreciably from the average. Secondly, the average figures do not all lie exactly upon the smoothed approximate curves here shown. Thirdly, the characteristics of a bore are not altogether constant. As already explained in Section 5.18, bore yields tend to decline in the course of time. In wet fields they often become drier (i.e. the enthalpy increases) as well as less productive. The curves shown in Fig. 8.3 are all more or less representative of bores tested fairly early in the development of each field.

When a bore is closed, the flow of both water and steam must of course fall to zero. The pressure instantaneously registered at the moment of closure is known as the 'shut-in' pressure, which may later decline as the fluid in the

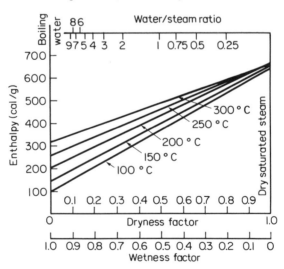

Figure 8.1 Relationship between enthalpy, temperature and quality for saturated water/steam mixtures.

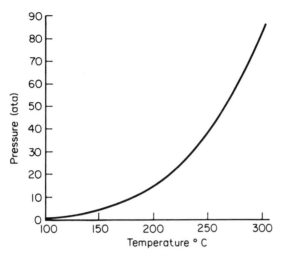

Figure 8.2 Temperature/pressure relationship for dry saturated steam.

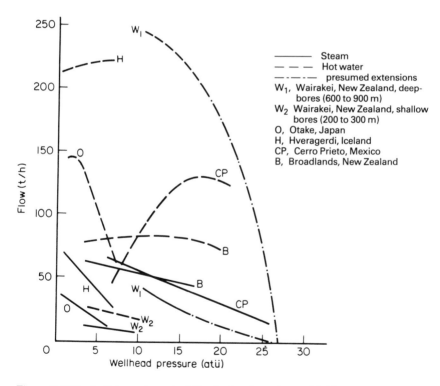

Figure 8.3 'Typical' bore characteristics for 'wet' geothermal fields.

bore cools off. In the case of the Wairakei deep bores (W_1 in Fig. 8.3) presumed extensions of the water and steam characteristics would suggest a shut-in pressure of about 27 atü, though curvatures sometimes fall off rather abruptly when a bore is gradually quenched by closure. It will be noted that steam yields tend to approximate to straight lines over the pressure range of greatest interest, while hot water characteristics are usually curved.

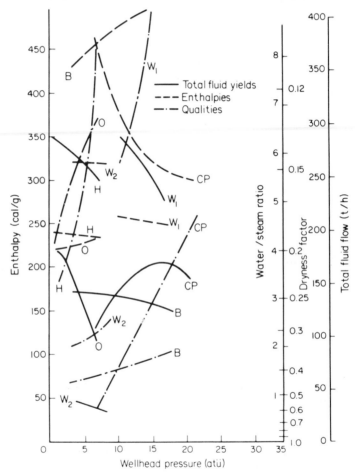

W₁, Wairakei, New Zealand, deep bores (600 to 900 m)
W₂ Wairakei, New Zealand, shallow bores (200 to 300 m)
O, Otake, Japan
H, Hveragerdi, Iceland
CP, Cerro Prieto, Mexico
B, Broadlands, New Zealand

Figure 8.4 Typical total bore yields, enthalpies and fluid qualities for 'wet' geothermal fields.

From the steam and hot water characteristics it is a simple matter to deduce the total fluid flow* (by adding the two), the quality (by comparing the two) and the enthalpy (from the steam tables or from Fig. 8.1). These further characteristics for the same fields as in Fig. 8.3 are shown in Fig. 8.4. They reveal the following features:

(i) The very great variety of bore characteristics found from field to field.
(ii) The relative constancy of the fluid enthalpy over a fair range of pressures at Wairakei (both shallow and deep bores), Hveragerdi and Otake.
(iii) The *rising* enthalpy at Broadlands and the *falling* enthalpy at Cerro Prieto, with increasing wellhead pressure.
(iv) In every case rising wellhead pressure causes increased wetness.
(v) The enthalpy of the shallow Wairakei bores is greater than that of the deep bores in the same field.

If a bore were to tap a zone of boiling water, the stagnant enthalpy of the flashing mixture (i.e. the enthalpy after allowing for the kinetic energy of the bore effluent) emitted by the bore would be expected to be constant, at the value of the enthalpy of the hot water tapped at depth. Thus the two classes of bore at Wairakei and also the bores at Hveragerdi and, to a lesser extent, Otake approximately conform with this expectation. The steeply rising and falling enthalpy curves for Broadlands and Cerro Prieto respectively suggest that the bores are fed from more than one 'horizon' in each case, each horizon contributing different proportions of the total heat with different pressure distributions down the bores. Such phenomena as these show that the true nature of fields must in fact be considerably more complicated than the simple stylized model of Fig. 5.2 would suggest. Increased wetness with rising pressure is only to be expected, especially if the enthalpy does not vary very greatly, since the higher pressure will tend to suppress flashing. The phenomenon of the higher enthalpy of the Wairakei shallow bores by comparison with that of the deep bores has already been commented upon in Section 5.10.

A point of interest to be noted in Fig. 8.4 is that whereas in every case but one the total discharge rises with falling wellhead pressure, at Cerro Prieto it reaches a peak and then declines. This is not altogether easy to explain. It will later be shown (Fig. 8.11) that a well tapping a source of constant fluid enthalpy would give a rising total fluid flow with falling wellhead pressure until it attains a maximum value at unrestricted discharge. Fig. 8.4, however, shows that the enthalpy at Cerro Prieto is far from constant: it rises sharply as the back-pressure falls, and the steam becomes much drier. It would seem that the bore simply tends to 'choke itself' as a result of friction

---

*The *total* fluid flow should of course include incondensible gases, but in the present context only the 'H_2O' fluids are being considered.

at high steam velocities. This explanation, however, rather begs the question; for it does not explain why the enthalpy should rise at low wellhead pressures. It could be, as suggested earlier, that there is more than one feeding 'horizon' at different depths. If, for example, there were a shallow infeed through a fracture system of fairly high impedance, and another infeed at greater depth from a zone of low impedance, then at low wellhead pressures the reduced fluid density in the bore would have a relatively greater effect on the back-pressure resisting the flow from the deeper infeed, which could then contribute a higher proportion of the total flow. Alternatively it is possible that there is only one infeed, but that at low back-pressures flashing occurs within the rock formation and not in the bore itself. The flow impedance to the denser fluid component (water) would thus be high, while the steam in the bore itself would be relatively unimpeded.

8.3 Dry hyperthermal fields

As the steam discharged from dry hyperthermal fields is usually somewhat superheated, there can clearly be no direct relationship between the variables (i) and (ii) of Section 8.1, while variable (v) will follow from these two and variable (vi) will have no meaning. At Larderello, in Italy, the degree of superheat ranges from about 83°C at a wellhead pressure of 6 ata to about 56°C at 16 ata. As low wellhead pressures are associated with high steam flows, there would be increased throttling action at low pressures as the steam passes at high velocity through the permeable formation, and this could account for the higher degree of superheat, as would be expected with isenthalpic expansion. At the Geysers field in California too the degree of

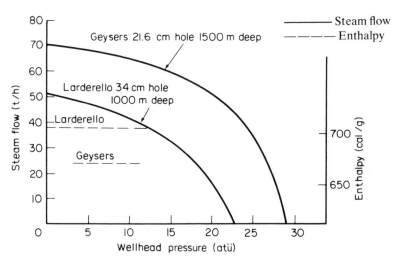

Figure 8.5 Typical bore characteristics in 'dry' geothermal fields.

superheat declines with rising wellhead pressure, ranging from about 28°C at 4 ata to about 10°C at 10 ata. The enthalpy at the Geysers varies somewhat from bore to bore, but is generally rather close to the maximum attainable enthalpy of rather less than 670 cal/g for saturated steam.

Typical steam flow and enthalpy characteristics, against a wellhead pressure base, are shown for Larderello and the Geysers – the two most famous 'dry' fields in the world – in Fig. 8.5. However, it is here advisable to repeat the warning already given about the word 'typical'. One bore feeding power unit 4 in the Geysers field was reported in 1980 as yielding 175 t/h of steam (pressure unspecified).

8.4 Hot water fields and low-grade aquifers

The characteristics of semithermal fields and of low-grade aquifers are far too idiosyncratic for generalizations to be made about them. Generally they will tend to yield water at fairly constant temperature, but the artesian pressure (if any) and the pumping pressure required are governed by the extent and permeability of the formation in which the hot water occurs, as these factors will govern the rate of drawdown.

8.5 Measurements

The measurements of wellhead temperatures and pressures offer no problems: they can simply be made by means of thermometers and pressure gauges. For wet bores, in fact, it is only necessary to measure one of these two, since they are interrelated in accordance with the curve of Fig. 8.2 or, for greater accuracy, in terms of the steam tables. It is the measurement of mass flows that requires particular care. Once these measurements have also been obtained – for each phase in a wet mixture – the derivation of enthalpies and of dryness or wetness factors can be quite simply deduced, either from the steam tables or, in the case of wet fields, from Fig. 8.1 approximately.

Mass flow measurements of dry saturated or of superheated steam can be made quite simply by means of sharp-edged orifice meters, after making sure that sufficient lengths of straight piping are placed upstream and downstream of the orifice to ensure non-turbulent flow (at least 25 pipe diameters upstream and at least 10 diameters downstream). The mass flow will then be a function of the manometric pressure drop across the orifice, of the upstream pressure and temperature and of the orifice restriction ratio: a calibration chart will be provided with each orifice flow-meter.

An alternative method of measuring dry steam is to discharge it at sonic velocity (so that the precise downstream pressure is immaterial) to the atmosphere through a cone of partly standardized dimensions, as shown in Fig. 8.6. The method is not very accurate owing to the number of rival formulae (quoted on Fig. 8.6) that can be used to interpret the results, but

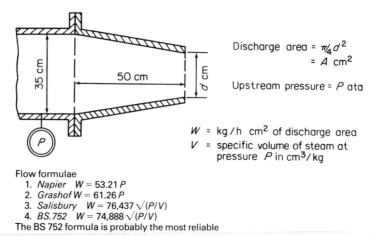

Discharge area = $\pi/_4 d^2$
= A cm^2

Upstream pressure = P ata

W = kg/h cm^2 of discharge area
V = specific volume of steam at
pressure P in cm^3/kg

Flow formulae
1. *Napier* $W = 53.21\, P$
2. *Grashof* $W = 61.26\, P$
3. *Salisbury* $W = 76,437\,\sqrt{(P/V)}$
4. *BS.752* $W = 74,888\,\sqrt{(P/V)}$
The BS 752 formula is probably the most reliable

Figure 8.6 Measurement of dry steam discharge to atmosphere through cones at sonic velocity. Correction for superhear: multiply W by 1.001 17 S, where S is the number of degrees of superheat in °C. The BS752 formula is probably the most reliable.

for approximate measurements its simplicity has something to commend it when the upstream pressure is at least 2 ata. Although the upstream pipe diameter and the cone length are standardized, the cone taper – and therefore the discharge area – may be varied so that suitably sized cones may be chosen for testing at convenient pressures. The accuracy of the various formulae will depend to some extent upon the degree of cone taper, but the BS 752 formula is probably as reliable as any.

The measurement of water/steam mixtures presents greater difficulties. The following are some of the methods used:

8.5.1 *Calorimetry* [96].   The whole of the water/steam output of a bore is discharged for a measured time into a tank containing a known mass of cold water at a known temperature. When the well discharge has ceased, the additional weight of water in the tank is a measure of the mass flow during the period of the test; (the water content is simply added while the steam content is condensed). From the rise in temperature of the water it is a simple matter to deduce the added heat, and by dividing this by the added mass the enthalpy of the bore fluid can be deduced and the dryness factor will be a derivative of this. Despite its simplicity, this method has its limitations. Unless a very large tank is used, the method is suited to bores of small output only; otherwise the duration of the test would have to be very short and the accuracy of the test would be sensitive to the speed of operating the test valves. A special rocking arm has been devised in New Zealand [96] to reduce this difficulty; but if valves are used it is necessary to *start opening* them at the beginning of the measured time and to *start closing* them at the end of the measured time, so that the periods of valve operation will tend to cancel one another out. The method can be used for bores of larger output by splitting the discharge in two at a T-joint, measuring the discharge from one branch only and doubling it. If a valve on the *unmeas-ured* branch is adjusted to give the same back-pressure as on the measured branch, it is not unreasonable to suppose that the flow resistance in each branch is the same, so that each accepts one-half of the total.

8.5.2 *Sampler measurement* [96].   By means of a travelling sampler tube, the flows at different points on the diameter of a discharge pipe can be measured by calorimetry and integrated, so as to arrive at the total discharge across the whole pipe. This is a very cheap method, as it requires only a small calorimeter and a light portable traversing device (Fig. 8.7), but it is not very accurate. The speed of traverse can be arranged so that the time in each position is proportional to the area of the annular ring being sampled. Ideally, the velocity of the fluid drawn off should be the same as the velocity of fluid in the main discharge pipe, so that the sampling tube causes a minimum of disturbance to the flow pattern.

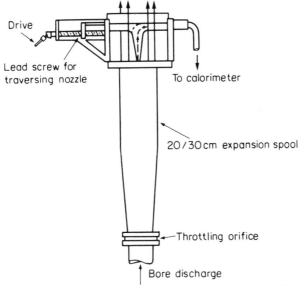

Figure 8.7 Bore enthalpy measurement by sampler. (After Fig. 2 of [96])

8.5.3 *Phase separation and the measurement of each phase separately.* This is the most accurate, but the most expensive method. Its reliability also depends upon the efficiency of the separator used, but as efficiencies of the order of 99.9% can usually be achieved this limitation should be insignifi-cant. The well fluid is passed through a cyclone separator, the steam flow

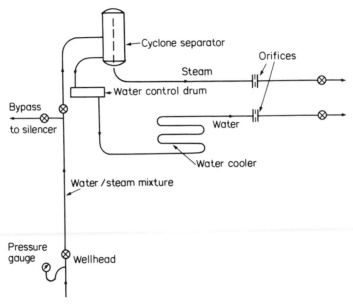

Figure 8.8 Well yield measurement by phase separation and orifices. (After Fig. 1 of [96])

is measured by calibrated orifice and manometer, and the hot water is measured in the same way. It is important that the water first be cooled sufficiently to ensure that boiling does not occur when passing through the orifice, otherwise the water flow readings will be quite unreliable. The arrangement is shown in Fig. 8.8.

8.5.4 *Critical lip pressure method* [97].   If hot water or a water/steam mixture is discharged at sonic velocity through an open-ended pipe, and if the pressure is measured at the discharge lip of the pipe, this pressure is a measure of the *heat flow* of the discharged fluid per unit sectional area of the pipe. James [97] has deduced the following formula, expressed in ft/s units:

$$G = \frac{11\,400\,P^{0.96}}{h^{1.102}}$$

where G = flow in lb/s ft^2 of discharge area.
        P = critical lip pressure in psia.
        h = fluid enthalpy in Btu/lb.

Where the two phases of a water/steam mixture are separated by means of a cyclone separator the water can be measured either by a cooled orifice (as in 8.5.3) or, after allowing for the flashed steam, by means of a V-notch or

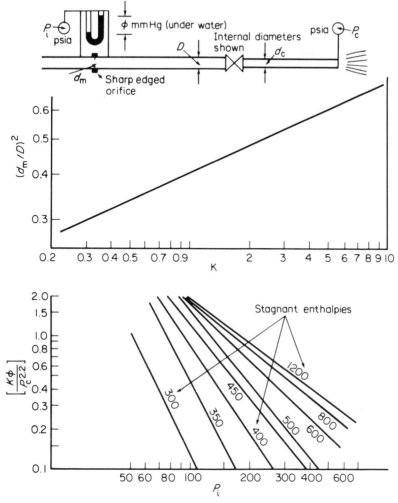

Figure 8.9 Derivation of enthalpy by means of orifice and critical lip pressure at sonic velocity discharge. (After Fig. 1 of [97])

Notes:

(i) d_c/D assumed to be 0.75 which gives convenient values of P_c. For other values of d_c/D use the correction formula $P_{c1}/P_{c2} = (d_{c2}/d_{c1})^{2.082}$

(ii) Recommended straight pipe lengths:
Upstream of orifice ≮ 25 D
Orifice to valve ≮ 10 D
Valve to discharge ≮ 25d_c

(iii) James original chart, giving pressures and enthalpies in British units is reproduced here. For conversions:
1 cal/g = 1.8 Btu/lb
1 psi = 0.068 03 atm.

weir. Alternatively, however, the water may simply be discharged at sonic velocity through an open-ended pipe and measured by applying the above formula. By measuring the temperature of the water leaving the separator its enthalpy is made known, and G can be deduced from P. It may be mentioned that the very existence of a positive gauge pressure for P is proof that the velocity is sonic or supersonic. If no such lip pressure is registered, then a smaller pipe must be used for the discharge until a lip pressure can be clearly measured. With water/steam mixtures the enthalpy of the mixture is *not* known; but the combination of an upstream orifice and a lip pressure gauge can provide all the information required for deducing the enthalpy and the mass flow in the manner shown in Fig. 8.9. The orifice factor K must first be determined by measurement, and the enthalpy then deduced from the lower graph after substituting the value K in the ordinate and observing the upstream pressure and the manometer pressure, as well as the critical lip pressure. This is a cheap method of measuring bore output characteristics, and is reasonably accurate within the idiosyncratic variations of a bore's performance from time to time.

8.5.5 *Method of cones.*   This is a relatively simple method of measuring the mass flow, quality and enthalpy characteristics of a wet bore by using a cyclone separator and a series of measuring cones of the type illustrated in Fig. 8.6. The method avoids the separate measurement of the water (though a lip pressure discharge pipe could provide useful corroborative evidence) and enables the dryness and enthalpy of the bore mixture to be deduced indirectly. It also dispenses with the need for long straight pipes required when a sharp-edged orifice is used for the steam measurement. The small sketch of Fig. 8.10 (a) shows a wet bore discharging directly into a cyclone separator after passing through a throttling valve V_1 which enables the wellhead pressure to be varied. The water is rejected (perhaps after corroborative measurement) and the steam discharge is connected to a measuring cone placed as closely as possible to the separator. Pressure measurements are taken just below the wellhead valve V_1 and also upstream of the cone. With various settings of the valve V_1 and with various sizes of cone, readings of both pressure-gauges are taken over a wide range of wellhead pressures. By using one of the cone formulae given in Fig. 8.6 – preferably BS 752 – the steam S_c discharged from the cone is deduced from the pressure P_c, and these steam flows are plotted against wellhead pressure in the manner shown in Fig. 8.10(b). However, none of these curves is the well steam yield characteristic because, owing to the pressure drop from P_w to P_c a fraction of the water phase will be flashed into steam, so that S_c will exceed the mass of steam issuing from the bore at the wellhead. The first task is to plot a graph of the pressure readings of both gauges (Fig. 8.10(c)) and to extrapolate the curves to meet the line of equal pressures. On that line,

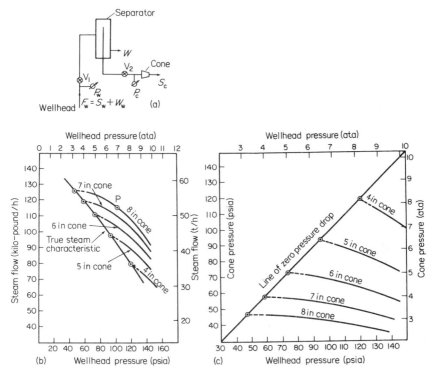

Figure 8.10 Measurement of well characteristics by means of separator and cones. (a) Test connections; (b) cone flow characteristics; (c) deduction by extrapolation of points on true steam characteristic.

where there is no pressure drop from the wellhead to the cone, there would be no flashing of the water. The points where the cone curves meet the line of equal pressure will therefore truly represent the well steam yield characteristic. These points are then transferred to Fig. 8.10(b) and joined up to form the steam characteristic of the well. To deduce the water characteristic a single example will illustrate the method. At a wellhead pressure of 100 psia the cone steam discharge (point P) is 116 500 lb/h, while the bore steam is 93 000 lb/h. The difference of 23 500 lb/h represents the steam flashed off from the water phase in dropping its pressure from 100 psia to 43 psia – the pressure at which an 8-in cone would discharge 116 500 lb/h of dry steam according to the BS 752 formula. The enthalpy of saturated hot water is 298.7 Btu/lb at 100 psia and 240.7 Btu/lb at 43 psia, while the latent heat at 43 psia is 931.2 Btu/lb. The proportion of water flashed will therefore be (298.7–240.7)/931.2, or 0.06228. Hence the water issuing from the wellhead will be 23 500/0.062 28, or 377 810 lb/h. Hence the mixture enthalpy at the wellhead can be deduced so:

$$
\begin{array}{lll}
\text{water heat} & 377810\times\ 297.7 = 112\,850\,000\ \text{Btu/h} \\
\text{steam heat} & \underline{93\,000\times1188.2 = 110\,500\,000\ \text{Btu/h}} \\
& \overline{470\,810\ \text{lb/h}} \qquad\qquad 223\,350\,000\ \text{Btu/h}
\end{array}
$$

$$
\text{Enthalpy} = \frac{223\,350\,000}{470\,810} = 474.4\ \text{Btu/lb}
$$

From this figure, with the help of steam tables, it is a simple matter to deduce the water content of the bore fluid at a wellhead pressure of 100 psia. The exercise may be repeated for several other points, thus enabling the enthalpy and water characteristics to be added to the steam characteristics. The accuracy of this method is limited by that of the cone formula used and by the efficiency of the separator, but the results should lie well within the vagaries of the bore itself. The method also ignores the very small pressure drop from the separator chamber to the cone entry, and if the separator were 100% efficient there could be a very small degree of superheat due to this small pressure drop if the pipe between the separator and the cone were lagged. The separator itself and the entry pipe should be lagged. If the water discharge is separately measured for corroborative evidence it would of course be necessary to add the quantity flashed, in order to arrive at the water quantity at the wellhead. As many cones may be used as are considered necessary to obtain consistent results, but if very small cones are used, care must be taken to avoid building up pressures in the separator higher than its design rating.

Cones can also be used for determining the characteristics of dry bores, without the intermediary of a separator, but thermometers must be used to ascertain the degree of superheat, so that a suitable correction may be applied to the cone discharge in accordance with the formula quoted in Fig. 8.6.

8.5.6 *Beta-rays and isotopes.*   Various suggestions have been made of how to determine the quality of fluid emitted from a wet bore by passing a stream of beta-rays through it and measuring the absorption rate, which would be a function of the fluid density. Unfortunately, as will be shown in Section 8.10, the velocities of the two fluids will not be the same, so the apparent and true densities will differ: also the result would be affected by the distribution of the water across the pipe cross-section. These problems are not easily solved. Alternatively, selective isotopes could be injected – one to be carried by the water and the other by the steam – so that selective measurement of radiation from each isotope at a chosen point over a fixed period would provide a means of determining the density which, in conjunction with the steam tables, would enable the enthalpy of the fluid mixture to be deduced. Some such method would enable a portable instrument to be taken from bore to bore for the analysis of flows without taking the bores out of service. Whether the obstacles are economic or technical, it would seem that no simple device has yet been produced, but the solution of the difficulties offers a challenge.

8.6 Heat capacity of a bore

Once the characteristics of a bore are known, it is a simple matter to deduce its heat capacity, provided that its application is known. An example will illustrate this point. Fig. 8.11 shows the flow characteristics of a hypothetical wet bore producing fluid at a constant enthalpy of 278 cal/g (500 Btu/lb). The figure is arbitrary, but not improbable. It will be seen that the total heat emitted by the bore is fairly constant at low pressures but declines rapidly at higher pressures. This feature follows from the shape of the total fluid flow curve and the constancy of the enthalpy. With rising pressure and temperature, the steam contributes less, and the hot water more, of the total heat. Fig. 8.11 is only illustrative: other bores would behave differently.

It will be noted that there are three heat-yield curves shown in Fig. 8.11. The upper one represents the heat yield at the wellhead, reckoned above 0° C, as taken from the steam tables. The middle curve also refers to the heat yield at the wellhead, but reckoned above an assumed local ambient temperature of river or lake water at about 19° C, or 67° F. It is, after all, only in the fact that the bore fluid exceeds some such temperature that it can be regarded as 'hot'. The lowest curve assumes a heat loss of 5% in transmission from the bore to the exploitation plant, and therefore represents a probable estimation of the *usable* heat. Now let it be assumed that this particular bore is to be used for a timber-drying plant which, according to Fig.2.1, needs heat at about 160° C; and let it further be assumed that a temperature loss of 3° C is incurred between the well and the plant. Fig. 8.11 shows that the particular well represented could supply about 59 million kgcal/h (net) at the required temperature. A paper factory, on the other hand, requiring heat at 180° C plus a further 3° C temperature drop from wellhead to plant, could obtain from the same well only about 52 Mkgcal/h at the required temperature.

If a typical Larderello type of dry bore were to be used, as illustrated in Fig. 8.12, it will be seen that the grade of available heat is higher at all wellhead pressures than required by either of the industries assumed in the above example for a wet bore. It would thus pay to operate the dry bore at the lowest practicable pressure – perhaps even sub-atmospheric – in order to obtain the maximum heat yield, and sometimes even to degrade that heat by dilution to the temperature required for the particular industrial application contemplated. With a 'Geysers' type of dry bore the temperatures are rather lower, though the steam flows are rather higher, than with a 'Larderello' type of bore. In either case, except for processes requiring the highest grades

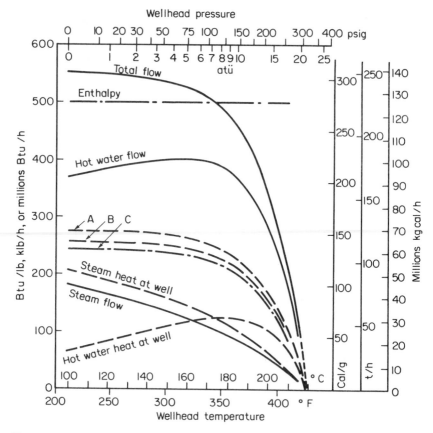

Figure 8.11 Typical heat flow pattern for a 'wet' bore, assuming a constant enthalpy of 500 Btu/lb or 278 cal/g.

Note: It is assumed that the heat both of the steam *and* of the water is used.

A, Total heat at well; B, net heat at well reckoned above 67°F (19.4°C) assumed ambient; C, usable heat assuming 5% transmission loss.

of heat, such as power generation, it would usually pay to operate the wells at the lowest practicable pressure (consistent with plant design requirements) at which heat yields are still rising with falling pressure. As with Fig. 8.11, three heat-yield curves are shown.

8.7 Power capacity of a bore

When it comes to the generation of electric power the situation is quite different from that of other heat-consuming industries, for the efficiency of a power plant declines with temperature owing to inescapable thermodynamic constraints. Thus the wellhead pressure of a bore at which

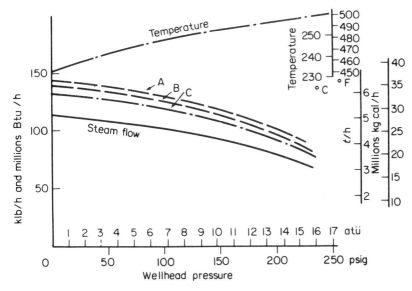

Figure 8.12 Typical heat flow pattern for a Larderello type 'dry' bore. A, Total heat at well; B, net heat at well reckoned above 67° F (19.4° C) assumed ambient; C, usable heat assuming 5% transmission loss.

maximum *power* potential is attained will not coincide with that at which maximum *heat* potential occurs. Rising wellhead pressure has two opposing effects: it reduces the steam yield but raises the extractable energy per kilogram of steam. The net result of these two opposing effects is to give a steam power potential curve that attains a maximum value at one particular wellhead pressure, falling off with either increasing or decreasing pressure on either side of that optimum. If consideration is also taken of the power extractable from the hot water phase of wet bores, a similar, though rather different, optimum pressure can be determined – despite the fact that over certain pressure ranges the hot water yield may sometimes *rise* with increased wellhead pressure (as in Fig. 8.11). Much will depend upon the cycle adopted, the fluid enthalpy and upon the turbine back-pressure. A typical pattern of power potential for a *wet* bore, supplying straight condensing and non-condensing turbines, is illustrated in Fig. 8.13. The use of better vacua would improve the power potential of the bore when supplying condensing turbines, but the gain would not necessarily be economically worthwhile because of the greater power required for gas exhaustion and the need for larger cooling facilities. (The power potentials of Fig. 8.13 are *gross,* and exclude deduction for auxiliary power consumption.) The question of single and double flashing will be covered in Chapter 10, but it may here simply be noted that by using the hot water and adopting two stages of flashing after the wellhead, in the example chosen for Fig. 8.13 a power gain of about 26% could be won by comparison with the use of steam alone in condensing turbines. With lower enthalpy fluids (i.e. having a higher proportion of hot water to steam) the proportional power gain could be substantially more.

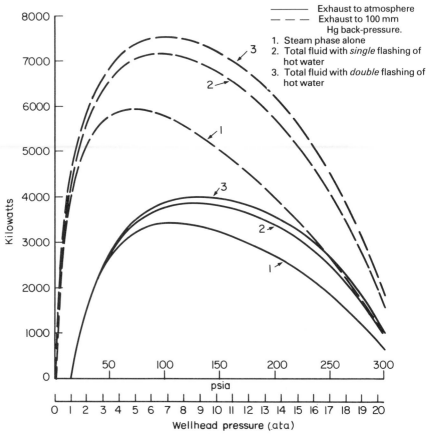

Figure 8.13 Power potential from typical 'wet' bore.
Note: The powers shown are at the generator terminals, without deduction for the power absorbed by gas exhausters or other auxiliaries.
Assumptions:
(i)   Same bore as for Fig. 8.11.
(ii) *Condensing* turbines sited at central power station fairly remote from bore. 25% of the absolute wellhead pressure and 6% of steam condensation assumed to be lost in steam transmission. In the case of flash steam a 2½% heat loss has been assumed in hot water transmission.
(iii) *Non-condensing* turbines sited fairly close to the bore. 10% of the absolute wellhead pressure and 1% of steam condensation assumed to be lost in steam transmission.

For a Larderello type of dry bore the pattern of power potential would be as illustrated in Fig. 8.14. The smaller steam yield, by comparison with the chosen example of wet bore, tends to be offset to a greater or lesser extent by the superheat, with the result that the steam phase alone of the chosen wet bore yields about 14% *less* power with non-condensing and about 5% *more* power with condensing plants at 100 mmHg back-pressure. If the hot water from the wet bore were also exploited for power by double flashing, both wells would yield about the same amount of maximum power when used with non-condensing plant. A Geysers type of dry bore (Fig. 8.5) could produce more power than either of the two bores considered above.

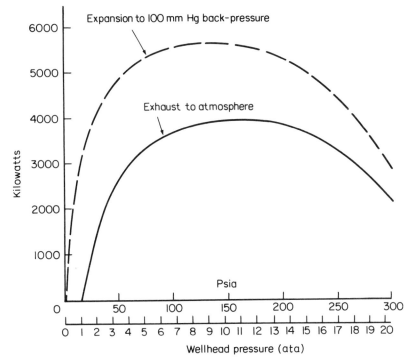

Figure 8.14 Power potential from Larderello type 'dry' bore.
Note: The powers shown are at the generator terminals, without deduction for the power absorbed by gas exhausters or other auxiliaries.
Assumptions:
(i) Same bore as for Fig. 8.12.
(ii) *Condensing* turbines sited at central power station fairly remote from bore. 25% of the absolute wellhead pressure and 5% of the heat content assumed to be lost in steam transmission.
(iii) *Non-condensing* turbines sited fairly close to bores. 10% of the absolute wellhead pressure and 1% of the heat content assumed to be lost in steam transmission.

It is emphasized that Figs 8.13 and 8.14 are examples only: every individual bore will have a different power potential curve, depending upon its mass flow and enthalpy characteristics and upon how the bore is to be used. But in every case there will be an optimum pressure at which the power potential will attain a maximum value. This optimum pressure, however, will not necessarily be the same as the *overall economic optimum pressure* (see Section 8.8).

James [98] has extended the notion of critical lip pressure (Section 8.5.4) to produce very simple, approximate, instantly usable formulae for estimating the electric power potential of a bore – wet or dry – by observing the pressure at the lip of a vertical open-ended pipe forming an extension to a blowing bore, as shown in Fig. 8.15. As already stated in Section 8.5.4, the lip pressure at sonic discharge velocity is a direct measure of heat flow per unit sectional area through a pipe; so by making reasonable assumptions as to pressure drops, overall conversion efficiencies and (for wet steam) the cycle to be adopted, it is possible to translate the lip pressure reading into electric power potential. The internal diameter of the discharge pipe must of course be known. James's formulae are based on a turbine entry pressure of

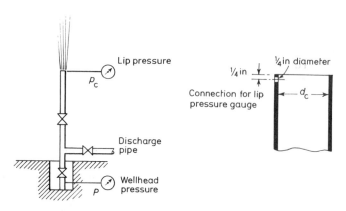

Figure 8.15 The James formulae for quick estimation of the electric power potential of a bore.

Formulae:

For wet bores (i) $\text{MWe} = p_c^{0.96} \left[\dfrac{d_c}{16.94} \right]^2$

(ii) $\text{MWe} = p_c^{0.96} \left[\dfrac{d_c}{11.92} \right]^2$

For dry bores (iii) $\text{MWe} = p_c^{0.96} \left[\dfrac{d_c}{14.48} \right]^2$

(iv) $\text{MWe} = p_c^{0.96} \left[\dfrac{d_c}{10.19} \right]^2$

For (i) and (iii)
$\begin{cases} p_c \text{ is in psia} \\ d_c \text{ is in inches} \end{cases}$

For (ii) and (iv)
$\begin{cases} p_c \text{ is in bars (abs.)} \\ d_c \text{ is in cm} \end{cases}$

50 psig and a wellhead pressure of 75 psig, and he recommends a high degree of throttling to attain the latter so as to keep in hand a reserve against future declining pressure: in fact he suggests testing at 175 psig at the wellhead, thus keeping 100 psi in hand. An accuracy of ±5% is claimed for wet bores of enthalpies from 400 to 600 Btu/lb (222 to 333 cal/g) and even better for dry bores: but these claims should be treated with reserve in view of the uncertainties as to pressure drop from well to turbine and of the conversion efficiency, especially the former. However, the method has the merit of cheapness and of speedy results. The method and the formulae could be modified to suit other assumptions as to turbine inlet pressure, pressure drop from well to turbine, wellhead pressure decline and conversion efficiency.

8.8 Economic optimum wellhead pressure for power production

Operating a bore at a pressure at which the power yield is a maximum will clearly save capital expenditure in drilling costs and in wellhead equipment and branch pipelines, in that fewer bores will be needed to produce a desired quantity of power. On the other hand there are other economic criteria to be considered. Low pressures imply high specific steam volumes and therefore bulky pipework and vessels. Low pressures will also tend to shorten the life of a finite reservoir of heat by drawing down the stored fluid too quickly. High pressures, on the other hand, involve the use of short turbine blades which are more conducive to blade gap losses than long blades and more vulnerable to the build-up of chemical deposits. High-pressure valves and fittings tend to be costlier, and the associated higher temperatures require thicker thermal insulation. The interplay of all these considerations has been carefully studied by James [99], who has concluded that for fluids of enthalpies exceeding 400 Btu/lb (about 220 cal/g) – and this covers most hyperthermal fields – the overall economic wellhead pressure both for dry *and* wet fields seems to lie in the neighbourhood of 6 ata. This comparatively precise figure is somewhat surprising in view of the fact that the optimum power yield pressures range from about 5 to 11 ata in Figs. 8.13 and 8.14. However, the curves in these figures are relatively 'flat-topped', and by choosing a wellhead pressure of about 6 ata the sacrifice in power from the theoretical optima would be within 3% for all condensing conditions. Only with atmospheric exhaust – rarely adopted for geothermal power generation – would the power sacrifice amount to about 11½% with a Larderello type bore, and much less in other cases. Hence there is plenty of scope for the other economic considerations.

8.9 Open discharge pipes

A certain amount of rough qualitative knowledge can be gained from observing a well freely discharging into the atmosphere through an open-ended pipe. Superheated steam will tend to emerge from the pipe in a parallel jet with a fairly long transparent gap between the pipe end and the point at which the steam starts to condense into an opaque stream of water droplets. After some distance the air friction dissipates the kinetic energy of the jet and the condensed steam forms a rising billowing cloud (Fig. 8.16(a)). Dry saturated steam produces a similar formation, but the transparent zone is quite short as the steam starts to condense very soon after leaving the pipe (Fig. 8.16(b)). Wet steam produces a characteristic 'tulip' discharge (Fig. 8.16(c) and Plate 9) due to the evaporative 'explosion' of the water particles as they flash as soon as the pressure falls to atmospheric, thus laterally distending the jet: there is no transparent zone. When a test cone is used, a slightly tulip-shaped formation is apt to occur, even with dry steam,

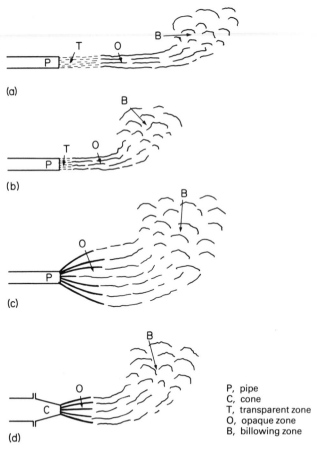

P, pipe
C, cone
T, transparent zone
O, opaque zone
B, billowing zone

Figure 8.16 Typical steam discharge patterns. (a) Superheated steam; (b) dry saturated steam; (c) wet steam, (d) dry steam from cone.

owing to the directional inertia of the fluid sliding along the tapered cone surface (Fig. 8.16(d)). Too much should not be read into these discharge patterns, as much will depend upon the humidity of the atmosphere and upon the wind; but very wet or highly superheated steam will usually betray their quality by the distinctiveness of the discharge jet pattern.

Plate 9 Typical 'tulip' discharge of wet bore fluid from open pipe at Cerro Prieto, Mexico [41]. (By courtesy of the Comisión Federal de Electricidad, Mexico)

8.10 Fluid velocities at wellheads

In a dry field, if the steam enters the lower uncased part of a bore at several different levels, the steam velocity will accelerate as it ascends from the bottom, partly because of the increasing number of feeding points and partly because of the falling pressure (and consequently the increasing specific volume) of the steam as it rises. At the point where the uncased zone ends and the steam enters the production casing, there may be an increase in the sectional flow area (see Fig. 7.2) which will cause a momentary check to the velocity; but after that the steam will again accelerate as it approaches the top of the bore. At the wellhead, it is a simple matter to calculate the average steam velocity if the mass flow, pressure and temperature are all known. From the last two the specific volume may be deduced and the mass flow converted to volumetric flow which, when divided by the cross-sectional area of the bore, will give the average steam velocity. Owing to skin friction, the velocity close to the bore walls will be somewhat less than in the middle of the bore. Obviously, for a given mass flow, the smaller the bore diameter the higher will be the steam velocity; and if this should be allowed to attain sonic velocity a maximum bore output will be imposed. The fact that the two bore characteristics shown in Fig. 8.5 are still rising at atmospheric wellhead pressure, with further pressure reduction, shows that subsonic flow still prevails (for the unspecified bore size used).

At Larderello most production casings are of 34 cm diameter – i.e. 0.0908 m^2 section. If a bore is normally operated at a wellhead pressure of, say, 9 ata (8 atü), the bore yield according to Fig. 8.5 would be about 44 t/h. At this pressure, at which the temperature is about 250°C, the specific volume of superheated steam would be 0.256 m^3/kg. Hence the mean steam velocity at the wellhead would be $(44\,000\times0.256)/(3600\times0.0908)$, or 34.46 m/s (113.1 ft/s).

In the case of a wet bore conditions are less simple. In the lower part of the bore the fluid would probably be hot water, possibly fed at different levels, and the upward velocity would increase as the fluid rises, with a check at the point of entry into the production casing. Owing to the high density of water, the velocities at depth would tend to be far lower than in a dry steam bore. But at some point in the bore, probably near the lower end of the production casing, flashing would develop at an increasing rate and thus reduce the specific gravity of the water/steam mixture, thereby imparting buoyancy to the fluid to assist the upward flow. The proportion of steam to water would steadily increase as the fluid rises, and the velocity would accelerate rapidly. The pressure drop over the last few metres before the wellhead valve will be considerable. At the wellhead the water phase will tend to cling to the bore walls while the steam phase will stay in the middle of the stream. Skin friction will thus ensure that the water velocity will always be appreciably less than the steam velocity. To take as an example the hypothetical wet bore of Fig. 8.11 and again assuming a working wellhead pressure of 9 ata, the bore yield would be 223 t/h, of which 177 t/h would be hot water and 46 t/h steam. At this pressure under saturated conditions the fluid specific volumes are 0.2124 and 0.00112 m³/kg for steam and water respectively, so that the *mean* fluid specific volume would be:

$$\frac{(177\times0.001\ 12)+(46\times0.2124)}{223}, \text{ or } 0.044\ 69 \text{ m}^3/\text{kg}.$$

With the same sized bore as assumed for Larderello – 34 cm diameter, 0.0908 m² – the *mean* fluid velocity at the wellhead would be:

$$\frac{223\,000\times0.044\ 69}{3600\times0.0908} = 30.43 \text{ m/s (99.83 ft/s)}.$$

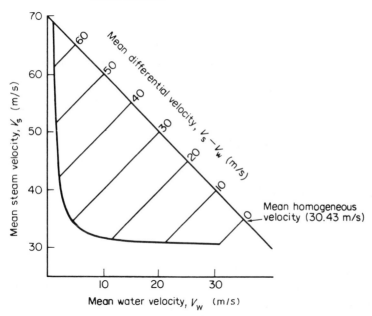

Figure 8.17 Relative fluid velocities at wellhead for typical 'wet' bore.
Assumptions: bore diameter = 34 cm (internal)
wellhead pressure = 9 ata
fluid flows { steam = 46 t/h
water = 177 t/h
total = 223 t/h
dryness factor = 20.63%
wetness factor = 79.37%
water/steam ratio = 3.85
enthalpy = 278 cal/g (500 Btu/lb)

The water velocity, however, is likely to be considerably lower, and the steam velocity higher than this mean value. There will even be velocity variations within each phase. The greater the mean velocity difference between the two fluids, the greater will be the proportional area of the bore section occupied by the water. The relationship between the mean velocities of each phase, for the chosen example, would be as shown in Fig. 8.17.

As already mentioned, if the fluid yield is sufficiently high and the bore diameter sufficiently small, sonic velocity will be attained at the wellhead and this will impose a lower limit to the wellhead pressure. If a wellhead valve be wide open so that the bore discharges freely to the atmosphere, the wellhead pressure will register atmospheric so long as the discharge is subsonic; but as soon as sonic value is reached a positive gauge pressure will start to build up at the wellhead. This will be equivalent to the 'lip pressure' referred to in Section 8.5.4. This means that with certain very powerful wells with slightly undersized bore diameters, as are to be found in the Cerro Prieto field in Mexico, it is impossible to reduce the wellhead pressure to atmospheric. Lower pressures above the wellhead valve could only be obtained by throttling, and increasing the pressure below the valve. Only the substitution of a larger diameter bore could enable atmospheric pressure to be attained at the wellhead when freely discharging with the valve fully open, but this would be unnecessary.

The total flow characteristic of the hypothetical wet bore of Fig. 8.11 is almost horizontal at atmospheric wellhead pressure, which shows that this bore, with the (unspecified) diameter in use, would just about attain critical flow – i.e. sonic velocity – when discharging freely to atmosphere.

The sonic velocity of a fluid is proportional to the square root of the elasticity/density ratio. With a perfect gas the elasticity is proportional to the absolute pressure. With wet steam mixtures the elasticity mainly depends upon that of the steam, while the density is very sensitive to the wetness. The sonic velocity of wet mixtures is therefore usually lower than that of dry steam at the same temperature, except at very low dryness factors when the high elasticity of water prevails and the sonic velocity rises rapidly towards that of water, which is very high.

9 Fluid collection and transmission

9.1 Wellhead gear

Except where pressurized or two-phase fluid transmission is adopted (see Sections 9.11 and 9.12) it is necessary for a quantity of equipment to be assembled at every wellhead in a wet hyperthermal field, to provide the means of controlling the fluids that emerge from the bore, of separating the water and grit from the steam, of disposing of unwanted fluids, of silencing the bore fluids when discharged to the atmosphere and of dissipating their energy, and of protecting the equipment and pipelines against excessive

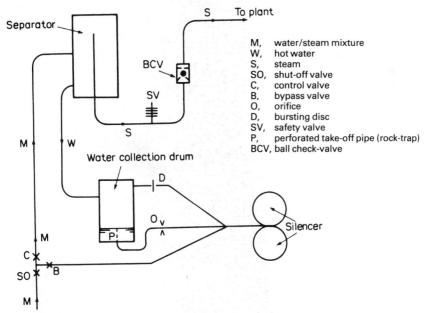

M,	water/steam mixture
W,	hot water
S,	steam
SO,	shut-off valve
C,	control valve
B,	bypass valve
O,	orifice
D,	bursting disc
SV,	safety valve
P,	perforated take-off pipe (rock-trap)
BCV,	ball check-valve

Figure 9.1 Diagrammatic arrangement of wellhead equipment and safety devices at a 'wet' bore. (After Fig. 5 of [101])

pressures. A typical assembly of wellhead equipment at such a bore is shown in Fig. 9.1 and Plate 10. The various components are connected together by means of pipework with conventional provision for thermal expansion – i.e. loops or bellows pieces – and most of them and their associated pipework will be thermally lagged. Where pressurized or two-phase transmission is adopted (see Sections 9.11 and 9.12) the wellhead equipment can be greatly simplified, as no separator, ball-check valve or water collection drum is required: some simplification is also possible at the wellheads in dry hyperthermal fields.

Plate 10 Typical wellhead gear assembly, Wairakei, New Zealand. Note the pipe expansion looping, the tall water/steam separator (centre), the water-collection drum (left centre), the grille-covered access steps into the cellar (left foreground) and the twin silencers (left background). Note also some trees (right) damaged by silicated spray (see Section 16.9). (By courtesy of the New Zealand Ministry of Works and Development)

9.2 Wellhead valving

Fig. 9.2(a) shows a typical wellhead valving arrangement as used at Wairakei. The valves are accommodated either in or just above the concrete cellar referred to in Section 7.2 and Fig. 7.1. The service valve is used to regulate the flow and pressure of the emergent fluids during testing, while

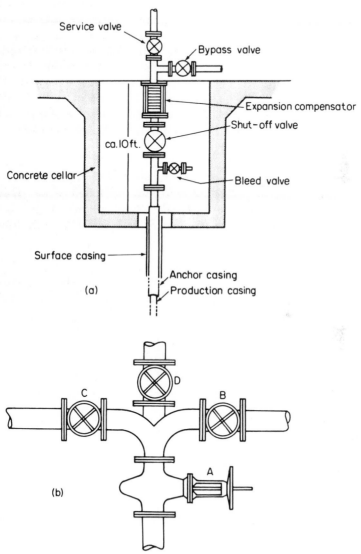

Fig. 9.2 Wellhead valving arrangements. (a) Typical wellhead valving arrangement at Wairakei (after Fig. 3 of [40]); (b) 'christmas-tree' valve assembly, commonly used in Italy, California and elsewhere (after Fig. 2 of [102]).

the shut-off valve enables the well to be isolated for maintenance purposes. A bleed valve permits the removal of non-condensible gases in a quenched bore, and a bypass valve permits the bore output to be discharged to waste. When a bore is taken out of service it is advisable always to bypass at least some of the fluid so as to keep the casings hot and thus avoid thermal shock due to alternating heating and cooling of the bore; and when a bore is in service the control and shut-off valves should normally be wide open so as to reduce wear and to eliminate the inefficient process of throttling; the control of pressure and flow is then exercised by the *system control* (see Chapter 13). An alternative valving arrangement, favoured in the dry fields of Larderello and the Geysers, is the 'Christmas-tree' assembly (Fig. 9.2(b)), where A is the shut-off valve, B the service valve, C a bypass control valve for discharging the steam to waste through a silencer when necessary, and D is to admit instruments or reamers to be lowered vertically into a bore.

9.3 Separators

The earliest crude separator tried out in a wet field was a simple 180° U-bend with a steam branch leading off from the inner wall of the downward discharging branch (Fig. 9.3). The water/steam mixture, in passing round the 180° bend, is subjected to very high centrifugal forces which throw the

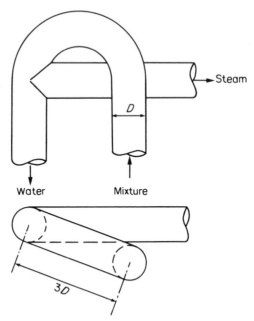

N.B. D is the internal diameter

Figure 9.3 Crude U-bend separator. (After Fig. 1 of [103])

water against the outer wall so that (theoretically) only dry steam can be withdrawn from the inner wall. If a mean water velocity of 20 m/s is taken (a not unreasonable figure according to Fig. 8.17), and a U-bend radius of 50 cm be assumed at the outer internal wall, the water acceleration would attain a value exceeding 80g. Under such conditions it might be thought that all the water would cling to the outer wall and that dry steam only would be drawn off from the inner wall, but owing to turbulence the device would remove only from 80 to 90% of the water. Some of the Wairakei bores yielded fluid containing as much as 8 parts of water to 1 of steam, so that with a U-bend efficiency of, say 85%, the steam lead-off pipe would still contain fluid of more than 50% wetness factor – a totally unacceptable figure. Various sophisticated separators have been devised, but one of the cheapest and most efficient is the Webre type [103], schematically illustrated in Fig. 9.4 and pictured in Plate 10 and capable of attaining an efficiency of 99.9% or more provided it is not over-loaded. The spiral inlet shown gives an improved performance over that of a simple tangential inlet that was tried earlier. At first, these cyclones were preceded (at Wairakei) by a U-bend so as to reduce the burden on the separator, but later it was found that this was not really necessary, so the water/steam mixture is now generally led through an expansion loop directly from the wellhead into the separator, thus greatly simplifying and cheapening the pipework.

Although the dimensional proportions shown for the Webre cyclone in Fig. 9.4 have usually been found satisfactory in several wet fields, it is of interest to record that at Cerro Prieto, Mexico, this was not so. Trouble was at first experienced there from the formation on the turbine blades of silica deposits that had to be removed periodically by sandblasting. The build-up of these deposits was reducing the effective power output by 5 or 6%, which could not be tolerated. Obviously the trouble was being caused by inefficient separation at the wellheads, resulting in the entry of unacceptable quantities of silica-saturated bore water into the steam mains. With long steam mains, as will be shown in Section 9.10, this might have been unimportant owing to the 'scrubbing' effect of pipe wall condensation; but at Cerro Prieto the wells are fairly close to the power station, so this effect is not very great. By raising the steam take-off pipe in the separators to a distance of only $D/4$ from the roof of the vessel (so that the cylindrical flow area for the steam became equal to the sectional area of the take-off pipe) the trouble was effectively overcome, though at the cost of a small increase in the pressure drop across the separators.

Another well-known separator, which avoids the unbalanced forces that operate in the Webre design, is the Peerless water/steam separator in which the mixture rises vertically and axially, impinging upon a domed plate that deflects most of the water downwards. Any residual droplets that may then be carried up by the rising steam pass through a labyrinthine mist-extractor

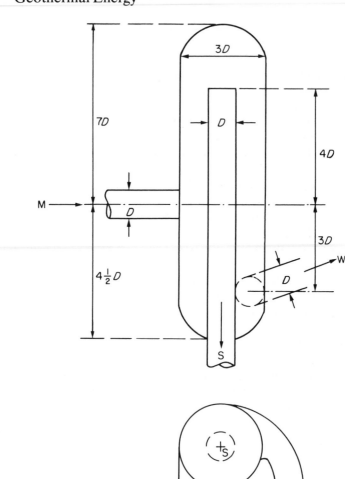

M,   Water/steam mixture at entry
W,   Water outlet
S,   Steam outlet

Figure 9.4 Schematic arrangement and recommended proportions of Webre type steam/water separator with spiral inlet. (After Figs 8 and 10 of [103])
Note: Internal diameters range from 76 to 137 cm according to the mass output of the bore.

before entering the steam outlet pipe. The efficiency of this design is about the same as that of the Webre design.

In dry fields the steam may contain a quantity of grit or dust, which is usually removed by means of a coaxial 'swirl' separator in the steam take-off pipe, the dust being discharged continuously through a 6 or 7 mm orifice permanently open to the atmosphere. The loss of steam through this orifice is negligible. Particles exceeding 10 μm in size can be eliminated in these dry separators. In wet fields fine pieces of rock and dust are washed away with the water discharge from the separators, but the occasional coarser fragments are trapped in the water collection drum by the perforated water take-off pipe (P in Fig. 9.1).

When considering separator performance it is essential to distinguish between mass wetness and volumetric wetness. Although this is obvious, the quantitative differences between the two figures can be surprising. For example, at a wellhead pressure of 8 ata, and with a water/steam mass ratio of 7:1, the *mass wetness* would be 87.5% while the *volumetric wetness* would be only 3.15%. This, of course, is due to the enormously differing specific volumes of the two fluids. At lower pressures the difference between the two figures would be greater, and at higher pressures it would be less.

Although the efficiency of the Webre type separator is very high under properly designed conditions, it deteriorates rapidly if the fluid inlet velocity is allowed to exceed about 50 m/s. The pressure drop absorbed by the separator is more or less proportional to the *volumetric* wetness at the inlet: except with very wet fluids it should be less than half an atmosphere.

9.4 Hot water discharge

Sometimes the hot water from a wet field is (regrettably) rejected to waste because the economics of extracting its inherent energy are not regarded as sufficiently attractive. Although the water/steam ratio is fairly steady at a constant pressure, it is nevertheless subject to short term fluctuations (gulping) and even to gradual long-term variations. The capacity of the water collection drum can take care of moderate short-term variations, but some cheap means of disposing of the hot water from this drum at the same average rate as that at which it is collected is necessary, if the water is to be discharged to waste. The use of float-controlled discharge valves would be costly and could sometimes be mechanically troublesome. So the problem has been solved in various wet fields simply by discharging the hot water through suitably sized bell-mouthed orifices. Such orifices, when passing boiling water, possess the convenient property of enabling a wide range of water flows (about 3:1 typically) to be discharged without upstream flooding on the one hand and without loss of seal on the other hand. The explanation of this phenomenon has been given in simple terms by Armstead and Shaw [101] and in more sophisticated terms in some of the bibliography references given by them. If the hot water is to be exploited for some useful purpose, it should either be pumped, or removed by gravity if sufficient fall is available, from the water collection drum to the exploitation plant.

Plate 11 Wairakei steam field, New Zealand. Discharge of superheated water through twin silencers. Despite the dramatic impact of this picture it represents waste and pollution. The steam is flashed off from high-pressure hot water on reduction to atmospheric pressure. (By courtesy of the New Zealand Ministry of Works and Development)

9.5 Silencers

When a bore is bypassed to waste, the noise can be ear-splitting and can even cause deafness to people subjected to it for too long without adequate protection. At Wairakei, even when the steam phase of a bore is being transmitted to the power plant, the hot water is normally rejected to waste, and the steam that flashes off from this hot water may sometimes be as much as, or more than, the useful steam (in mass). When the whole of a bore is being bypassed to waste, the volume of discharged fluids can be enormous (see Plate 11). Thus there is need at Wairakei for silencers at all times, whether the bores be in service or bypassed. The same applies to other wet fields where the hot water is rejected to waste. Even where the hot water from a wet field is usefully exploited, and also for dry bores, it is sometimes necessary to bypass to waste the full fluid output during emergencies or maintenance periods. Hence it is essential to provide silencers at *all* geothermal bores, even though in some cases they may be needed only occasionally.

A simple design of silencer, devised some years ago in New Zealand, is used at Wairakei and has found favour in other fields also [104]. It is diagrammatically illustrated in Fig. 9.5. The fluid impinges upon a cusped

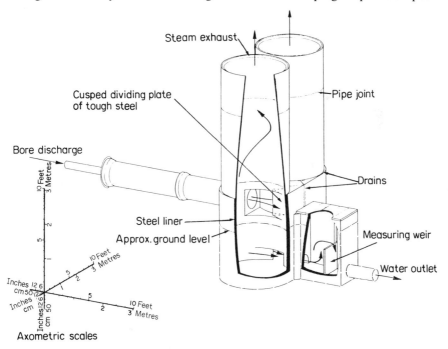

Figure 9.5 Cut-away axometric sketch of twin cyclone silencer as used at Wairakei, New Zealand. (After Fig. 6 of [104])

Plate 12 Typical branch steam line, Wairakei, New Zealand [90]. Note the stiff right-angled expansion loops, the concrete 'apron' to prevent erosion of the pumice sloping ground surface, and the twin silencers from a bore at the top of the slope. (By courtesy of the New Zealand Ministry of Works and Development)

Plate 13 General view of bore field, Wairakei, New Zealand. Note the three lagged 30-in steam pipes on the left and the five 20-in steam pipes to the right. On the extreme right is the 19-in experimental hot water transmission pipe, also lagged. To the far right can be seen one of the control vent-valve houses discharging a moderate amount of steam, while in the background can be seen several well-head silencers, steam pipe loops and branch steam lines [90]. (By courtesy of the New Zealand Ministry of Works and Development)

dividing plate of toughened steel at the tangential junction of twin cylindrical towers. The water swirls round the cylinders, losing its kinetic energy in friction, and is passed out over a measuring weir to a waste channel. (When using this weir for measurement purposes it is of course necessary to make due allowance for flashed steam in order to arrive at the quantity of water issuing from a wet bore). The steam escapes to the air from the top of the cylinders. This type of silencer deflects most of the noise skywards and, more important, lowers the pitch of the noise from a high frequency scream to a tolerable deep-noted roar. Silencers of this type can be seen in Plates 10–15. The cylindrical silencer towers were originally made of reinforced concrete, but some of them failed in service from the combined effects of high temperature, erosion and mechanical vibration. A far more durable material for this duty, already proved capable of giving up to 12 years service, is treated timber. *Radiata* pine, treated with 5% pentachlor-

Plate 14 Vertical steam pipe expansion loop, allowing vehicle access beneath, Wairakei, New Zealand [90]. See Section 9.8 and Fig. 9.6(a). Note the twin silencers in the background. (By courtesy of the New Zealand Ministry of Works and Development)

Plate 15 Aerial view of the Wairakei geothermal development, New Zealand. Note the bore field in the middle distance, the steam transmission pipelines, the A and B power stations, the cooling water pump-house, the Waikato River in the foreground, the cooling water outfall (right), the step-up substation and high voltage transmission lines (behind the B station) and the Wairakei Hotel (left). The equipment between the A and B buildings is the flash vessels and scrubbers used for the experimental hot water transmission scheme (see Sections 9.9 and 13.3; see also Fig. 13.3). (By courtesy of the New Zealand Ministry of Works and Development)

phenol (by weight) in Shell industrial oil No. 4 has been found very satisfactory.

In Iceland a rock-filled silencer in a steel container has been successfully used.

9.6 Safety devices

The working pressure of a steam collection system will usually be a fraction of the shut-in pressure of the bore, and although the bores and wellhead valves will be designed to withstand that shut-in pressure, all equipment beyond will be designed for a lower pressure than the bore and its valving. A safety-valve should therefore be provided at each wellhead as a statutory requirement. But as the lifting of a safety-valve is an occurrence to be avoided, except as a last resort (because of the erosion damage to seatings and because of the difficulty in getting them to re-seat without appreciably lowering the pressure), a cheaper safety device is also often provided at each wellhead in the form of a bursting disc. Such discs can be easily replaced and are relatively cheap, but as they are not usually accepted by Boiler Inspectors they have to be backed up by conventional spring- or weight-loaded safety-valves (see Fig. 9.1). Conventional safety-valves have the advantages that the lifting pressure can be varied at will and set precisely, and that they will re-seat when the pressure has fallen sufficiently; but they have the disadvantages of high cost and of the difficulty of maintaining an absolutely tight seal. In geothermal conditions this may result in the precipitation of dissolved solids in the area around the seatings, which could necessitate increased maintenance and sometimes even prevent the valve from discharging its full rated output. A further disadvantage of the safety-valve is that its capacity when passing a steam/water mixture cannot be precisely known. Bursting discs, on the other hand, are cheap, simple and require no maintenance. They have the disadvantages that they cannot be set precisely, that their bursting pressure depends upon temperature, that once burst the well cannot be used until the maintenance staff have replaced the broken disc with a new one, and that the material of the disc can be weakened by age, vibration and chemical attack, so that a disc may burst at a lower pressure than intended. If this should occur during normal operation, the system may become disturbed. A further safety device is sometimes provided at wet wells in the form of a steel ball-float which lifts to shut off the steam supply to the steam collection system in the event of a separator flooding or failing to perform its duty properly. The entry of a gulp of water into a steam main, and perhaps ultimately reaching a turbine, could be disastrous; hence the need for the ball check-valve. The lifting of the ball against its upper seating would tend to induce the shut-in well pressure in all the wellhead equipment; hence the bursting disc and safety-valve should be placed between the top of the bore and the ball-valve at the entry to the steam collection system, as in Fig. 9.1. Normally, the disc will burst on the lifting of the ball and the entire bore fluid will be discharged to waste through the silencer. But if it should fail to act, the safety-valve will lift and relieve the pressure in the wellhead equipment.

9.7 Steam branch pipes (see Plate 12)

From each well a steam branch pipe, sized to the output of the well and to allow for the pressure-drop over its length, will conduct the almost dry steam from the wellhead to the nearest convenient point in the steam mains that carry the steam from the field to the power station or other utilization plant. These branch pipes must be anchored at certain points by means of solid ground supports, and provision must be made for thermal expansion and contraction between anchor points either by 'zig-zagging' or 'snaking' the pipes – usually in a horizontal plane, though sometimes by means of vertical loops where road access must be provided from one side of the branch pipe to the other. Pipe supports between anchors may take the form of simple sliding pads on concrete blocks, of flexible upright metal rods capable of bending one way or another to take up pipe movement, or of suspension rods hanging from light portal or 'gallows' frames.

9.8 Steam mains (Plate 13)

The carrying capacity of the steam mains will normally increase as they progress from the remoter parts of the bore field towards the exploitation plant, gathering additional steam in the process. The increased capacity may simply be provided by enlarging the diameter of the pipe(s), or alternatively by adding more pipes in parallel. Wherever there is a change in the number of pipes it is convenient to provide a valved manifold to enable individual pipes to be isolated for servicing. All such manifolds must be at anchor points. Junctions with well branch-pipes should also be at anchor points, where the pipe movement is zero; and anchors must of course also be provided wherever the pipeline changes direction, so as to take up the large lateral components of the expansion forces. Compensation for thermal expansion and contraction can be provided in a number of ways:

9.8.1 *Zig-zagging or lyre bends.*   These methods are suitable for pipes of moderate diameter. As with branch pipes adequate anchoring is necessary, and suitable sliding or suspension supports must be provided between anchors.

9.8.2 *Axial bellows pieces.*   These, of the type shown on a wellhead in Fig. 9.2(a), transmit high compressive or tensile forces to the anchors, which must in consequence be sufficiently robust to resist these forces. They provide a rather expensive means of compensation, and are therefore used only where space restrictions prohibit the use of any form of looping – e.g. within a power house building, or in a hot water transmission pipe where vertical looping (for reasons that will be explained in Section 9.9) cannot be adopted. Internal metal sleeves greatly reduce steam friction in corrugated axial bellows.

9.8.3 *Hinged bellows pieces.* These are capable of angular flexure without net axial yield. They transmit only minimal axial thrusts when used in three-pin arch loops, as shown in Fig. 9.6(a). Loops of this kind are commonly adopted at Wairakei (Plate 14), where they provide the additional asset of road access to motor vehicles beneath. The loops are supported by means of light steel frames. Each loop is capable of absorbing at least 0.75 m of pipe movement, and at Wairakei they are spaced at intervals of about 300 m or so. The relatively high cost of hinged bellows pieces tends to be offset by the lightness and wide spacing of anchors. The straight pipe lengths between loops and anchors are normally supported on rollers carried on concrete pads let into the ground. Welded onto the pipes are short steel contact pieces long enough to penetrate the lagging so as to present a metal surface to the rollers and so prevent damage to the lagging when pipe movement occurs.

9.8.4 *Articulated bellows pieces.* These are fairly long pieces of pipe provided with flexible corrugations and with flanges at either end. The flanges are interconnected by means of a pair of long hinged rods which permit the assembly to flex sideways while keeping the flanges parallel to one another. Such units, in conjunction with hinged bellows pieces can absorb complicated expansion movements such as occur where a pipeline passes down a steep hillside (see Fig. 9.6(b) and Plate 16).

9.8.5 *Slip joints.* Though extremely simple in conception these are not often entirely satisfactory, as they need careful maintenance and are vulnerable to chemical deposition. They consist simply of one tube sliding concentrically within another and provided with a stuffing box and gland packing.

Different designers of geothermal installations favour different solutions to the problem of pipeline expansion. In New Zealand, zig-zagging is generally used for branch lines while hinged and articulated bellows pieces are commonly used for steam mains. In the dry fields of Larderello and the Geysers, bellows pieces are also used to a limited degree, but greater reliance is placed upon zig-zagging and upon vertical loops. With all forms of expansion compensation it is usual to take up half of the pipe expansion movement when the system is cold, so that the initial pipe stresses are more or less equal (but reversed) to the final stresses when the system is hot.

Where necessary, intermediate supports may be provided with lateral or overhead steel guides to prevent possible buckling of pipes under stress; this can apply both to branch lines and to mains. A most erudite analysis of pipeline construction problems and of stressing under thermal changes has been prepared by Pollastri [105].

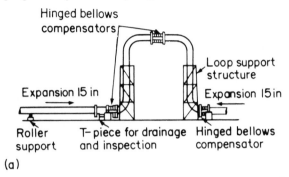

(a)

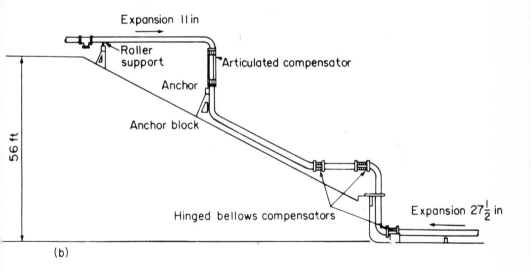

(b)

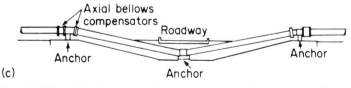

(c)

Figure 9.6 Use of bellows expansion compensators of various types for steam and hot water transmission at Wairakei, New Zealand. (a) Steam line at loop; (b) steam line at bluff; (c) hot water line under road. (After Fig. 7 of [40])

Plate 16 Steam transmission over difficult terrain, Wairakei, New Zealand. Note the variety of expansion arrangements where the steam pipes descend a steep bluff (see Section 9.8 and Fig. 9.6(b)). (By courtesy of the New Zealand Ministry of Works and Development)

As already mentioned, mild steel – fortunately a cheap material – is adequate for pipelines and pressure vessels filled with geothermal fluids. This is because of the virtual immunity of this metal against H_2S attack in the absence of oxygen – which is never present in bore fluids except in combination with other elements (as in CO_2). The immunity derives from a self-protective skin of iron sulphide which rapidly coats the inside walls (see Section 15.3.4). In the early days at Wairakei, in order to minimize the amount of welding (which could possibly be a seat of chemical or physical weakness) *seamless* Mannesmann tubes were used for the main pipelines. These, however, were expensive; and after it had been clearly shown that welding, if very carefully performed, constituted no danger, it was decided thereafter to use *seam-welded* tubes for the steam mains, though seamless tubes continued to be used on branch lines. Tubes of this type are now almost universally used for steam mains in all geothermal installations throughout the world. Seam-welded tubes are normally machine-bent to a cylindrical shape, and the edges are machined to a suitable profile and held securely together after bending. The joint is then welded longitudinally along the pipe. The seam welding may most cheaply be done at the site, as this enables the steel to be shipped in compact plate form instead of in cylinders, The longitudinal seam welds may be done manually or automatically, and after tubes of convenient length have been produced they are mounted in position and joined together by means of manual circumferential welds to form long sections of pipeline. Another form of pipe sometimes used is the *spiral-seamed* pipe, formed by winding long strips of fairly narrow steel plate helically, so that one edge abuts the other. The strip edges have been previously shaped, and the spiral butt joint is then welded to form a solid tube. Carefully prepared welds can possess almost the same strength as the plate itself. Welding should be undertaken only by highly skilled and qualified welders who are periodically subjected to exacting tests to ensure that their skill has not deteriorated. Welds should be X-ray tested (50% of longitudinal and 100% of circumferential were so tested at Wairakei), and sample test pieces should be cut off for physical testing. With thick plates of more than ½ to ¾-in thickness (say 12.5 to 19 mm) it is necessary to relieve the welding stresses by heat treatment. Bolted flanges, screwed and subsequently welded circumferentially, are obligatory wherever dismantling may be needed for maintenance purposes; elsewhere consecutive pipe lengths are circumferentially welded together – this process being known as 'stove-pipe' welding. At Wairakei, the steel plate used for the welded steam mains is to BS1501-151B, and the finished tube is to BS806 standard. Other countries may call for different standards.

9.9 Hot water transmission

Although the transmission of moderately hot water for district heating and other purposes (see Chapter 11) at temperatures near or below atmospheric boiling point is commonly practised in Iceland and elsewhere, only once (hitherto) has the transmission of very hot geothermal water at temperatures of 200°C or more been attempted over a distance of at least 1.5 km. This was for an experimental installation at Wairakei for the extraction of power from hot water that would otherwise have been wasted. The theoretical aspects of extracting power from hot water will be dealt with in

Chapter 10, but the *transmission* of this water raises special problems that fall within the scope of the present chapter.

The primary problem of transmitting very hot water is to prevent boiling during transmission, as the formation of steam pockets and their subsequent collapse could give rise to very high and unpredictable stresses which might cause a burst pipe. It is necessary to recognize that superheated water (i.e. above 100°C and pressurized) is an explosive fluid many times as dangerous as steam at the same temperature. For example, although hot water at 200°C has an isentropic heat drop to atmospheric pressure of only about 11% of that of steam at the same temperature, its density is 110 times as great. Hence a pipe or vessel containing water at 200°C has about 12 times the potential for doing mechanical work as the same pipe or vessel containing steam at 200°C; and this may be taken as a measure of its explosive capacity. A burst pipe transmitting very hot water could have disastrous consequences. The only way of preventing boiling in a hot water pipe is to ensure that at all points the *hydraulic* pressure exceeds the *vapour* pressure. The hydraulic pressure is a complex function of the initial vapour pressure at the point of collection, pipe friction, pipe gradients and dynamic heads arising from changes of water velocity due to valve movements. In the Wairakei scheme a safe margin between hydraulic and vapour pressure was assured by a combination of pumping, attemperation (by injecting lower temperature water from second grade wells) and of carefully controlled speeds of valve operation. A full description of this scheme may be found in the bibliography [106]: technically it was entirely successful, but unfortunately it had to be abandoned after about one year of operation because the wells selected for the experiment, which for obvious reasons were those closest to the power plant, proved to be near the edge of the field – a fact not suspected when the project was launched – and their output of hot water dried up. The hot water pipeline was subsequently converted into a subsidiary steam line. Owing to a fear that the whole field might ultimately yield dry steam, the hot water transmission scheme was not extended further into the heart of the bore field where there was (and still is) a large quantity of hot water available, with the result that several tens of megawatts of potential power from that hot water has been going to waste ever since the Wairakei installation was first initiated late in the 1950s. (This at least was the situation late in 1980; but at that time reinjection tests had been in progress for some time, and the possibility was then under consideration of reinjecting the spent bore water at Wairakei in the light of the experience gained from the tests.) Prevention of boiling in the pipeline could, of course, be ensured by pumping alone, so long as a generous pressure margin be provided over and above the vapour pressure; but this might be needlessly extravagant in pumping energy. The combination of pumping and attemperation, supported by a carefully designed control system, ensured that at Wairakei the pumping energy amounted to no more than about 1/200th part of the power potential of the transmitted hot water.

Thermal expansion of the hot water mains in the experimental scheme was taken up by means of internally sleeved axial bellows pieces so as to reduce pipe friction to a minimum and so as to avoid the unsparable loss of hydraulic head that would result from the adoption of vertical expansion loops (of about 6 m in height) as with the steam mains. At road crossings, therefore, the pipes were led beneath the road (Fig. 9.6(c)). Fairly heavy anchors had to be provided to take the high axial thrusts.

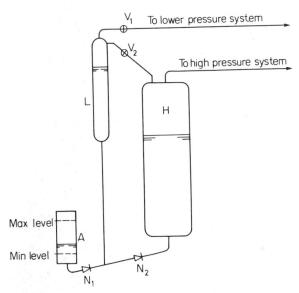

Fig. 9.7 The author's proposal for vapour-pumping (or 'pumpless' pumping) for imparting static head to a hot water transmission system. (After Fig. 5 of [106])

It has been mentioned that attemperation formed a feature of the experimental hot water transmission scheme at Wairakei. This was achieved by taking advantage of the fact that in that field there were two bore systems operating at different pressures and temperatures [40, 106]. The hot water from the higher pressure system was lifted by low head pumps into a head tank (Plate 17) about 12 or 13 m above ground level so as to provide the necessary surplus pressure. This head of water 'floated' on the upper end of the hot water transmission pipe (see Fig. 13.3 later). Hot water from the lower pressure bores was injected into the pipeline by means of high head pumps. The injected water reduced the mean temperature of the transmitted water and thus lowered the vapour pressure, at the same time making it possible to recover the energy of some of the lower temperature water, the

Plate 17 Head tank for experimental hot water transmission scheme, Wairakei, New Zealand. See sections 9.9 and 14.3; see also Fig. 13.3. (By courtesy of the New Zealand Ministry of Works and Development)

separate transmission of which would have proved uneconomic. As an alternative to the use of pumps, with their attendant problems of maintenance and of using special materials, the author devised a system of vapour pumping, or, to use a more graphic if less scientific term, 'pumpless pumping', which would have worked on the principle illustrated in Fig. 9.7 and described more fully in the references [106]. Although the system nominally relied upon the dual pressure system of Wairakei, it could equally be introduced into a single pressure system at reduced efficiency simply by using the atmosphere as the lower pressure. As with the actual Wairakei system a collection tank, A, is provided for each (high pressure) well and a common head tank, H, would serve a whole group of wells. Associated with each collection tank is a lift cylinder, L, connected as shown in Fig. 9.7. The water in A is allowed to rise and fall alternately between pre-designed maximum and minimum levels. When it rises to its maximum height, a level-sensitive device opens a valve, V_1, thus connecting the associated lift cylinder to the lower pressure system (valve V_2 at this time being closed). The pressure difference between the two systems impels the water from A into L, passing on its way through a non-return valve, N_1. As soon as the water in A falls to the prescribed minimum level, a signal causes valve V_1 to close and V_2 to open. This equalizes the pressure between L and H (being that of the high pressure system) and the water drains from L into H. Meanwhile A starts to refill and the cycle is repeated. The net effect of each cycle is to raise into the head tank, without the use of pumps, a quantity of high pressure water equal to the capacity of A between maximum and minimum levels. A similar cycle in meanwhile operating with other high pressure collection tanks and their associated lift cylinders. Since the operation of the various control valves is actuated solely by the water levels in the collection tanks, there is nothing to prevent a group of wells thus being connected to the same lift system even if they are at different ground levels. (With pumps, differences of ground level would require different pump characteristics for each well.) Standard collection tanks would be provided for each well: those wells having a high water yield would operate at a higher cycle frequency than those of low yield. Although there are certain secondary problems which need not here be described, 'pumpless pumping' would have the advantage of extreme mechanical simplicity. The lifting of water in this way would be effected at the cost of some high pressure steam and of some condensation on the water surfaces of the lift cylinders when the equalizing valves V_2, are opened. These losses provide the necessary lifting energy, but the temperature reduction in the lifting cylinder would provide useful attemperation. Although the efficiency of the process is only about one-third of that obtainable by mechanical pumping, the energy is still quite small – about $1\frac{1}{2}\%$ of the transmitted energy as against $\frac{1}{2}\%$ at Wairakei for lifting high pressure water.

The method of controlling the rate of movement of the valve admitting the hot water to the plant – the third ingredient principle of hot water transmission – will be described in Chapter 13.

It is possible that some future projects in wet fields may require the transmission of very hot water over fairly long distances, and that the principles briefly described here may be put into practice. However, it is also possible that the adoption of two-phase transmission or of submersible pumps (see Sections 9.11 and 9.12) may relegate hot water transmission, as such, to the level of historic and academic interest only, unless perhaps for non-power uses.

9.10 Thermal lagging: condensation in steam pipes and heat losses

During its passage from the wellheads to the exploitation plant, both steam and hot water will lose a certain amount of heat. All vessels and pipes, other than waste pipes and silencers should therefore be lagged with any of the thermal insulating materials (e.g. magnesia) commonly used in power station practice, or with other materials such as vermiculite which may be locally available, cheap and effective. The cost of lagging can be very high, and it is necessary to strike an economic balance between its cost and the value of the heat saved. In dry fields, which usually deliver superheated steam at the wells, it is advisable to ensure that the temperature does not fall during transmission to condensation point. With hot water transmission, quite a modest amount of lagging can ensure a temperature drop of much less than 1°C/km. In fact Bodvarsson quotes a 14-in pipeline in Iceland transmitting hot water over 15.3 km at 87°C (sending end) with a loss of only 3°C, or about 0.2°C/km. With dry saturated steam, to which the steam leaving the wellhead separators in wet fields approximates, it would probably not be economic to reduce condensation on the pipe walls to less than about 5–7% during transmission, depending upon the length of the pipelines. Moreover, a certain amount of condensation is useful as a 'scrubbing device' for purifying the steam in transit. Even the most efficient of wellhead separators will let through small quantities of bore water with the steam, and these waters are likely to contain chlorides which, even in very small proportions, in the presence of hydrogen sulphide (almost invariably a constituent of geothermal steam) can do great damage to turbine blades. The scrubbing is effected by providing downward facing T-pieces at intervals of perhaps 150 m with drainage facilities (see Fig. 9.6(a)). Condensed steam, forming on the pipe walls, will mostly tend to collect along the bottom of the pipe, so that at each T-piece much of the water will fall into the downward pointing branch, from which it will be drained away. Originally, mechanical traps were provided to open intermittently whenever sufficient water had collected. More recently, at some geothermal installations, it has been decided that a simple open orifice is cheaper, entails no greater fluid loss and is more reliable. The scrubbing process, which may be likened to rinsing, results from the repeated dilution of the condensate by further condensation, and its repeated partial removal through the drains. The degree of purification attainable can be astonishing, as will be illustrated by a hypothetical example.

Table 9.1 Scrubbing effect of imperfectly lagged pipeline carrying slightly wet saturated steam. (S is the salt concentration in ppm at the exit from the wellhead separator.)

Location		Steam (kg/h)	Water (kg/h)	Total salts (kg/h $\times 10^{-6}$)	Salt concentration (ppm)
At exit from wellhead separator		20 000	100	100.S	S
1st trap	entry	19 900	200	100.S	0.5.S
	exit	19 900	50	25.S	0.5.S
2nd trap	entry	19 800	150	25.S	0.166 6667.S
	exit	19 800	37.5	6.25.S	0.166 6667.S
3rd trap	entry	19 700	137.5	6.25.S	0.045 4545.S
	exit	19 700	34.375	1.5625.S	0.045 4545.S
4th trap	entry	19 600	134.375	1.5625.S	0.011 628.S
	exit	19 600	33.594	0.3906.S	0.011 628.S
5th trap	entry	19 500	133.594	0.3906.S	0.002 924.S
	exit	19 500	33.398	0.0977.S	0.002 924.S
6th trap	entry	19 400	133.398	0.0977.S	0.000 732.S
	exit	19 400	33.349	0.0244.S	0.000 723.S
7th trap	entry	19 300	133.349	0.0244.S	0.000 183.S
	exit	19 300	33.337	0.0061.S	0.000 183.S
8th trap	entry	19 200	133.337	0.0061.S	0.000 046.S
	exit	19 200	33.334	0.001 53.S	0.000 046.S
9th trap	entry	19 100	133.334	0.001 53.S	0.000 011.S
	exit	19 100	33.333	0.000 38.S	0.000 011.S
10th trap	entry	19 000	133.333	0.000 38.S	0.000 003.S
	exit	19 000	33.333	0.000 09.S	0.000 003.S

Total condensation $= \dfrac{20\,000 - 19\,000}{20\,000} = 5\%$

Initial wetness $\dfrac{100}{20\,100} = 0.498\%$. Final wetness $= \dfrac{33.333}{19\,033} = 0.175\%$

Salt reduction $\dfrac{100}{0.000\,09} = 1.11 \times 10^{6}$ times

Salt concentration reduction $\dfrac{1}{0.000\,003} = 333\,333$ times

Let it be assumed that the fluid issuing from a wellhead separator consists of 20 t/h of steam and 100 kg/h of hot water. This implies a wetness factor of nearly 0.5% – a very cautious assumption. Let it further be assumed that drains and traps are spaced at intervals of 150 m and that each is 75% efficient – i.e. that three-quarters of the accumulated water in the pipeline is removed at each drain and that one-quarter remains behind to be carried along with the steam. Finally let it be assumed that 100 kg/h (or 0.5% of the initial dry steam flow) is condensed in each 150 m section of pipeline between traps. The figures of Table 9.1 clearly reveal the enormous degree of purification that can result from the scrubbing effect of repeated dilution and partial draining. It will be seen that even after the fourth drain (600 m) 99.61% (100–0.39) of the solubles have been removed, while after the tenth drain (1500 m) less than one-millionth part of the original solubles remain in the pipeline. At Wairakei, the chloride content of the bore water may sometimes be as high as 3500 ppm, so that after the fifth trap the chloride concentration would only slightly exceed 10 ppm, while after the tenth trap it would have fallen to 0.0105 ppm only. Even the first figure would just be about acceptable to turbines, while the second figure virtually implies distilled water. It will also be seen from Table 9.1 that the condensation loss in 1500 m would be 5% – a fairly realistic figure – and that the steam enters the plant in a *drier* condition than it leaves the wellhead separator. (This latter feature would not be true, however, if the separator were much more efficient than has been assumed in this example). The figures in Table 9.1 are, of course, illustrative only, and a lower removal efficiency at the drain points would reduce the degree of purification considerably. The following points may be noted from them:

(i) The total residual salts are a function of the number and efficiency of the traps: they are independent of the rate of condensation in the pipeline.
(ii) The salt concentration in the water phase *does* depend on the rate of condensation in the pipeline.
(iii) The residual wetness depends upon the rate of condensation in the pipeline.

A problem may sometimes arise where wells are situated within a short distance of the power station, and the length of pipework between them is not enough to ensure the efficient removal of most of the salts carried over from the bores. (The example of Cerro Prieto has been cited in Section 9.3.) It might at first be thought that a reduction in the thickness of the pipe lagging would provide the answer by increasing the quantity of diluting condensation and at the same time cheapening the lagging costs; but point (i) above shows that this would be ineffectual in reducing the total amount of salts entering the turbines, though according to point (ii) the salt concentration in the droplets would indeed be improved. Spacing the drain pots more closely to one another would also be ineffective because much turbulence is set up at each pot and it requires a long stretch of downstream pipework for the flow pattern to quieten sufficiently to ensure that most of the condensate travels in a rivulet along the bottom of the pipe. Without this rivulet the next downstream drain pot would fail to remove much liquid. In theory the pipework could be artificially lengthened by following a devious route: this could be a conceivable, though inelegant, solution. Raising the steam take-off pipe in the separators, as effectively tried at Cerro Prieto (see Section 9.3) could well be the answer, but in extreme cases it may be unavoidable that the steam be washed in a scrubber at the entry to the power station (see Section 10.7 and Fig. 10.7). Thus, very short steam transmission distances do not necessarily make for economy.

A very curious phenomenon has been observed at Wairakei, relating to the pipe wall condensation formed in the steam mains. It has been found that the lower internal surface of the pipes has sometimes been eaten away, occasionally to more than half-way through the wall thickness. The surprising thing is that this has occurred only towards the receiving end of the pipelines, remote from the wells – i.e. where the condensate has attained a high degree of purification, is practically free of dissolved salts and has attained an almost neutral pH value. No such attack on the pipe walls has been observed nearer the wells, where salt concentrations are much higher. A makeshift remedy is to rotate the vulnerable lengths of pipeline from time to time so that different parts of the inside surface are brought into the lowest position; but this is clearly no proper remedy. Success has been attained, however, by injecting small quantities of bore water into the affected lengths of steam pipe [107]: the attack immediately stopped. (See also Section 15.3.4.)

It would seem that steam pipelines can be both too short and too long for trouble-free operation!

9.11 Pressurized water transmission

It has long been realized that fluid transmission could be enormously simplified and cheapened if it were possible to install submerged pumps at a suitable depth in wet bores, capable of raising to the surface and transmitting from the wellhead to the exploitation plant very hot pressurized water extracted from depth in a water-dominated field. The advantages of so doing would be as follows:

(i) Hot water is economically transmittable over longer distances than steam.

(ii) No wellhead equipment other than a simple shut-off valve, and perhaps an emergency or testing bypass, would be needed.

(iii) By the suppression of boiling, no steam – with its inconveniently large specific volume – would be formed until after the pressurized water has reached the exploitation plant.

(iv) Little or no chemical precipitation would occur in transit if the heat losses are kept low by means of good lagging. For example, with silica present in the form that permits supersaturation, the water would probably be clear of the plant and rejected, having yielded up its energy, before any precipitation occurs. For the exploitation of geothermal brines this could also be of immense value.

(v) Transmission pipes would be far smaller and cheaper than for steam.

(vi) Transmission velocities could be kept fairly low without undue extravagance in pipe diameters.

(vii) Heat losses in transmission would be much less than for steam.

(viii) On arrival at the exploitation plant the pressurized water could be flashed in any convenient number of stages, and steam could then be made available at pressures only slightly above convenient plant admission pressures, with virtually no steam transmission problems (owing to the short distances involved) except for scrubbing – see Section 10.7.

(ix) A submerged pump is capable of delivering a greater fluid output than that which a bore can yield unaided. Hence for a given required quantity of heat to be delivered, fewer bores would be needed.

Until recently no-one had succeeded in producing a commercially acceptable pump capable of standing up to the arduous conditions of geothermal service. However, three types of submerged pump are now available on the market or are in an advanced state of development. It is too early yet to say with certainty that any or all of them are of proven reliability beyond all doubt, as it must take several years of service in the field to establish complete confidence. Nor is it yet possible to pronounce as to which of three types will give the best overall value. However, several submerged pumps are already in use in geothermal installations, and some of them are reported to be performing excellently. There would seem to be no reason why such pumps should not before long become accepted as a reliable and valuable innovation, even if a few 'teething troubles' may first have to be experienced and overcome. Each type has a potential 'heel of Achilles' that is under close observation.

The three types of down-hole submerged pump are as follows.

9.11.1 *Line shaft pump.*   This is driven by a motor, turbine or engine at ground level. The drive is transmitted to the deep pump through a length of line shafting contained in tubing and supported by bearings – usually at about 5 ft spacing throughout its length. The bearings are force-lubricated

Plate 18 A 23-stage line-shaft submerged geothermal pump in the process of installation. This pump is about 38 ft in length and is rated to produce 2100 USgal/min at a head of 1300 ft. The pump is to be set at a depth of about 1200 ft and will operate at about 400° F (204° C). As the longest shaft sections are 21 ft long, this pump is made up of two units coupled together by means of a tandem adapter that can be seen above the lower pair of elevator clamps. (By courtesy of Peerless Pump, an Indian Head Company)

from ground level; and it is on the reliability of these bearings, their effective lubrication and immunity from chemical attack or physical deterioration, that users require absolute assurance. Down-hole pumps may have to be positioned at depths of 1000–2000 ft or more – so some hundreds of bearings may be needed to support the line shaft. Pumps of this type are available for working at 482° F (250° C) or even more. Boiling water at this temperature has an enthalpy of 467 Btu/lb (259 cal/g) and a saturation pressure of 562 psig (39.25 atü). The makers are aiming to produce a pump capable of operating at much higher temperatures – perhaps as much as 560° F (293° C), corresponding to boiling water of 562 Btu/lb (312 cal/g) and a saturation pressure of 1118 psig (78.3 atü). Time will show to what extent success will be achieved.

Plate 18 shows a line shaft pump about to be inserted down a 12-in bore. Its sheer size may cause surprise to those who have never before seen a down-hole pump.

9.11.2 *Down-hole turbine pump.*   This pump, at the time of writing, is still under development but has undergone some promising tests. Its principle is to dispense with a ground level drive so that a long line shaft is no longer necessary. Instead, a submerged down-hole boiler generates steam by extracting a small fraction of the heat from the bore water. The steam drives a small, compact turbine unit directly coupled to the pump impeller which is placed at the bottom of the assembly. The pumped bore fluid flows past the boiler/turbine/pump assembly through the annular space surrounding it. Fresh feed water is led from the surface to the boiler through a tube passing down the bore, and the exhaust steam from the turbine passes up the bore through another tube to the surface, where it may be exhausted to atmosphere or alternatively condensed and recycled to the boiler. An alternative to the steam turbine is being developed, using a binary fluid which is condensed at the surface and pumped down to the boiler. This type of pump could well prove to be reliable, but it could perhaps be the most highly priced of the three types, though the overall 'dollar efficiency' might still be favourable.

9.11.3 *Down-hole electric pump.*   With this design a down-hole submerged electric motor is directly coupled to a multi-stage pump with an annular bore water intake and vertical delivery pipe. The motor is of a very sophisticated design, being restricted in diameter to less than that of the bore, so that the torque radius is very small: in consequence it is very long. The electric down-hole pump is relatively cheap, costing far less than the line-shaft pump; but its vulnerability lies in the difficulty of sealing a submerged motor and in the ability of insulation to withstand the high temperatures.

Down-hole pumps of all types must be set at a depth in the well below the level at which flashing would occur if the bore were discharging without the aid of a pump. Boiling at depth is prevented by allowing the gases emitted from the bore water to accumulate in the sealed annular space surrounding the uprising pump delivery pipe or the line shaft tube. These gases maintain a pressure slightly in excess of the vapour pressure at the free surface of the bore water. Should the gases accumulate to excess, and tend to depress the water level to a point where the pump might start to cavitate, the trapped gases may be vented at the surface to the extent required. Venting may be necessary only once every few days, as the emission of gases from bore water, when not boiling, is moderate. A bubble pipe may be used to detect the water level at depth.

Although a pumped bore may yield more fluid than the same bore operating unaided, it is necessary to ensure that the pump capacity is not so great that the water delivery becomes starved through inadequate permeability in the formation; otherwise the pump will cavitate and chemical deposition may occur in the formation.

9.12 Two-phase fluid transmission

An alternative to the separate transmission of steam and hot water in wet hyperthermal fields is to carry both fluids together in the same pipe. This greatly simplifies the wellhead equipment and pipework, and reduces costs. The method was tried out experimentally in New Zealand and Japan in the late 1960s with considerable success: it has now been adopted for regular service in several modern geothermal installations – notably in Japan and Iceland. The method enables wellhead separation to be dispensed with altogether. Separation, and perhaps subsequent further flashing, is done at the utilization plant.

For some time, engineers were blinded by the undesirability of steam formation in hot water mains and of hot water in steam mains; the common bogey being water-hammer. Certain investigators, notably James [108], argued that as two-phase fluids rose up the bore casings without any disastrous effects, a pipeline forming a continuation of the bore might equally survive the rough treatment of transmitting flashing, boiling water/steam mixtures over the ground surface. Test rigs in New Zealand and Japan have shown that with two-phase fluid transmission no water hammer has been experienced, even with fairly rapid valve movements; no noticeable corrosion or erosion has occurred in the pipelines; and although pressure drops are higher than when transmitting steam alone, they are not unacceptably high. Pressure drops have been found to conform approximately to the following formula [108]:

$$\Delta P_m = \frac{\Delta P_s}{d^{0.5}}$$

where ΔP_m = pressure drop through a given pipe passing water/steam mixture;

ΔP_s = pressure drop through the same pipeline passing dry saturated steam;

d = gravimetric dryness fraction of the mixture.

(The initial vapour pressure is assumed to be the same for both fluids.)

Thus with a water/steam mixture of, say 4:1, d would be 0.2 and the pressure drop for the mixture would be 2.236 times as great as for dry steam. But as the pressure drop is also inversely proportional to the fifth power of the diameter (see Section 9.13), this could be compensated by increasing the diameter by about 17½%. In any case, pressure drop is not without a modest 'dividend' in the form of some additional flash steam formed in transit. Furthermore, the pressure drop incurred in wellhead separators would be avoided.

The claimed advantage of two-phase transmission are as follows:

(i) The savings in wellhead gear can outweigh the extra cost of the rather larger pipework.

(ii) The adoption of relatively large separators close to the utilization plant has an economic 'scale effect' advantage by comparison with the use of many smaller individual wellhead separators, and their maintenance can be more easily and cheaply handled.

(iii) Significant amounts of additional power can be extracted from the hot water by multiple flashing at the plant without the use of a costly separate hot water collection and transmission scheme with its attendant controls (see Chapter 13). Under Wairakei conditions the power capacity could have been raised by about 50% if power had been extracted from all the hot water available.

(iv) The wasteful rejection of hot bore water at the wellheads can be avoided, thereby not only conserving energy and reducing heat pollution but also saving the very costly drainage channels that would otherwise be necessary: it also eliminates the emission of vast quantities of flash steam into the air from the wellhead silencers. With multi-stage flashing at the plant, the water may be rejected at about atmospheric pressure, with very little consequent flashing, and the effluent water can be discharged from a single point only – near the exploitation plant.

(v) Two-phase flow pipes can negotiate moderate uphill gradients.

(vi) The relative levels of different bores is of minor importance. With separate hot water transmission, unless 'pumpless pumping' is adopted, each pump must be designed to suit the level of the associated wellhead (above datum).

(vii) Fluids discharged from many bores may simply be merged together by joining wellhead branches to a single main pipeline (or more, if required).

(viii) Aesthetic gains would result from the elimination of many elaborate wellhead equipment assemblies, and from the large reduction in the amount of escaping steam.

In short, the system would appear to be both economically and aesthetically advantageous. It is, however, necessary to heat and pressurize a pipeline with steam before admitting the mixed fluids, so as to avoid water hammer; and steep gradients must be avoided. It is also well to point out that two-phase fluid transmission, requires the use of scrubbers after separation at the receiving end (see Section 10.7) for the removal of chlorides – normally effected in steam transmission pipelines by repeated condensation and drainage (Section 9.10). Scrubbers are large and costly pieces of equipment that are not altogether free from operating troubles.

It is of interest to consider why water-hammer, which can be troublesome if water is present in a steam line or if steam is present in a water line, apparently does not usually occur in pipes transmitting very wet mixtures such as are emitted from wet geothermal bores. This is rather an obscure problem that has been studied by several investigators [109–112] and appears to be closely related to the flow pattern adopted by the two fluids within the transmission pipe. It would seem that two-phase flow, between certain maximum and minimum proportions of water to steam, tends to assume the annular flow pattern where the water mainly clings to the pipe walls while the steam passes through the middle – a relatively stable pattern in which each fluid interferes but little with the other. The unwanted presence of steam in a hot water pipe favours 'bubble flow', which can result in water-hammer if a rise in pressure should cause the bubbles to collapse so that the displacing water strikes the pipe roof forcibly. The unwanted presence of water in a steam pipe will usually start as 'rivulet flow', which is quite stable provided the steam velocity is 'moderate' as defined in Section 9.14; but at high steam velocities there is a danger that the rivulet may be swept up by the 'wind' of steam and hurled downstream in slugs, which would also cause water-hammer. The transition from moderate to high steam velocity is probably related to the fact that the tendency for water to be swept up is greater for a large surface than for a small, and less for deep water than for shallow. Near the edges of a rivulet the ratio of water surface to depth is high; and at some critical steam velocity the edges of the rivulet will become entrained with the steam. The rivulet will immediately respread itself and the process will be repeated at an accelerating rate because the ratio of surface to depth increases as the rivulet becomes smaller. Thus, above some critical steam velocity rivulet flow could not survive. With annular flow there are no 'edges' at which this process could start, so a much higher steam velocity would be permissible. At Wairakei it has been reported that two-phase flow has been successful over distances of about 1 km but that water-hammer has occurred with greater lengths. A suggested explanation is that the pressure drop induced by the flow would not only increase the specific volume of the steam but would also add to the mass of steam by evaporation from the water annulus. Hence the steam velocity would become higher with increasing length. Thus at some critical length it seems probable that the steam velocity would become too great even for annular flow to remain stable. A possible cure would be to introduce a gradually tapered pipe enlargement where the length becomes too great for stability. This is a subject to which further studies will doubtless be devoted; but it has been established beyond doubt that two-phase flow is often practically and economically feasible for such transmission distances as are frequently involved in the development of a wet geothermal field.

9.13 Pressure drops in steam pipes

The well-known 'Babcock' or 'Gutermuth and Fischer' formula, usually expressed in ft/s units is

$$\Delta p = 0.4716 \left(1 + \frac{3.6}{d} \right) \frac{LVw^2}{d^5}$$

where Δp = pressure drop (psi),
L = pipe length (ft),
V = specific volume of steam (ft^3/lb),
d = internal diameter of pipe (inches),
w = mass flow (lb/s).

Thus, for example, the pressure drop in 20-in internal diameter straight pipe carrying dry saturated steam at a pressure of 8 ata and at a flow rate of 100 tons/h would be 2.56 psi per 1000 ft. To be accurate, since the specific volume continuously increases as the pressure falls, it would be necessary to assume a series of short lengths and adjust for the changing value of V each time, integrating the whole; and even to allow for condensation. A simpler, and sufficiently accurate method is to do the sum for the initial condition, deduce V at the delivery end of the pipe, and then to repeat the calculation using the mean value of V at the two ends. Actual pressure drops are invariably complicated by the presence of bends, fittings, elbows etc., each of which is equivalent to larger lengths of pipe than that which they occupy. For example a bend of radius equal to four times the pipe radius is equivalent in flow resistance to a straight length of pipe about four times the actual length (along the curve) of the bend. In a long and complicated pipeline with bends, loops, valves, bellows pieces, etc., the effective pressure drop may be a considerable percentage higher than that calculated for a plain straight pipe. Various handbooks are available for providing the necessary data for calculation.

The presence of small quantities of water in the steam does not greatly affect the formula, so long as corrections are made for V and w, but with very wet water/steam mixtures the formula gives too high a result – which is fortunate for two-phase flow transmission.

9.14 Limiting fluid transmission velocities

With steam and two-phase fluid transmission, the limiting velocity of the fluids in pipelines is imposed by the following considerations:

(i) The permissible pressure drop between the assumed economic well-head pressure and the designed inlet pressure to the exploitation plant must not be exceeded. Where a wellhead separator is installed, it would be wise to allow a pressure drop of about 10% of the absolute wellhead pressure to be absorbed in the separator and its associated pipework. Pressure drops incurred in the branch lines and mains should be separately calculated.

(ii) Steam velocities should not exceed the values determined by the formula:

$$v = \frac{1300}{P^{0.54}}$$

where v is the steam velocity in ft/s
P is the steam pressure in psia.

This formula is recommended by James [113] as a definition of a 'moderate' velocity below which the rivulet of condensate flowing along the bottom of a steam pipe is considered to be more or less immune from being swept along in gulps by the faster moving steam. Apart from the risk of water-hammer, the re-entrainment of water caused by excessive steam velocities is apt to carry the water past the next downstream collection pot and so escape removal. The reduced removal efficiency would mean that the degree of purification would be far less than suggested in Table 9.1, Section 9.10. So long as the velocity is restricted to the value determined by the formula, a removal efficiency of at least 70% should be ensured.

(iii) Notwithstanding (ii) above, steam velocities should never be allowed to exceed about 150 ft/s owing to the risks of erosion and vibration in fittings such as expansion compensators and valves. This limitation would only become influential at pressures below about 55 psia.

(iv) Hot water velocities must be limited by considerations of induced frictional pressure drops, in accordance with the principles set out in Section 9.9. Apart from that, and as an overriding consideration, water velocities should never exceed 8 ft/s because small roughnesses on pipe surfaces could cause local concentrations of friction that could lead to the formation of steam bubbles.

10 *Electric power generation from geothermal energy*

10.1 Evolution and situation at the end of 1980

The historical evolution and planned growth of geothermal power generation have been shown graphically in Fig. 0.1. The supporting figures are set out in Table 10.1, and are summarized below:

Installed capacity at the end of 1980	2585.8 MWe
Definitely planned additions	6196.6 MWe
Installed capacity in the year 2000 according to curve A, Fig. 0.1	8782.4 MWe
Tentatively planned additions	8094.3 MWe
Installed capacity in the year 2000 according to curve B, Fig. 0.1	16876.7 MWe

It cannot be claimed that these figures are precise. The fact that decimals of megawatts are shown may therefore suggest a lack of proportion in the presentation of Table 10.1; but the aim is to show even those countries where geothermal development has started on the humblest scale. Once decimals have been admitted to a double entry table, the need for preserving an arithmetical balance both vertically and horizontally compels their retention, even in totals where they lack significance. In the preparation of Table 10.1 the following points have been taken into consideration:

(i) Plant capacities have been ascertained partly from published data and partly fromn questionnaires sent by the author directly to the developers. In a few instances there have been inconsistencies in the information available, in which cases the author has been compelled to decide upon what he considers to be the most probable figures. It must be borne in mind that plans are continually changing, and that the most carefully prepared estimates need constant revision.

(ii) Gross plant capacities have been quoted where known, without deduction for power station auxiliaries such as pumps, exhausters, etc.

162

Where net figures only are known, reasonable additions have been made in order to convert them to the gross equivalent, according to the types of plant involved–condensing, non-condensing, binary, etc. Where there is doubt, quoted capacities have been assumed to be gross.

(iii) As Table 10.1 forms an assessment of installed capacity only, and not of generated energy, no deductions have been made for certain plants that are known to be under-loaded – e.g. Krafla (Iceland) and Wairakei (New Zealand). Allowance for this under-loading will be made when energy estimates are made later.

(iv) The fact that Japan has taken 14 yeas to develop 224.5 MWe of geothermal power plant suggests that the estimate of 7000 MWe (total) to be installed by the mid-1990s is somewhat optimistic: it nevertheless originates from the Japan Geothermal Energy Association, who seem confident of achieving this goal. By contrast, the USA estimates appear to be rather over-cautious, to judge from various published estimates quoted below:

Date and source of estimate	MWe installed in the year 2000	
1975 F. Maslan *et al.*, p. 2414, Proc. UN Geothermal Symposium, San Francisco, 1975 [114]	Normal development Crash programme (Hydrothermal systems only)	108 000 515 900
1975 Document ERDA-86, p. L-12 [115]	39 000	
1979 Document DOE/ET-00900 IGCC-4, p. 14 [116]	20 000–40 000	

The USA estimate of 3933.8 MWe (total) for the year 2000, deduced from Table 10.1, falls very much short of these figures. Admittedly some of the published estimates have been more in the nature of goals than of expectations; but for the purpose of preparing Table 10.1 it has been deemed advisable to take into account only those plants in the USA that are now either definitely planned or are now under active consideration. The caution exercised in the case of the American figures is likely to have offset any undue optimism that may perhaps apply to the figures for Japan. On balance, it is suggested that imprecise though Table 10.1 must surely be, it is probably as close an estimate as can be made in 1981 in the light of the prevailing circumstances and uncertainties.

In so far as generated electrical energy is concerned, it is estimated that the world's total installed geothermal power plant capacity of 2586 MWe was producing about 18.3 TWh p.a. in 1980. This implies a mean annual plant factor of 80.9% – a rather low figure for geothermal power generation, caused partly by the underloading of certain stations and partly by the temporary reduction of plant factor at the Geysers, California, owing to

Table 10.1. Geothermal power plants, in service and planned, at the end of 1980 (MWe gross)

Country	In service at end of 1980	Planned additions	1981	1982	1983	1984	1985/89	1990/94	1995/2000	Totals
Chile	3	Definite	—	—	—	—	—	15	—	15
		Tentative	—	—	—	—	—	—	—	—
		Total	—	—	—	—	—	15	—	15
China, People's Republic	2.1	Definite	—	6	—	—	—	—	—	6
		Tentative	—	—	—	—	18	—	—	18
		Total	—	6	—	—	18	—	—	24
Costa Rica	—	Definite	—	—	52	—	—	—	—	52
		Tentative	—	—	—	—	52	—	—	52
		Total	—	—	52	—	52	—	—	104
El Salvador	111.3	Definite	—	—	—	—	—	—	—	—
		Tentative	—	—	—	—	55	110	275	440
		Total	—	—	—	—	55	110	275	440
Ethiopia	—	Definite	—	—	—	—	15	—	—	15
		Tentative	—	—	—	—	—	—	—	—
		Total	—	—	—	—	15	—	—	15
Greece	—	Definite	—	—	—	—	—	—	—	—
		Tentative	—	10	—	—	—	—	—	10
		Total	—	10	—	—	—	—	—	10
Guadeloupe	—	Definite	—	—	4.9	—	—	—	—	4.9
		Tentative	—	—	—	—	3.1	—	42	45.1
		Total	—	—	4.9	—	3.1	—	42	50
Guatemala	—	Definite	—	—	—	—	—	—	—	—
		Tentative	—	—	—	—	—	—	—	—
		Total	—	—	—	—	—	—	—	—
Honduras	—	Definite	—	—	—	—	55	55	220	330
		Tentative	—	—	—	—	—	—	—	—
		Total	—	—	—	—	55	55	220	330
Iceland	35	Definite	—	—	—	—	50	50	—	100
		Tentative	6	30	—	—	—	—	—	36
		Total	6	30	—	—	50	50	—	136
Indonesia	0.2	Definite	2	30	—	—	—	—	—	32
		Tentative	—	—	—	—	—	60	—	60
		Total	2	30	—	—	—	60	—	92
Italy	439.6	Definite	2	—	15	15	3	—	—	44
		Tentative	—	—	—	15	—	—	—	18
		Total	2.4	11.6	15	30	3	—	—	62
Japan	224.5	Definite	—	50	—	350	2875.5	—	—	3275.5
		Tentative	—	—	—	—	—	3500	—	3500
		Total	—	50	—	350	2875.5	3500	—	6775.5
Kenya	—	Definite	15	15	—	—	—	—	—	30
		Tentative	—	—	15	—	—	—	—	15
		Total	15	15	15	—	—	—	—	45

Country										
Mexico	Definite	150	30	—	220	220	—	—	—	470
	Tentative	—	—	—	—	—	100	230	450	780
	Total	—	—	—	220	220	100	230	450	1250
New Zealand	Definite	202.2	—	—	—	—	80	80	—	80
	Tentative	—	—	—	—	—	—	—	80	80
	Total	—	—	—	—	—	80	80	80	160
Nicaragua	Definite	—	—	—	—	—	35	—	—	35
	Tentative	—	—	—	—	—	—	—	65	65
	Total	—	—	—	—	—	35	—	65	100
Panama	Definite	—	—	—	—	—	—	—	—	—
	Tentative	—	—	—	—	—	75	—	—	75
	Total	—	—	—	—	—	75	—	—	75
Philippines	Definite	446	—	165	—	165	—	—	—	555
	Tentative	—	—	—	225	660	—	—	—	660
	Total	—	—	165	225	825	—	—	—	1215
Portugal (Azores)	Definite	—	3	—	—	15	—	—	—	18
	Tentative	—	—	—	—	—	—	—	—	—
	Total	—	3	—	—	15	—	—	—	18
Santa Lucia	Definite	—	3	—	—	—	—	—	—	—
	Tentative	—	—	—	—	1.5	—	—	—	1.5
	Total	—	3	—	—	1.5	—	—	—	1.5
Turkey	Definite	0.5	—	—	—	—	—	—	—	—
	Tentative	—	—	—	—	—	—	99.5	50	149.5
	Total	—	—	—	—	—	—	99.5	50	149.5
USA	Definite	960.4	8	379.2	287	231.9	440.1	—	—	1346.2
	Tentative	—	—	52	—	127.4	491.4	852.4	104	1627.2
	Total	—	8	431.2	287	359.3	931.5	852.4	104	2973.4
USSR	Definite	11	—	—	—	—	50	—	—	50
	Tentative	—	—	—	—	—	200	—	—	200
	Total	—	—	—	—	—	250	—	—	250
Grand totals	Definite	2585.8	66.4	716.8	803.9	996.9	3612.6	—	—	6196.6
	Tentative	—	—	52	15	803.9	1015.5	5001.9	1206	8094.3
	Total	2585.8	66.4	768.8	818.9	1800.8	4628.1	5001.9	1206	14 290.9

Cumulative totals

Curve A, Fig. 0.1	2585.8	2652.2	3369.0	4172.9	5164.8	8782.4	8782.4	8782.4	With definitely planned additions
Curve B, Fig. 0.1	2585.8	2652.2	3421.0	4239.9	6040.7	10668.8	15670.7	16876.7	With all planned additions

environmental problems (see Chapter 16). By the year 2000 it should be possible to attain a mean annual plant factor of at least 85% for the aggregate of installed geothermal power capacity, in which case the generated electrical energy should amount to the following approximate values, according to the total geothermal plant capacity installed by then:

Installed capacity (MWe)	Energy generated (gross) (TWh p.a.)	Comments
8782	65.4	See curve A, Fig. 0.1
16 877	125.7	See curve B, Fig. 0.1
63 950	476.2	See curve C, Fig. 0.1
40 000	297.8	Compromise suggested for Table 0.1

Despite the inefficiency inherent in the process of converting heat into electricity, imposed by inescapable thermodynamic constraints, and despite the desirability of encouraging direct applications of earth heat, electric power generation is likely to remain for some time the commonest form of geothermal development. This is partly for the reasons given in the early part of Chapter 2 and partly because high-temperature fields, though limited in number and location, are likely to be developed more rapidly than lower temperature geothermal resources as they tend to offer greater and quicker commercial rewards and are less dependent on location for ensuring a market for their production. Although geothermal power generation is likely gradually to yield some of its present pre-eminence to other applications of earth heat, it will almost certainly retain great importance for a long time.

10.2 Cycles

The various practical cycles used for geothermal power generation are illustrated diagrammatically in Fig. 10.1.

10.2.1 *Indirect condensing cycle* (Fig. 10.1(a)). In the early days of power development at Larderello in Italy, the corrosive nature of the steam was thought to be too aggressive for its direct admission to the steam turbines. Accordingly an 'indirect system' was at first adopted, which involved the use of a heat-exchanger by means of which 'clean' steam was raised from contaminated natural steam. The system included the means of disposal of non-condensible gases accompanying the steam, and also enabled the (then) valuable boron and ammonia to be recovered from the concentrated liquor in the heat-exchanger and from the gas discharges. Before the metallurgy of turbine materials had advanced very far, and when the recoverable minerals and gases commanded good prices, this cycle was economic despite the fact that from 15 to 20% of the power potential of the steam had to be sacrificed in the heat-exchanger. Examples of this cycle were still in operation in Italy until about the middle of the twentieth century, but improved metallurgy on the one hand and the declining economic attractions of mineral extraction (by comparison with other production methods) on the other hand have now

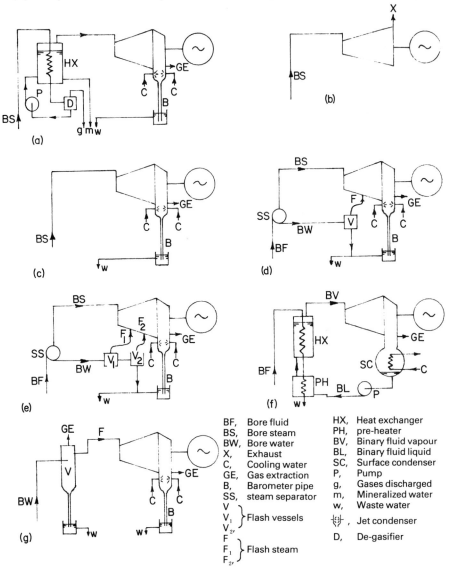

Fig. 10.1 Geothermal power generation cycles. (a) Indirect condensing; (b) non-condensing; (c) straight condensing; (d) single flash; (e) double flash; (f) binary fluid cycle; (g) hot water sub-atmospheric.

rendered this cycle virtually obsolete, though it could still serve a purpose if exceptionally valuable minerals in high concentrations were present in the fluid emitted from a field. It could also perhaps be used where the geothermal steam contains a fairly high proportion by weight of non-condensible gases. Such steam is often emitted from new bores in the Monte Amiata district in Italy, though the gas content declines considerably in the course of time.

10.2.2 *Direct non-condensing cycle* (Fig. 10.1(b)).   This is the simplest and, in capital cost, the cheapest of all geothermal cycles. Bore steam, either direct from dry bores or after separation from wet bores, is simply passed through a turbine and exhausted to atmosphere. Such machines may consume about twice as much steam (for the same inlet pressure) as condensing plants per kilowatt of output and are therefore wasteful of energy and costly in bores. Nevertheless, they have their uses as pilot plants, as standby plants, for small local supplies from isolated bores, and even perhaps (though to a very limited extent) for supplying peak loads (see Section 10.16 (iv)). Non-condensing machines must also be used if the content of non-condensible gases in the steam is very high (greater than 50%), and will usually be used in preference to the condensing cycles for gas contents exceeding about 10%, because of the high power required to extract these gases from a condenser.

10.2.3 *Straight condensing cycle* (Fig. 10.1(c)).   Since there is no need to recover the condensate for 'feed' purposes, as in a conventional thermal plant, direct contact jet condensers with barometric discharge pipes can usually be used in place of costlier surface condensers. However, it may sometimes be necessary to use surface condensers if the nature and concentration of the non-condensible gases are considered to offer a threat to the local environment (see Chapter 16).

10.2.4 *Single flash cycle* (Fig. 10.1(d))*.   In wet fields it may be possible to extract quite a lot of supplementary power from the hot water phase by passing it to a flash vessel operating at a lower pressure than that at which the

---

* *Notes concerning flash cycles:* First, it is emphasized that although in Figs 10.1(d) and (e) a single turbine has been shown admitting pass-in steam, it would be equally possible for the plant to take the form of separate turbines connected in cascade, with the exhaust from an upstream turbine mingling with flash steam and passing on to the entry of a downstream turbine. Secondly, it is well to point out at an ambiguity of *nomenclature* with flash cycles. Some people prefer to call Fig. 10.1(d) 'double flash', arguing that the first stage of flashing occurs in the bore itself. If the boiling water deep in the aquifer is regarded as the original source of energy, there is logic in this argument. The same people would regard Fig. 10.1(e) as 'treble flash'. To the author it seems more reasonable to treat the *wellhead fluid* as the starting point of the thermal cycle from the engineering viewpoint: otherwise Figs 10.1(b) and (c) would have to be called 'single flash' in wet fields although no surface flash vessel is involved.

main steam is admitted to the turbine(s). The flash steam may then be used to pass through the lower pressure stages of the prime mover(s). Ideally, the maximum power yield from the hot water can be obtained if the flash vessel operates at a temperature midway between that of the collected hot water and that of the condenser. The unflashed fraction of the hot water is then rejected to waste or put to some industrial or other use.

10.2.5 *Double flash cycle* (Fig. 10.1(e))*.   Ideally, the maximum power would be extracted from the hot water phase in a wet field by using an infinite number of flash vessels connected in cascade – or at any rate by using one flash vessel for every turbine stage; but such a procedure would be an economic absurdity. It *can*, however, sometimes pay to provide two flash vessels as shown in Fig. 10.1(e), operating as closely as possible at temperatures of one-third and two-thirds way between the collected hot water temperature and the condenser temperature. This was the cycle adopted for the experimental hot water transmission scheme at Wairakei, described in Section 9.9. Approximately 68 to 70% of the supplementary power from the hot water, under Wairakei conditions, was supplied by the first flash, F_1: this shows that there would be diminishing returns from further stages of multiple flash which would render yet another stage uneconomic.

10.2.6 *Binary fluid cycle*.   Much thought has been, and is being, given to the use of refrigerant fluids of very low boiling point, such as Freons, isobutane, propane and others in a closed turbine-feed-boiler cycle as shown in Fig. 10.1(f). The theoretical advantages of the binary cycle are:

 (i) It enables more heat to be extracted from geothermal fluids by rejecting them at a lower temperature.
 (ii) It can make use of geothermal fluids that occur at much lower temperatures than would be economic for flash utilizations.
(iii) It uses higher vapour pressures that enable a very compact self-starting turbine to be used, and avoids the occurrence of sub-atmospheric pressures at any point in the cycle.
(iv) It confines chemical problems to the heat exchanger alone.
 (v) It enables use to be made of geothermal fluids that are chemically hostile or that contain high proportions of non-condensible gases.
(vi) It can accept water/steam mixtures without separation.

There are, however, the following disadvantages:

 (i) It necessitates the use of heat exchangers which are costly, wasteful in temperature drop and can be the focus of scaling.
 (ii) It requires costly surface condensers instead of the cheaper jet-type condenser that can usually be used when steam is the working fluid.

*See footnote on p. 168.

(iii) It needs a feed pump, which costs money and absorbs a substantial amount of the generated power.
(iv) Binary fluids are volatile, sometimes toxic and sometimes flammable; and must be very carefully contained by sealing.
 (v) Makers are generally inexperienced, and high development costs are likely to be reflected in high plant prices – at any rate unless and until the cycle becomes commonly adopted in practice.
(vi) Large quantities of cooling water are needed.

The first operational example of a binary cycle geothermal plant, as distinct from laboratory prototypes, was the small 750 kW (nominal) unit at Paratunka in Kamchatka, USSR. This unit used geothermal hot water at 81°C as the primary heat source, Freon 12 as the binary fluid and cooling water at 6–8°C. High parasitic losses reduced the net output to 450 kW only [117]. It is understood that this machine is now closed down.

More recently other binary cycle plants have been put into service, or are in the process of development. Several very small units (50 to 200 kWe) are already operating in the People's Republic of China, and a 20 kW experimental unit is being installed in Central France. In the USA a 5 MWe unit is in service on the Raft River, Idaho, where a source temperature of only 150°C is available; and a 10/12.5 MWe unit is being operated at East Mesa in the Imperial Valley, California. At Los Alamos, New Mexico, a small 60 kW unit was put into service in 1980 to demonstrate that power could be generated from heat extracted from hot dry rock. More ambitious plans include a 45 MWe unit to be commissioned in 1984 and a 65 MWe unit for 1986: both will be at Heber in the Imperial Valley, and will generate power for the San Diego Gas and Electric Company. Other binary projects are under contemplation.

The Magma Power Company's 10/12.5 MWe plant at East Mesa deserves more detailed description. It is a *dual* binary plant, using separate propane and isobutane circuits (Fig. 10.2). Well water at about 182°C heats an isobutane cycle supplying a 10 MWe turbine working at 500–600 psig. The rejected heat from the isobutane enters a propane cycle working at 58 psig and about 94°C with a 2.5 MWe turbine that supplies all the auxiliaries, so as to leave a net output of about 10 MWe from the gross capacity of the two turbines. The system makes use of down-hole bore pumps and a propane superheater. All the plant is outdoors except the main alternator and electrics.

A fairly new innovation which may influence the development of the binary cycle is the direct contact heat-exchanger, in which the immiscible binary and bore fluids are brought into direct contact with one another in the same vessel, thus permitting heat to be exchanged between them without the need of any conducting metallic separating barrier, and with a consequent reduction in the terminal temperature difference. A 500 kW demonstration

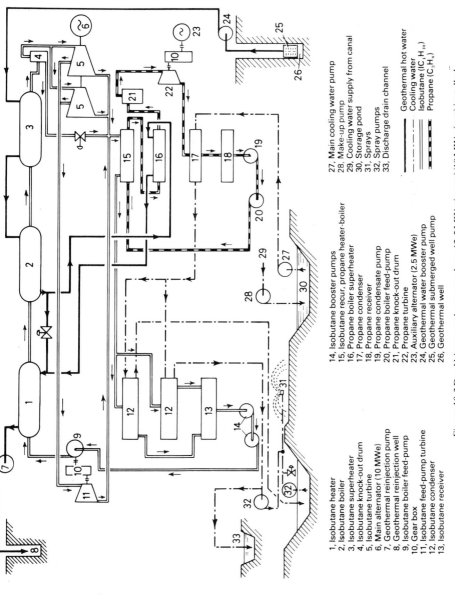

1, Isobutane heater
2, Isobutane boiler
3, Isobutane superheater
4, Isobutane knock-out drum
5, Isobutane turbine
6, Main alternator (10 MWe)
7, Geothermal reinjection pump
8, Geothermal reinjection well
9, Isobutane boiler feed-pump
10, Gear box
11, Isobutane feed-pump turbine
12, Isobutane condenser
13, Isobutane receiver

14, Isobutane booster pumps
15, Isobutane recur, propane heater-boiler
16, Propane boiler superheater
17, Propane condenser
18, Propane receiver
19, Propane condensate pump
20, Propane boiler feed-pump
21, Propane knock-out drum
22, Propane turbine
23, Auxiliary alternator (2.5 MWe)
24, Geothermal water booster pump
25, Geothermal submerged well pump
26, Geothermal well

27, Main cooling water pump
28, Make-up pump
29, Cooling water supply from canal
30, Storage pond
31, Sprays
32, Spray pumps
33, Discharge drain channel

—— Geothermal hot water
—·—·— Cooling water
—··—··— Isobutane (IC₄H₁₀)
▬▬▬ Propane (C₃H₈)

Figure 10.2 Dual binary cycle power plant, 12.5 MWe (gross)/10 MWe (net) installed at East Mesa, Imperial Valley, California, USA. (By courtesy of the Magma Power Company)

pilot unit is undergoing test at East Mesa in the Imperial Valley, California, in order to study this device: it is under the joint supervision of the US Department of Energy, the University of California and the Lawrence Livermore Laboratory.

The proposal [44] mentioned in Section 5.10 to inject a binary fluid into a bore would amount to using the bore itself as a direct contact heat-exchanger. At surface level the two fluids would have to be separated; the binary fluid would be expanded in a turbine and, after condensation, would be reinjected down the bore. The proposal has not yet been commercialized, and there are some doubts as to the effectiveness of separation of the fluids; but the idea has interesting possibilities.

The binary cycle has its enthusiasts and its critics. Almost certainly it is likely to have net economic attractions in certain circumstances – e.g. for exploiting low-enthalpy geothermal fluids – and to be of no advantage in others. The recent appearance of so many binary cycle plants in service should soon provide enough practical evidence for a balanced judgment to be made. It is well to remember that as the practice of bore fluid reinjection becomes more prevalent (see Chapter 16), thus enabling hot discharged geothermal waters to serve as useful 'boiler feed' to the underground heating cycle, the force of advantage (i) above will be weakened.

10.2.7 *Hot water sub-atmospheric cycle* (Fig. 10.1(g)).   Only one case of this cycle is known to the author, namely at Kiabukwa, in the province of Katanga, Zaire [118] where a small British-made 220 kW power plant was put into service as long ago as 1953. It is believed to be no longer in use. This plant used hot spring water at 91°C which was sprayed into a chamber at a low pressure of 0.3 ata, and the resulting flash steam passed through a three-stage low-pressure turbine exhausting to a condenser supplied with cooling water at 24°C. The plant was not self-starting, and the vacua in the flash chamber and condenser had first to be established by means of an ejector supplied with steam from an auxiliary wood-fired boiler. The natural occurrence of a hot spring enabled drilling costs to be avoided, but it is difficult to understand how such a plant could have been economic. Nevertheless, the plant was of great technical interest and is believed to have given good service for several years. It is doubtful whether this cycle will ever be repeated. An interesting phenomenon occurs at Kiabukwa, in that the discharge of hot water from the hot spring fluctuates in a tidal cycle of 12½ hours, although the site is about 1300 km from the sea. This is probably due to the action of 'land tides' opening and closing horizontal subterranean fissures by very small movements which are nevertheless significant in relation to the width of what are probably very narrow cracks.

10.2.8 *The hybrid cycle.*   This is still under development. The principle is first to bleed off the non-condensible gases from the pressurized bore water and to use these gases to pre-heat a binary fluid. The pressure of the bore water is then dropped and the flash steam passed through a turbine. The residual hot water is fed to a direct contact heat-exchanger (see Section 10.2.6) to vaporize the binary fluid for a separate turbine.

10.3 Power potential of geothermal fluids

The fact that geothermal fluids occur in nature at relatively low temperatures by comparison with those adopted in conventional thermal power plants means that the cycle efficiency of a geothermal power plant must always, of thermodynamical necessity, be low. Such plants are economic solely by virtue of the extreme cheapness of the *heat* (see Chapter 14) by comparison with fuels: they are also desirable from the aspects of energy conservation and minimal pollution.

The actual power potential of steam delivered to a power plant will depend not only upon its pressure, temperature and quality, and upon the back-pressure at the turbine exhaust, but also upon the configurations of pipework, separators and valving within the power station, in all of which will be induced losses of pressure and energy. It will also depend upon the quality of the generating plant. Considerable differences in power potential will therefore be experienced from plant to plant, but the curves shown in Fig. 10.3 may be regarded as fairly representative for dry saturated steam. They allow for reasonable turbine and alternator efficiencies and are expressed in terms of energy delivered at the alternator terminals; no allowance being made for the power consumed by gas-exhausting equipment, cooling water pumping and miscellaneous station auxiliaries. These parasitic demands can vary widely according to the quantity of non-condensible gases present in the steam and the vacuum maintained in the condenser and method of cooling. They will vary from virtually zero for a non-condensing turbo-generator unit to as much as 47 or 48% for a condensing plant where the non-condensible gas content of the steam is as high as about 20%. (With binary plants the parasitic losses may be very high indeed – e.g. about 40% at Paratunka (USSR) down to 20% for the East Mesa plant shown in Fig. 10.1; but this fact is irrelevant to Fig. 10.3, which relates to steam.)

Where dry superheated stream is available, its power potential cannot be represented by a simple series of curves as in Fig. 10.3 because much will depend upon the degree of superheat. To calculate the power potential in such cases, the available heat drop must be deduced from a Mollier chart and allowance made for the combined efficiencies of the turbine, the alternator

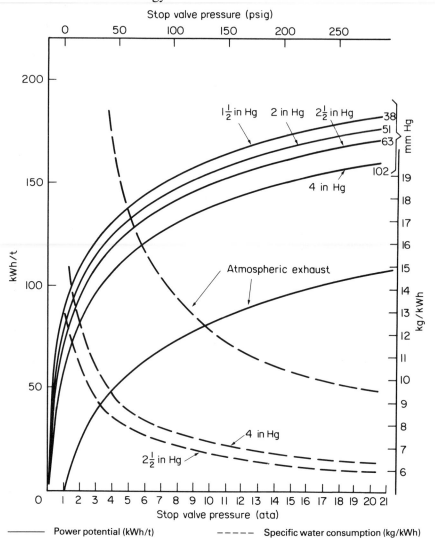

Figure 10.3 Typical power potentials of dry saturated steam at various turbine admission pressures, exhausting to atmosphere or to different vacua.
Note: The broken lines are, of course, a multiple of the reciprocals of the full lines.

and the station pipework, etc. These efficiencies would probably range from about 70 to 76% according to the admission pressure and the size and arrangement of the plant units.

The power potential of boiling water, as recoverable by the use of single or multiple flashing is approximately as shown in Fig. 10.4(a) and (b). (The expressions 'single flash' and 'double flash' here relate to the number of

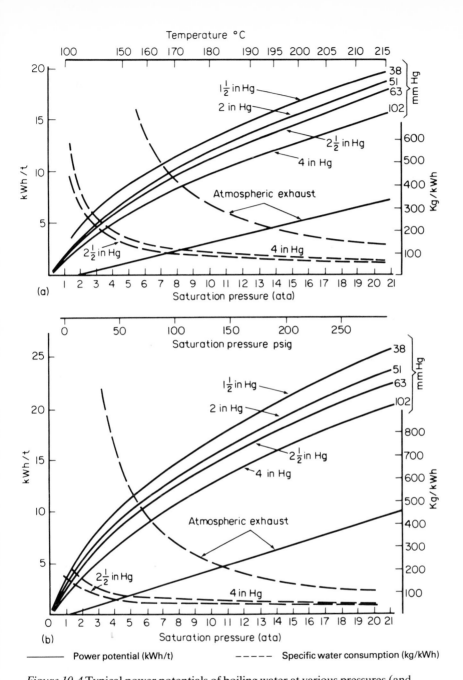

— Power potential (kWh/t)   - - - - - Specific water consumption (kg/kWh)

Figure 10.4 Typical power potentials of boiling water at various pressures (and temperatures – see top scale) flashing at optimum pressures and exhausting to atmosphere or to different vacua. (a) Single flash; (b) double flash.
Note: The broken lines are, of course, a multiple of the reciprocals of the full lines. Although the flash potential for non-condensing turbines has here been shown, this is really of academic interest only: it would almost certainly never be economically worthwhile. The curves are, of course, based upon the adoption of the economic flashing temperatures mentioned in Sections 10.2.4 and 5.

flashing operations *above ground* and exclude the bores – see footnote p. 168.) Although this potential, when expressed per kilogram of fluid, is only a few percent of that of dry saturated steam at the same pressure, it should be remembered that the water steam ratio is sometimes very high – perhaps 6 or 7 to 1 – so that the *total* power potential of the hot water may form a very substantial fraction of that of the steam in a wet field. It has already been mentioned in Section 9.12(iii) that the power generated at Wairakei when the field was first exploited could have been raised by about 50% if full use had been made of the hot water. With the passing of the years there have been changes in the pressures, temperatures and enthalpies of the bore fluid at Wairakei, so this proportion may well have fallen; but it could still be impressive. In general the use of the hot water in a wet field for power generation could sometimes effect a large saving per kilowatt, particularly if two-phase transmission or down-hole pumping is adopted. The whole question, however, should be carefully studied in relationship to the alternative of bore water reinjection (see Chapter 16).

It is emphasized that when using Figs 10.3 and 10.4 the chosen point on the base line should be the pressure at the power plant – *not* at the wells. Figs 8.13 and 8.14, which attempted to show the typical power potentials of *wells*, made reasonable assumptions as to transmission losses while embodying the characteristics of Figs 10.3 and 10.4.

10.4 Efficiencies of geothermal power plants

The poor thermal efficiencies of geothermal power plants, mainly due to the low fluid temperatures and pressures, are not helped by the absence of a closed feed cycle, though the cheapness of direct contact jet condensers, normally used with geothermal plants, by comparison with the cost of surface condensers, would largely offset any economic gain attainable from feed-heating, with its attendant heat-exchangers. The thermal efficiencies of geothermal power plants corresponding with the *empirical* power potential curves of Figs 10.3 and 10.4 and reckoned above an assumed natural ambient water temperature of 15°C, are shown in Fig. 10.5. They are not impressive. If James's theory (Section 8.8) of an optimum wellhead pressure of about 6 ata is correct, with perhaps 4½ ata at turbine entry, a thermal efficiency of only about 15% could be expected with a 4 inch (102 mm) Hg back-pressure, if steam only were used. When it is remembered that a modern conventional thermal plant should have an efficiency of well over 30%, this shows how extremely cheap earth heat must be for geothermal power to be economically competitive (as it undoubtedly can be) with fuel-fired conventional thermal power.

If a wet field is being exploited, and use made of the hot water by double flashing above ground, the generating efficiency would be reduced although the amount of power generated would be increased. For example, let it be

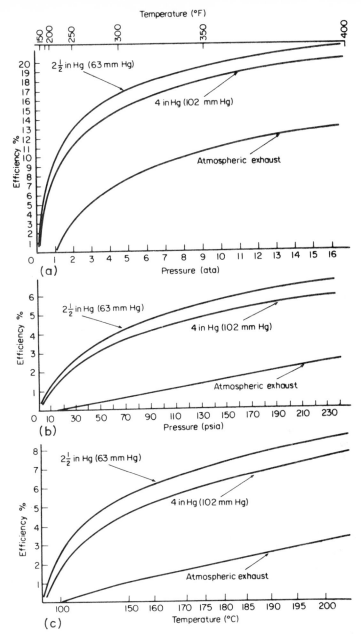

Figure 10.5 Typical geothermal generation efficiencies. (a) Dry saturated steam; (b) boiling water (single flash); (c) boiling water (double flash).
Notes:
  (i) Pressures and temperatures are at turbine entry.
 (ii) All horizontal scales apply to all three figures.
(iii) No allowance is made for gas exhausters or other auxiliaries.
(iv) The lowest curves in (b) and (c) explain the remark in the note below Fig. 10.4 that flashing would not be economic in conjunction with non-condensing sets.

assumed that a 4:1 water/steam mixture be delivered (separately) at a power plant with steam at 4½ ata and hot water at 158°C (1.4°C below the wellhead temperature, assuming a 25% absolute steam pressure drop in transmission), and that the hot water is double-flashed at the power plant and that the back-pressure is 4 inch Hg (102 mm Hg). Then it can be shown that the mean efficiency would be only 10.05%. At first sight this reduction of efficiency from almost 15% to about 10% despite the increased use of heat may seem surprising. But the explanation lies in the fact that low though the efficiencies of Fig. 10.5 may be, they are not nearly so low as the *overall* efficiencies if reckoned on the total fluid heat yielded at the wells. In the example cited above, the efficiency of using the steam phase alone, though 14.9% intrinsically, is only 7.27% when compared with the total heat yielded up by the wells. This large difference is explained partly by the rejection of the hot water and partly by the energy losses in transmission. Countless other examples could be cited to show the very low overall generation efficiencies in terms of total well heat yields in wet fields. Although the use of the hot water will raise this overall efficiency, it will still be considerably less than the efficiency of power generation by steam alone owing to the very low efficiency at which power can be generated from flashed hot water. This, however, does not imply that the use of the hot water for power generation cannot be economically justified. There is more to the economics of power generation than the efficiency of just one part of the process.

With dry bores efficiencies should be slightly better than indicated by Fig. 10.5(a) owing to the higher average dryness of the steam during its passage through the turbine resulting from the use of slightly superheated steam. But as the degree of superheat is usually moderate by the time the steam reaches the plant, the differences are likely to be small and may well be swamped by the quality differences from plant to plant. For example, at the Geysers the makers of one of the 55 MW (gross) units specify a steam consumption of 907 530 lb/h at 100 psig/355°F entry with a 4 inch Hg back-pressure. This is equivalent to an efficiency of 17.3%, which lies almost exactly on the appropriate curve of Fig. 10.5. As there should be no condensation losses in transmission, owing to the superheated condition of the steam, and only very small dry drain losses; and as there is no hot water to be rejected, the overall plant efficiency reckoned on the total bore heat yield could probably be arrived at approximately by allowing about 5% heat loss in transmission – i.e. 17.3×0.95 or 16.43%.

If two-phase fluid transmission is adopted, use is sure to be made of the hot water. The absence of rejected condensate in transmission, and the gain in steam quantity from flashing as a result of the pressure drop should help to make generation rather more efficient (as well as cheaper in capital outlay) than with separate fluid transmission. Likewise, pressurized transmission effected by down-hole pumps should enable power to be generated more efficiently from wet bores than with any of the alternative means of transmission, because the temperature loss with hot water transmission is very small and virtually the full wellhead fluid potential becomes available at the plant; it being necessary only to deduct the power absorbed by the down-hole pump(s).

It is emphasized that all the empirical curves of Figs 10.3, 10.4 and 10.5 must be viewed in the light of the fact that the non-condensible gas content in geothermal fluids vary widely from field to field, and that this variation will affect overall plant performances and efficiencies. Hence the figures, for this reason also, can be no more than indicative of conditions that have broadly been experienced in exploited fields.

Although low efficiencies are undoubtedly a feature of geothermal power generation at present, there is no reason why this should always be so. If we should ever succeed in penetrating to great depths at acceptable cost, or in tapping magmatic pockets at moderate depth, it may become possible to attain far higher fluid temperatures and pressures, which, as can be seen from the trends of the curves in Fig. 10.5, and as must also follow from thermodynamic considerations, will certainly greatly raise power generating efficiencies. Again, there is the possibility of improving the efficiency of field use as such, by means of reinjection (see Section 16.4) which could perhaps offer an alternative – and possibly a cheaper one – to the direct use of the hot water issuing from a wet field by adding to the field life rather than by extracting more heat from the bore fluids. The whole question of the efficiency of geothermal exploitation is undoubtedly very complex.

10.5 Use of hot water: general

Since wet hyperthermal fields are more common than dry, very careful thought should be given to the various ways in which energy can be extracted from the hot water phase, bearing in mind what has been said in the preceding Section. Much of this problem has been discussed and analysed elsehwere [106]. It is instructive to list the various methods:

(i) *Reinjection.* See Sections 10.4 and 16.4.
(ii) *Use of binary fluids.* See Section 10.2.6. The use of binary fluids need not of course be limited only to the water phase of wet fluids: it could equally be applied to both the steam and hot water phases simultaneously.
(iii) *Use for some non-power purpose* in a dual or multi-purpose project. (See Chapters 2 and 12.)
(iv) *Use of hydraulic pressure.* Except in the case of geopressurized fields (See Section 5.16) this is seldom of much value, as it is usually insignificant in comparison with the *thermal* pressure resulting from the temperature of the water.

(v) *Thermal use in conjunction with conventional power plant.* In 1961 Hansen [119] suggested that hot geothermal waters could be used as a means of feed-heating in a conventional fuel-fired plant in place of the usual bled steam feed-heating system, thus allowing *all* the steam to pass right through the turbine into the condenser and so generate more power. He claimed that the performance of a 150 MW conventional steam station (100 kg/cm^2/538°C/3.825 mm Hg) could be improved by a 7% reduction in heat rate if about 450 kg/h of geothermal hot water at 197°C were used in place of bled steam in all but the highest pressure feed-heater. It might be objected that, where high-temperature fields are concerned, this interesting proposal presupposes just the right combination of fuel requirements and geothermal availability, and a nice balance of costs. Nevertheless, the broad idea was again raised in 1975 by Harrowell [120], and it was very nearly adopted when a low-grade aquifer was discovered near Southampton in southern England and a borehole was sunk in 1980 adjacent to the 480 MWe oil-fired Marchwood power station. This plant then held a rather low position in the 'merit order' of the national power stations, as it was aging; but it was estimated that by using the 67°C bore water for feed heating, the station efficiency could be substantially improved, the plant's position in the national merit order raised, and about 2800 tons/year of oil fuel could be saved. Unfortunately, as a result of a change in the relative prices of oil and coal, Marchwood had to be demoted by the Generating Board and assigned to very low load factor duty; so nothing came of the proposal as it ceased to be economic. Nevertheless, in another situation the use of hot geothermal water might well prove to be economically attractive if used in this way.

(vi) *Isentropic expansion,* by converting the heat energy of the hot water into kinetic energy to be harnessed in a turbine. This proposal is fraught with difficulties, but it will be discussed in Section 10.15.

(vii) *Isenthalpic expansion,* or flashing. This is the conventional method of using the heat energy of geothermal hot water, as proposed in Sections 10.2.4 and 5, and illustrated in Figs. 10.1(d) and (e). Flashing is sufficiently important to warrant a separate section (Section 10.6).

The fact that the intrinsic energy of hot water per kilogram is far less than that of steam at the same temperature is sometimes offset by the far higher gravimetric proportion of water to steam, which may mean that the energy content of the water phase may be comparable with that of the steam phase of a mixed fluid issuing from a geothermal bore. Indeed, the *heat* content of the water may exceed that of the steam, though its lower grade may make it less valuable than the steam for conversion into mechanical and electrical energy.

10.6 Flashing

The quantity of flash steam emitted from boiling water when its pressure is reduced to a value below saturation level, and when expansion takes place isenthalpically – i.e. when all kinetic energy temporarily generated is re-converted by friction into heat – can be deduced from the formula:

$$f = \frac{H_w - h_w}{L},$$

where f is the fraction of the water flashed into steam;
  H_w is the enthalpy of the hot water before flashing;
  h_w is the enthalpy of the residual water after flashing; and
  L is the latent heat of evaporation at the lower pressure.

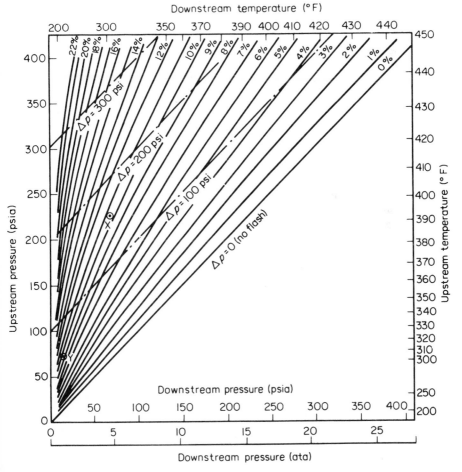

Figure 10.6 Isenthalpic flash chart.

The interpretation of this formula can be effected from the steam tables, but for approximate ready reckoning Fig. 10.6 may be useful. The flash steam quantities, when multiplied by the power potentials of dry saturated steam deduced from Fig. 10.3 will give the power potential of the hot water as deduced from Figs. 10.4(a) or (b), provided that the flashing temperatures are taken at the optimum values between the initial and 'sink' temperatures of the cycle.

Example. 1 t/h of boiling water at 200°C (15.34 ata) is to be double-flashed into a turbine, the condenser of which is maintained at a temperature of 52°C (102 mm Hg). The optimum flashing temperatures will be:

1st stage $200-\frac{1}{3}(200-52) = 150.7°C$ (equivalent to 4.77 ata)
2nd stage $200-\frac{2}{3}(200-52) = 101.3°C$ (equivalent to 1.04 ata)
First state flash $H_w = 203.6$ cal/g $h_w = 151.6$ cal/g $L = 504.9$ cal/g

Hence $f = \dfrac{203.6-151.6}{504.9} = 0.103$ t/h

Residual water $= 1-0.103$, or 0.897 t/h

Second stage flash $H_w = 151.6$ cal/g $h_w = 101.2$ cal/g $L = 538.5$ cal/g

Hence $f = \dfrac{151.6-101.2}{538.5} \times 0.897 = 0.0839$ t/h.

Power from 1st stage flash at 4.77 ata $= 0.103$ t$\times$114 kWh/t (Fig. 9.3)
Power from 2nd stage flash at 1.04 ata $= 0.084$ t$\times$58 kWht/t (Fig. 9.3)
i.e. 11.74+4.87, or *16.61 kW.*

Check. From Fig. 10.4(b), power potential of boiling water at 15.34 ata
$= 16.6$ kWh/t.

According to these figures, the first and second stage flash proportions amounted to 10.3% and 9.36% respectively, and are represented by points X and Y in Fig. 10.6.

Despite the optimum flashing temperatures determined by dividing the total temperature range by an integral number of flashing stages (2 for single- and 3 for double-flashing), it may often be convenient to choose rather different flashing temperatures (and pressures) in order to take advantage of existing pressure systems that have been established for other reasons. For example, in the experimental hot water scheme at Wairakei, where the temperature of the hot water arriving at the plant was normally about 193°C and the vacuum temperature was 33.3°C, the optimum flashing temperatures would have been 140°C and 86.5°C for two-stage flashing. Apart from the undesirability of adopting a sub-atmospheric pressure for the second stage, it was convenient to flash into two established steam systems at the power station operating at about 148°C and 102°C. This involved a sacrifice of only about 4.3% of the ideal power obtainable from the hot water, because the curves of power potential against various intermediate temperatures are fairly 'flat'.

If flashing is allowed to take place by simply admitting boiling water into a flash vessel at a lower pressure, the operation passes momentarily through an *isentropic* stage in which the heat energy of the boiling water is at first converted into kinetic energy. Although this kinetic energy is subsequently destroyed by friction, so that the overall process becomes *isenthalpic,* high water jet velocities are temporarily achieved. For pure isentropic expansion, boiling water could gain a jet velocity of $91.4\sqrt{\Delta H}$ m/s, where ΔH is the heat drop in cal/g. This figure would be modified in practice by imperfect conversion of heat into kinetic energy, but the actual jet velocities can nevertheless be substantial. Fears were felt at Wairakei, before the hot water flashing experiment was undertaken, that the wear and tear on flash vessels might be excessive. For example, in expanding from 193°C to 148°C the isentropic heat drop is 2.683 cal/g and the theoretical jet velocity would approach 150 m/s. Various ways were considered for minimizing the isentropic factor so that isenthalpic expansion could be effected as gently as possible. These included a gang-operated group of valves in series which would reduce the heat drop (and therefore the velocity) across each valve, a re-boiler, a resistance pack approximating to an ideal 'porous plug' and a thermo-compressor arrangement. However, it was found that by admitting the flashing water tangentially into a cylindrical vessel – i.e. by adopting in effect a simple cyclone separator – the wear and tear from the flashing process was negligible.

Although the experimental hot water transmission scheme at Wairakei (see Section 9.9) had to be abandoned through lack of hot water, great ingenuity has since been exercised in salvaging much of the energy content of the hot water by the process of 'inter-flashing' – that is, flashing in the bore field instead of at the power station. Owing to the adoption of two separate steam systems operating at different pressures it was possible to pass the discarded hot water from some of the higher pressure wellhead separators through other (flashing) separators situated close to the bores which discharged their flash steam into the lower pressure steam pipework system, discarding their hot water at a lower temperature than would have been possible had the water been rejected from the higher pressure separators alone [121]. Although this enabled about two-thirds of the power potential of some of the hot water from the higher pressure wells to be recovered, and avoided the expensive hot water transmission system, it involved the transmission of low-pressure steam from the field to the plant, and this is the least efficient method of heat transmission (by comparison with high-pressure steam and hot water), as will be seen from Section 14.12. Nevertheless, the arrangement possessed overall economic attraction and it was later extended by the installation of yet a *third* steam transmission pipe system operating at a still lower pressure – about 2 atü in the bore area and slightly above atmospheric pressure at the receiving end, for admission into the 30

MWe pass-in turbines at the low pressure stop valves. The economics of transmitting steam at such a low pressure over a distance of 2 to 3 km would usually be unattractive; but at Wairakei this third steam system has enabled the power plant to maintain a high output despite the falling pressure at entry to the high-pressure turbines and has thus put to good use capital expenditure incurred 25 years ago when prices were low.

10.7 Flash steam scrubbing

As already mentioned in Section 9.10, the presence of even very small quantities of chlorides can do severe damage to turbine blades in the presence of bore gases. In the case of remote bore steam it has been shown (again in Section 9.10) how the scrubbing effect of long pipelines can remove virtually all chlorides that may be carried into the pipelines as a result of imperfect separation at the wellheads. Interflashed steam would also be purified in this way, but when flash vessels are located close to the plant it is vitally necessary to 'scrub' the flash steam before admitting it into the turbines, because with flash vessels a small amount of highly concentrated chloridic water is unavoidably carried over.

Fig. 10.7 illustrates the type of scrubber used at Wairakei for the experimental hot water scheme. The flash steam with its small entrainment of chloridic water droplets enters at the side near the bottom of the cylindrical scrubber and passes vertically upwards through three mesh separator packs and two impingement 'bubble-plates', leaving at the top for admission to the turbines. Stainless steel wire meshes were first used at Wairakei for the separator packs, but these were rapidly eaten away; which proved the need for scrubbing, especially as the H_2S content of flash steam is very small by comparison with that of the bore steam with which the former would mix when passing through the turbines. Inert plastic meshes were successfully substituted for stainless steel. Chloride-free wash water is flushed over the impingement plates, falling by gravity to a discharge point lower down the scrubber, as shown. The turbine makers had stipulated a maximum chloride content of 10 ppm in the *water phase* of any final carry-over entrained in the scrubbed flash steam – not 10 ppm related to the total mass flow of the slightly wet steam. The chloride content from the entrained water droplets from the flash vessels was about 4550 ppm. Thus the scrubbers had to reduce the chloride concentration by 455 times. The assumed efficiency of the mesh separators was 97.5% – allowing 2.5% of the liquid phase to pass through and rejecting 97.5% downwards. The Wairakei low pressure scrubbers were rated for a steam flow of 63500 kg/h and a carry-over of 3855 kg/h (about 6% of the steam flow) of water was assumed to be lifted from each impingement plate, most of it being captured and thrown back in the next mesh separator above. It was also assumed tha the flash steam entering the scrubber contained 0.5% of entrained water (317.5 kg/h). Under the

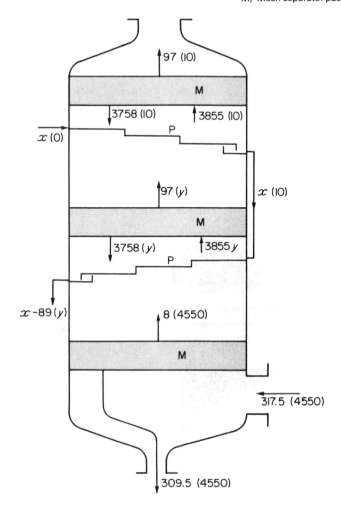

P, Impingement or bubble plates
M, Mesh separator packs

Figure 10.7 Diagram of low pressure flash steam scrubber as used at Wairakei for the experimental hot water transmission scheme.
Notes:
(i) Figures show water flows in kg/h at various points, with the chloride content in ppm in brackets.
(ii) The wash-water x is taken from the station separators and is virtually chloride free. Its temperature will approximate to that of the entry steam (about 100° C), so the steam throughput (63.5 tons/h) suffers no appreciable loss. In fact by draining the wash-water from the intermediate pressure scrubber a slight gain is won from flashing.
(iii) Entrainment of water from bubble plates is approximately proportional to the volumetric steam flow.

assumptions made it is estimated that about 590 kg/h of wash-water would be necessary to ensure that the entrained water (estimated at only about 0.15% of the total fluid) would contain chloride not exceeding 10 ppm in concentration. A larger quantity of wash-water would reduce the chloride content of the final carry-over still further.

In Fig. 10.7 water quantities are shown in plain numbers (kg/h) while chloride concentrations in ppm are shown in brackets. The calculations are as follows:

Upper impingement plate

Water balance $\quad x + 97 + 3758 - 3855 = \text{overflow} = x$

Chloride balance $\quad 10(3855 - 3758) + 10x = 97y$

$$970 + 10x = 97y$$

$$x = 9.7y - 97 \tag{1}$$

Lower impingement plate

Water balance $\quad x + 8 + 3758 - 3855 = \text{overflow} = x - 89$

Chloride balance $\quad (8 + 4550) + 3758y + 10x = 3855y + (x - 89)y$

$$36400 - 8y - xy + 10x = 0 \tag{2}$$

Substituting Equation 1 in Equation 2

$$36400 - 8y - 9.7y^2 + 97y + 97y - 970 = 0$$
$$y^2 - 19.175y - 3652.5 = 0$$

Hence $\quad y = 70.778$

and $\quad x = (9.7 \times 70.778) - 97 = 589.55$.

Thus the wash-water flow-rate would be about 590 kg/h and the chloride concentration on the lower impingement plate would be rather less than 71 ppm. Not only has the chloride concentration been reduced to the required level of 10 ppm but the flash steam has also been made drier, by reducing the wetness from 0.5% to about 0.15% by weight.

10.8 Condensers and vacuum

Since no boiler feed is needed with geothermal power plants, condensing turbines need not usually be provided with costly surface condensers. Relatively cheap jet condensers may be used instead, in which the turbine exhaust directly impinges upon, and mixes with, the cooling water so that the two fluids are discharged together. The cost saving arises mainly from the absence of tubes, while a smaller terminal temperature difference can be achieved by the direct contact. With condensing geothermal turbines the vacuum can be maintained either by direct once-through river or lake cooling or by closed loop cooling through a cooling tower. If there is an abundant supply of natural cooling water available as, for example, is provided by the Waikato River at Wairakei, direct cooling has much to commend it, but it may give rise to pollution troubles (see Chapter 16). Cooling towers have the advantage of requiring no make-up water, other than for starting-up purposes, and can therefore serve a geothermal power plant in an arid location. The reason for this lies in the wet exhaust that results from using steam at low pressures and temperatures. For example, let it be supposed that a geothermal turbine is supplied with dry saturated steam at 7 ata and exhausts to a vacuum of 102 mm Hg (4 in Hg). Assuming a turbine efficiency of 75%, it can be seen from the Mollier diagram that the wetness at exhaust would be about 12%. The total heat of the exhaust, at this wetness, would therefore be $[(0.88 \times 619.88) + (0.12 \times 51.65)] = (545.49 + 6.20) = 551.69$ kcal/kg. Let it be supposed that the cooling water enters the condenser at 28°C and leaves at 48°C, at which its latent heat is 570.3 kcal/kg. To evaporate 1 kg of cooling water would require $(48 - 28) + 570.3$, or 590.3 kcal, which could be supplied by 590.3/551.69 or 1.07 kg of turbine exhaust. Thus there would be a mass gain of 7% on the cooling water circulated, and no extra water would be needed – only a purge. At part load, however, there may be insufficient condensate. The turbines at the Geysers cannot operate at less than half load for this reason, without external make-up. The quantity of cooling water required will depend upon the desired vacuum, the temperature of the cooling water (either from a natural direct source or from a cooling tower), the wetness of the steam at the turbine exhaust and the quantity of steam 'swallowed' by the turbine; but the cooling water may be 30 to 40 times the steam consumption of the turbine.

All the major existing geothermal power plants use cooling towers, except Wairakei where there is a large river available for direct cooling. The choice of a riverside site was also originally influenced by the flatness of the terrain and the absence of hot springs beneath the foundations, and also partly by the belief that the centre of the bore field was only about 1.5 to 2 km distant. Subsequent development of the field showed that the more productive bores were rather more distant than had at first been supposed, so that possibly a power plant sited in the middle of the bore field might have been a better choice (despite the poorer vacuum obtainable with cooling towers) since due to the proximity of the bores there would be shorter pipelines and higher available turbine entry pressures (or alternatively bigger steam yields). However, the foundation conditions in the bore field would have been difficult, so the choice of a riverside site may well have been a wise one after all.

There are various possible arrangements of condenser/turbine assemblies that are favoured by different designers, as illustrated in Fig. 10.8. Arrangement (a) is used at Wairakei: it has the advantage of easy accessibility but the disadvantage of great building height which, in a seismic area, can be very costly in steel. Arrangement (b) seems to be the most popular one: it is adopted at the Geysers (units 1 to 4), in Japan (Plates 19 and 20), Mexico

Plate 19 Onuma geothermal power plant, Japan (10 MWe). Note the low level turbine house and high level external jet condenser. See Fig. 10.8(b). (By courtesy of the Japan Geothermal Energy Association)

Plate 20 Onikobe geothermal power plant Japan (25 MWe). (By courtesy of the Japan Geothermal Energy Association)

and the USSR. It has the merits of cheaper building costs, with the turbine at ground level; but the disadvantages of a tall external structure for the condenser and barometer pipe, a large and expensive vapour duct (of corrosion-resistant metal which must be kept drained (by means of an extraction pump) and may be vulnerable to frost) and a less easily accessible condenser. Arrangement (c) which is used at Larderello, is a hybrid solution that is fairly economical in building costs but which requires a deep pit for the barometer pipe and the use of a warm water extraction pump. Another hybrid arrangement, (d), is favoured at some of the later units at the Geysers: it avoids the deep pit of arrangement (c) but requires a larger condensate extraction pump. Fig. 10.8 is an over-simplification of the problem, which can be influenced by ground contours and the level at which cooling towers (if used) can conveniently be placed. Considerations such as these will affect the lifting capacity of pumps and the degree to which the vacuum suction can be used to induce the cooling water to flow into the condenser. All arrangements can be used either with direct river cooling or with cooling towers.

At the beginning of this section it was stated that with geothermal power plants it is not *usually* necessary to provide surface condensers. There are, however, exceptions. Where atmospheric pollution is experienced from the excessive emission of hydrogen sulphide, it may be necessary to use surface condensers despite their higher cost by comparison with direct contact jet condensers. This is because it is difficult to remove the offending gas after it has entered into solution with the condensate/cooling water mixture in a jet condenser: the problem is reduced if the relatively small quantity of condensate is kept entirely separate from the much larger quantity of cooling water. The newer power plants in the Geysers field in California are, for this reason, provided with surface condensers. The problem of H_2S abatement will be dealt with in Chapter 16.

Condenser vacuum can be maintained by means of mechanical exhausters or by ejectors activated by steam or high-pressure water. Mechanical exhausters are the most economical in energy consumption, but ejectors are usually favoured because of their low maintenance costs and trouble-free operation. At Wairakei, rotary exhausters were at first installed for regular use, with standby steam ejectors. Although the latter consumed about 4½ times as much steam as that required to supply the electricity for the motors of the mechanical exhausters, they soon came to be used in preference to the rotary exhausters, which were subject to a good deal of maintenance trouble. Water ejectors are more efficient than steam ejectors as they involve no loss of latent heat, but they are costlier than steam ejectors and require large quantities of water [122]. The use of boiling water ejectors has been suggested [123] but it is not known if this has yet been tried. Two useful studies of mechanical exhausters are given in [124, 125]. The capacity of any

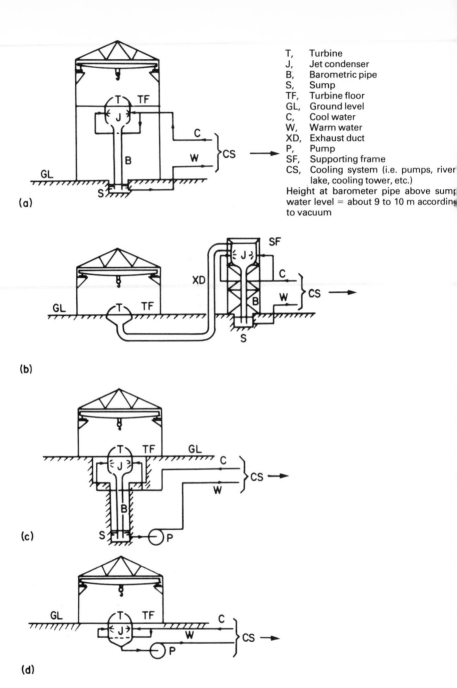

T,   Turbine
J,   Jet condenser
B,   Barometric pipe
S,   Sump
TF,   Turbine floor
GL,   Ground level
C,   Cool water
W,   Warm water
XD,   Exhaust duct
P,   Pump
SF,   Supporting frame
CS,   Cooling system (i.e. pumps, river lake, cooling tower, etc.)
Height at barometer pipe above sump water level = about 9 to 10 m according to vacuum

(a)

(b)

(c)

(d)

Figure 10.8 Different arrangements of jet condensers and geothermal turbines. (a) High level turbine; (b) low level turbine, external condenser; (c) condenser pit; (d) use of extraction pump to save height.

exhausting equipment must of course depend not only upon the condenser vacuum but also upon the proportion of non-condensible gases contained in the steam. If this proportion exceeds about 10% or so by weight, condensing machines become uneconomic owing to the high power absorbed by the gas exhaustion process. For gas contents exceeding about 10% by weight it becomes necessary to use cycles (a) or (b) of Fig. 10.1. Chierici [45] suggests that cycle (a) is suitable for gas contents of 10–15%. For very high gas contents, non-condensing turbines (cycle (b)) are the only practicable choice.

James [126] makes a case for adopting an economic vacuum of 127 mm Hg (5 in Hg) as an universal optimum for geothermal condensing plants, but this theory has not gained widespread acceptance. In general, except where large quantities of river, lake or sea-water are available, it is not worth striving for too low a back-pressure, as this can sometimes necessitate very costly cooling towers and high energy-consuming gas extraction equipment.

By comparison with conventional fuel-fired thermal power plants, the amount of heat rejected to waste by condensers is very great. A modern fuel-fired station of, say 33% efficiency rejects about 2 kWh (thermal) for every kWh of electricity generated; whereas a geothermal power plant of, say only about 16% efficiency, would reject 5.25 kWh (thermal) for every kWh generated. Hence, the cooling towers installed at geothermal power stations are enormously larger than those at conventional thermal stations of the same capacity.

10.9 Fuelled superheating

The efficiency limitations of geothermal turbines imposed by the relatively low pressures and temperatures normally encountered in hyperthermal fields could theoretically be improved by using supplementary fuel to impart a high degree of superheat to the steam before entering the turbines. This would have the further advantage of giving drier steam in the low pressure stages with consequent reduction of blade erosion, and could enable higher tip speeds to be adopted (see Section 10.10) and larger outputs per turbine of any given type to be achieved. This was considered at the design stage of Wairakei, but under the conditions then prevailing only a very doubtful economic case could be argued in favour of fuelled superheating – and then only for overload purposes. The idea was therefore not pursued. In 1970 the whole question was re-examined [127] with a view to the possibilities of supplying peak loads geothermally. It was observed that steam at 7.8 ata could yield about 25% more power per kilogram when superheated to 343°C then when saturated – at 170°C. A verdict of *non-proven* was returned because the economics of superheating were so dependent upon the cost of fuel, the shape of the load curve, the size of plant and various other factors that no general conclusion could be drawn. It is only fair, however, to say that James [126] took a more optimistic view in 1970. With well reasoned argument he deduced that, under Wairakei conditions at any rate, fuelled superheating would probably have paid at the time when he was writing, though he was not dogmatic in asserting a general case for its adoption. But the spectacular increases in fuel prices since 1973, *vis-à-vis* other forms of inflation, will now almost certainly have ruled out permanently all ideas of fuelled superheating.

A conceivable application of the superheat principle might be to dilute the wet steam part-way down the expansion process in a turbine with a proportion of superheated infeed of additional steam so as to reduce the wetness and consequent blade erosion, raise the efficiency of the lower pressure stages of the turbine, and enable the adoption of higher tip speeds for the blades of the lowest pressure stage. Much would depend upon the pressures involved and on the cost of fuel; but the possibility might be worth investigation in certain circumstances although the market trend in fuel prices suggests that the notion is not very likely to prove attractive.

In one sense this proposal to use fuel to supplement geothermal energy could be regarded as the converse of the notion advanced in Section 10.5 – to use geothermal heat to boost a conventional fuel-fired power station. A possible variant would be to sustain an aging thermal field (see Section 5.18) by means of supplementary fuel in order to get the maximum use from a geothermal power plant that has not yet completed its serviceable life. There could be circumstances where the cost of fuel would be more than covered by the extended use of capital assets that would otherwise be incapable of earning their keep.

10.10 General points of geothermal turbine design

Wood [122] has brought out some factors influencing the design of geothermal turbines. One is the permissible tip speed of the low pressure blades which, for metallurgical reasons (see Chapter 15), must be limited to a moderate value – about 275 m/s – in order to avoid excessive erosion (which is proportional to the square of the tip speed) in the wet environment of the low-pressure steam. This in turn fixes a limit to the area of the annular exhaust ring if the blade length is not to be excessive, and thus to the steam flow-rate if the leaving losses are to be kept within reasonable limits. The power capacity of a geothermal turbine will therefore depend upon the pressure drop from inlet to exhaust, the number of exhausts and the rotational speed. Until 1971 the largest geothermal turbines in the world were the 30 MW units at Wairakei, each supplied with mixed flow steam entry at 4.4 and 1.05 ata, with double exhausts to 38 mm Hg back-pressure, running at 1500 rev/min; but from that year several 55 MW units were put into service at the Geysers, California, with steam entry at 7.7 ata/179°C, 102 mm Hg back-pressure, double exhausts and a speed of 3600 rev/min. With quadruple exhausts still larger capacities become practicable. By 1975 a 110 MW unit had been commissioned at the Geysers, several more units of that rating are now in service, and one unit of 135 MW (gross) was commissioned in 1980.

The high wetness factors encountered in geothermal turbines, owing to the low steam entry temperatures by comparison with those prevailing in modern conventional thermal plants may necessitate the installation of water separators after partway expansion. At Wairakei, where the turbines are arranged in 'cascade' in three pressure ranges – 13/4.5, 4.5/1.05, 1.05/0.05 ata – separators are incorporated within the station pipework system between each pressure range so that the steam entering the next downstream turbines will be reasonably dry; while internal inter-stage separating facilities are provided within the mixed pressure turbines. In this way the wetness of the steam entering the condensers is limited to about 10 to 10.5% under the worst conditions from the mixed pressure sets and about 9% or slightly above from the 11 MW low pressure sets.

In parts of the Italian fields – particularly in the Monte Amiata region – newly sunk bores are characterized by very high gas content – sometimes as much as 80% or even more. The only way of economically extracting power from such bores is by means of non-condensing turbines. But in the course of two or three years the gas content falls to perhaps 10% or even less by weight, under which conditions a condensing plant becomes far more attractive economically. Dal Secco [124] has suggested that for gas contents in the 10–50% range the indirect cycle (Fig. 10.1(a)) would be the most economical; but for what is likely to be a transitory phase only it would clearly not be worth introducing this cycle specially. The Italians have therefore devised an ingenious two-stage development plant [128]. At first a non-condensing 15 MW unit is installed, which may give perhaps two or three years' useful service. Later, when the gas content has dropped sufficiently, they mount a second low pressure double-flow turbine coaxially with the non-condensing unit and directly coupled to a rotary gas-exhauster. This second unit is provided with a condenser. The non-condensing unit may run continuously while the second unit is under erection, but when the latter is ready for service it is necessary to shut down the first unit for about two or three days only while the two machines are mechanically coupled together, the exhaust from the high pressure unit diverted to the inlet of the low pressure unit, and the lubricating system arranged to serve both units. The result is a 15 MW 2-cylinder condensing machine with greatly reduced steam consumption which may give many years of good service. In this evolutionary way a new field or region may be developed logically and economically in a practical manner.

Blade failures have occurred at Wairakei and Matsukawa due to resonant vibrations. This is a form of trouble by no means confined to geothermal plants: it has also occurred extensively in conventional thermal power

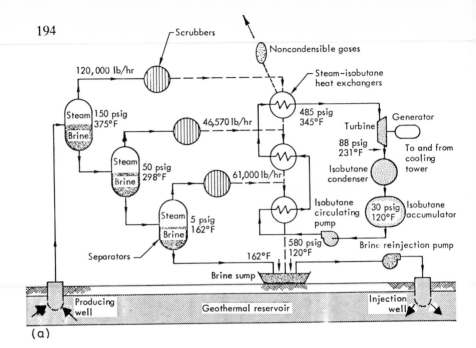

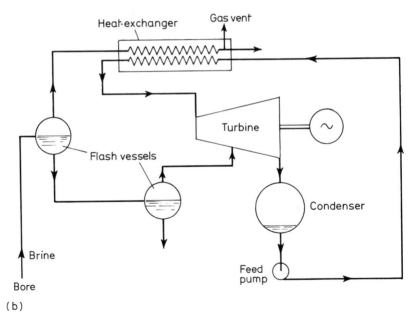

Figure 10.9 Considered methods of extracting power from hot brines. (a) Schematic arrangement of proposed three-stage flash system for use with geothermal brines in conjunction with an isobutane binary cycle (after Fig. 36 of [132], by courtesy of the San Diego Gas and Electric Company); (b) dual admission indirect steam cycle (after Fig. 4–17 of [133]).

stations in many parts of the world where the natural frequencies of blades or blade systems, vibrating in some particular mode, respond to a harmonic of some exciting frequency caused, for example, by the nozzle-passing frequency of the first stage blades. The resulting resonance can cause blade fractures through metal fatigue. Although this is not a specific weakness of geothermal turbines, matters may sometimes be aggravated in such turbines by the build-up of chemical deposits on the blades, thus altering their natural resonant vibration frequency. This is a hazard to which designers should be very much alive.

The question of turbine blade materials will be dealt with in Chapter 15.

10.11 Geothermal power plant maintenance

It is important that a systematic scheme of preventive maintenance be observed at geothermal power plants, together with rigidly planned periodic overhauls and the stocking of adequate spare parts – particularly turbine blade assemblies, both fixed and rotating. In this way high plant availability may be assured so that a geothermal power plant may operate at very high annual plant factor – often 90% or more [129, 130]. (See also Chapter 13, Section 6.)

10.12 Extraction of power from geothermal brines

In certain parts of the world – e.g. the Imperial Valley, California – the geothermal fluids consist of boiling brines of high saline concentrations of up to 300 000 ppm and sometimes having caustic properties. Whilst such brines may offer industrial possibilities for the recovery of salts and other minerals, they can also be used for other purposes including power production. The power potential of these hot brines is believed to be very great – e.g. a single 30 km^2 area in the Imperial Valley overlying such brines has been thought capable of supporting an electrical output of 15 000 MWy [131] – but their chemically aggressive nature, particularly due to the combination of chlorides and H_2S, raises serious utilization problems because of the risk of saline carry-over into the turbines. In the 1970s the San Diego Gas & Electric Company set up a test plant at Niland in the Imperial Valley with a view to establishing a 5 MWe binary cycle power plant as illustrated in Fig. 10.9(a), using isobutane as the working fluid. As the plant was for experimental purposes only, and mainly intended to discover whether the scrubbing facilities would be adequate, the turbine was omitted. The thrice repeated flashing process had the effect of raising the salinity of the rejected brine very considerably, as also the concentration of silica. It is not therefore altogether surprising that scaling problems were experienced. The test plant has since been abandoned and dismantled.

More recently, the direct transfer of heat from highly saline bore fluids to a binary fluid by means of heat exchangers, after the manner of Fig. 10.2, has proved to be more successful. Also, some of the special prime movers described in Section 10.14 could be used to extract mechanical energy from hot pressurized brines. The indirect cycle shown in Fig. 10.9(b) would seem to have certain advantages in that the gases are removed in the first-stage flash, but it might be necessary to scrub the second-stage flash.

10.13 Extraction of power from geopressurized fields

Although the exploitation of geopressurized fields has not yet become commercially established, the resource is believed to be so vast that much thought has been devoted to the question of how best the energy could be extracted when access to these fields becomes economically acceptable. Fig. 10.10 shows a schematic plant arrangement that would probably provide a good and relatively inexpensive method of exploitation. The hot mixture of water and gas, after rising up a production bore, would pass into a separator, from which most of the methane would be discharged at the top into a gas pipeline. The water, still under very high pressure, would be passed through a water turbine whereby the *hydraulic* energy would be converted into electricity. The spent water would still contain quite a lot of gas in solution, in addition to the heat content. This residual gas would be removed in a second separator, and the hot water would pass through a heat-exchanger in which it would surrender its heat to a binary fluid cycle embodying a turbo-generator and condenser. The bore water, having given up most of its heat in the exchangers, would be reinjected below ground. Fig. 10.10 is very

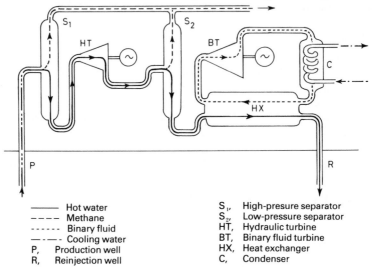

——— Hot water	S_1,	High-presure separator
– – – Methane	S_2,	Low-pressure separator
- - - - - - Binary fluid	HT,	Hydraulic turbine
— · — · Cooling water	BT,	Binary fluid turbine
P, Production well	HX,	Heat exchanger
R, Reinjection well	C,	Condenser

Figure 10.10 Extraction of power from geopressurized fields. (After Fig. 8 of [134])

much simplified, with such auxiliary equipment as reducing valves and feed pumps omitted; but it suffices to give a clear picture of what is broadly intended.

10.14 Special geothermal power generators

Although geothermal turbines are usually fairly orthodox in broad conception, a number of special power generation devices have been conceived for geothermal application. Some of these have only reached the laboratory or experimental prototype stage, while others are mere paper conceptions: but it is of interest to consider them.

10.14.1 *The Sprankle HPC (Hydrothermal Power Company Ltd) prime mover.*   A prototype of this machine has actually been working on trial in several locations since 1975. It is illustrated in Fig. 10.11 and consists basically of two intermeshed helical 'gears' [136]. Such a machine, when driven by external mechanical power, can serve as a compressor: when used in reverse, it can act as a positive displacement prime mover. Its principle intended use is for the extraction of power from boiling brines. The hot brine is admitted to the rotors at point A and passed through the helical passages B, C and D, falling in pressure and flashing off vapour as it proceeds, being finally exhausted from E. The pressure exerted by the fluid upon the helical vanes produces a rotary driving force. It is thus a two-phase flow machine. If the brine concentration is at or near saturation, salts will be deposited on the helical and stator surfaces, and the rubbing action is alleged to form a hard, smooth surface of salts which effectively seals the machine against leakage and short-circuiting of the fluids past their intended path through the machine. The makers are offering units of up to 5 MW capacity, and claim that 'the process approximates to an isentropic expansion from a saturated liquid. It is theoretically the simplest conversion process and is thermodynamically optimum'. This is tantamount to a claim of very high efficiency. Actual tests have shown *engine* efficiencies of 49 to 55% at different outputs and speeds and with a range of entry fluid pressures from 24.4 to 76.5 psia with vapour factors from 11.6% to 33.3% – all exhausting to atmosphere [137]. The expander has a high weight/power ratio and is not really suitable for use as a condensing machine. It could be useful for extracting small quantities of power from hot brines (see Section 10.12), but it will be necessary to establish whether the long term effects of internal flashing will be damaging to the machine surfaces in the way that cavitation can damage marine propellers.

10.14.2 *The Robertson engine.*   This device was not specifically designed with geothermal application in view: it was intended to operate on a variety of working fluids, but the inventor considers that it should also be suitable for extracting energy from geothermal fluids – water and/or steam. The machine is illustrated in Fig. 10.12. It has a tapered rotor revolving within a tapered stator. The rotor has a 'single-start' helix, while the stator has a 'double-start' helix of exactly twice the rotor pitch. The rotor is mounted on a crank, and its rotation on the crank-pin is governed by gears so that each cross-section of the rotor oscillates within the stator as it revolves. This

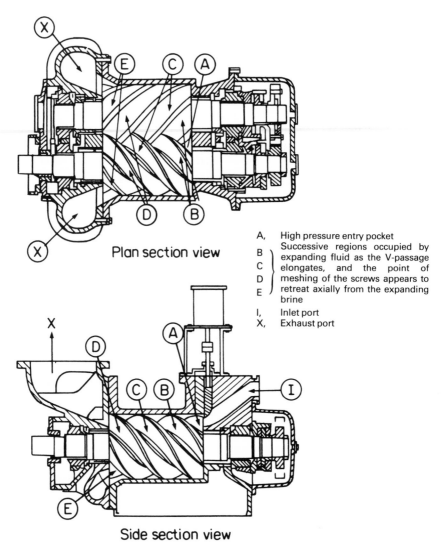

Plan section view

A, High pressure entry pocket
B ⎫ Successive regions occupied by
C ⎬ expanding fluid as the V-passage
D ⎪ elongates, and the point of
E ⎭ meshing of the screws appears to retreat axially from the expanding brine

I, Inlet port
X, Exhaust port

Side section view

Figure 10.11 The Sprankle (Hydrothermal Power Company Ltd) double helical hot brine expander. (From HPC Brochure [135])

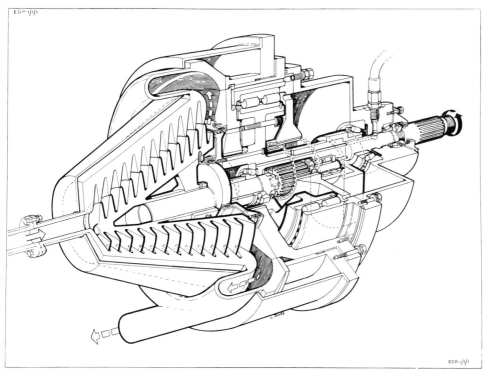

Figure 10.12 The Robertson engine. (By courtesy of the inventor)

arrangement gives rise to a series of trapped volumes which expand progressively from one end of the machine to the other as the rotor rotates. The
number of trapped volumes is half the number of rotor pitches. The taper
shape allows the rotor and stator to be lapped by means of grinding paste so
as to provide a near-perfect fit, and it allows for progressive adjustment to
compensate for wear. As with the Sprankle machine, chemical deposits,
precipitated from the water phase as it progressively boils during its passage
through the engine, would be wiped so as to give a self-cleaning action.
Power loss due to leakage past the helical surfaces would be small, partly
because of this self-cleaning action and partly because of the low pressure
drops between successive trapped volumes. Gases released from solution
would assist the steam in exerting useful work-creating forces on the rotor.
One of the main merits of the engine is that it virtually has only one moving
part, and as the flow is unidirectional there is no need for valves or timing
ports. The volume occupied by the fluid forms a higher proportion of the
stator volume than in the Sprankle machine: the Robertson machine should
therefore be lighter and more compact per kilowatt of output, though
insufficient information has been made available for the relative efficiencies

of the two devices to be compared. As with the Sprankle machine, the long term effects of flashing have yet to be proved. The designer, Mr T.J.M Robertson of the Atomic Energy Research Establishment, Harwell, England, has made and tested a small fractional horse-power model which ran on compressed air at 2500 rev/min and showed a remarkable sustenance of torque at all speeds down to the stalling point. He plans to make a 40 shaft horse-power prototype to rotate at 3000 rev/min with an inlet pressure of 100 psia saturated steam, exhausting to atmosphere – an expansion ratio of 6.8:1. The overall size of the machine is not expected to exceed about 53 cm diameter $\times$ 61 cm long, but the designer hopes to improve on this somewhat by optimization.

10.14.3 *The bladeless turbine* [136].   An assembly of plain metallic discs is mounted on a shaft, with narrow spaces between the discs. Geothermal fluid – hot water and/of steam – is admitted to the spaces between the discs at the periphery, and is directed to follow an inward spiral path towards a central exhaust port. The driving force is derived from frictional boundary layer drag between the fluid(s) and the disc surfaces. Such a machine would be very cheap to construct and to maintain: the discs could easily be dismantled, cleaned of chemical deposits and reassembled. Nevertheless, its efficiency is unlikely to be high as much energy would be wasted in frictional turbulence, and as centrifugal forces would act in opposition to the centripetal flow induced by the pressure drop between inlet and exhaust.

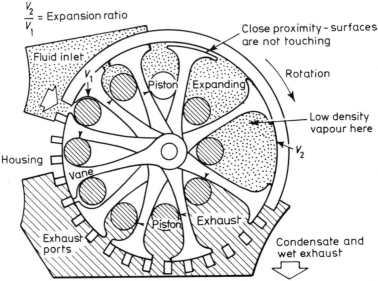

Figure 10.13 Diagrammatic representation of the KROV (Keller Rotor Oscillating Vane) positive displacement machine. (After Fig. 6 of [136])

10.14.4 *The KROV (Keller Rotor Oscillating Vane) machine* [136].   This is a positive displacement machine (as in fact also are the Sprankle and Robertson machines) designed to deal with geothermal fluids of all qualities. Its somewhat fanciful action is self-evident from Fig. 10.13. Although it should be capable of handling a wide range of fluid conditions over high expansion ratios it is mechanically very complex and its efficiency is unlikely to be high. Nor is the sealing between inlet and exhaust, which would be dependent upon contact between the roller-pistons and vanes, likely to be very sound.

10.14.5 *The Armstead–Hero turbine* [138].   The intended purpose of this concept is to provide an extremely cheap turbine capable of handling wet or dry geothermal fluids at the wellheads of newly sunk productive bores, so as to put to immediate use the fluid that would normally be blown to waste. It could thus serve as a pilot plant. It could also be used as an emergency standby unit, for which high efficiency is unimportant, and could even make small contributions to system peak loads. As a pilot plant it could be moved from field to field as early bores in one field are brought into permanent service and as new fields are explored. In principle the machine reverts to the 2000 year old Hero engine, in that it is a reaction turbine of the simplest type. For use with dry steam the arrangement shown in Fig. 10.14(a) would be used; steam being admitted through a hollow shaft and discharged through tangential peripheral nozzles provided with a small axial 'rake' to clear the exhaust away from the flat cylindrical rotor. For wet steam, the two phases would be separated within the rotor itself by centrifugal action and the use of a scoop (Fig. 10.14(b)) which would enable the water to be discharged at a velocity appropriate to its intrinsic heat energy (which would of course be less than that of the steam discharged at the periphery).

It is recognized that boiling water cannot be expanded isentropically because at the moment of passing through a nozzle it becomes a water/steam mixture and, as will be shown in Section 10.15, such mixtures cannot be efficiently converted into mechanical work. However, by discharging the water at short radius it is at least possible to avoid the drag that would result if the water were carried to the periphery; and it is possible that some very small contributory torque could be supplied by the hot water, even though it would be far below the theoretical intensity. The water jet quantity would be self-adjusting owing to the behaviour of boiling water when passing through bell-mouthed orifices (see Section 9.4). Advantage is taken of the moderate sacrifice of efficiency resulting from a large drop in rotational speed to limit the tip speed to a practicable maximum value. Steam friction would probably limit the efficiency to a modest level, but the extreme cheapness of the machine, with its absence of blades and its adaptability to different fluid conditions simply by changing fixed nozzles, should make it economic for limited purposes.

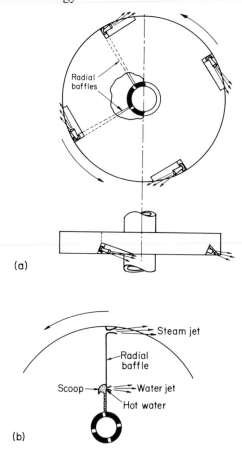

Figure 10.14 The Armstead–Hero turbine for use at dry or wet geothermal wellheads. (a) As used with dry steam. The recessed nozzles, the number of which is arbitrary, are 'raked' so as to discharge the exhaust with an axial component of velocity. (b) As used with water/steam mixtures. The water discharge is controlled by a self-regulating orifice and is placed at a radius suitable to its isentropic discharge velocity. (After Figs 5 and 7 of [138])

10.14.6 *The gravimetric loop machine* [139].   This is essentially for use in a binary fluid circuit. Fig. 10.15 shows the principle. Three parallel fluid columns of considerable height (64 m in a protoype) are arranged to form two loops, the central column being common to both. In the left-hand loop a refrigerant (e.g. Freon 11) is pumped counter-clockwise round the loop: in the right-hand loop, water is circulated clockwise by gravimetric forces. The bottom of the water loop is heated by geothermal fluid. The heated water, on contact with the Freon at the bottom of the central column, causes the latter to boil, and the released Freon bubbles give buoyancy to the fluid in this column so that the water and boiling Freon rise to the top where the two fluids, being immiscible, are separated. The Freon is liquefied by means of a condenser and is recirculated round its loop by a pump, while the water crosses over to the right-hand column and descends, passing through a water turbine near the bottom of that column. The different densities of the fluids in the right-hand and central columns provide the driving head for the turbine.

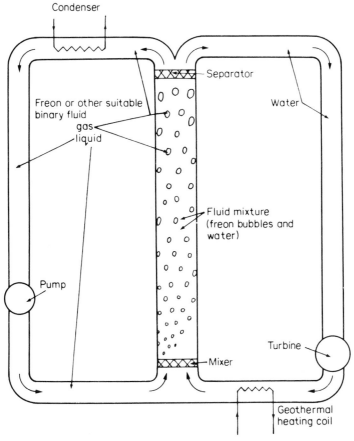

Figure 10.15 The gravimetric loop binary system. (After Fig. 1 of [139])

10.14.7 *EGD (Electro-gas-dynamics)* [140].   This is a concept developed in the USA with a view to the direct conversion of combustion energy into electricity without the intermediary of an engine. The system has certain similarities to MHD (magneto-hydro-dynamics) except that it lacks a magnetic circuit and produces electricity from a blast of ionized particles through an electrostatically charged field. In 1966 the author suggested to

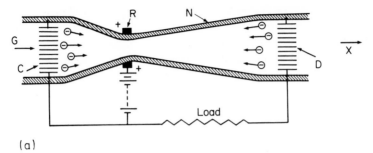

(a)

G, high pressure geothermal water/steam mixture with saline water phase
C, charging grille-cum-atomizer
R, charging ring
N, expansion nozzle of electrically insulating material
D, Discharging grille
X, low pressure exhaust
⊖→, charged particle with direction of *pull* (not motion)
       Although d.c. has here been shown to demonstrate the principle, a.c. could be used instead

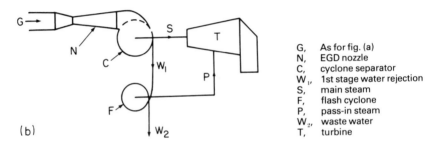

G,	As for fig. (a)
N,	EGD nozzle
C,	cyclone separator
W_1,	1st stage water rejection
S,	main steam
F,	flash cyclone
P,	pass-in steam
W_2,	waste water
T,	turbine

(b)

Figure 10.16 Electro-gas-dynamics, or EGD. (a) Principle of EGD; (b) possible use of EGD in connection with 'bottoming turbine'.

the patentees that their principle could perhaps be adapted to the production of power from geothermal fluids, somewhat in the manner shown in Fig. 10.16(a). It depended upon the fact that all known geothermal waters are to some degree saline, and therefore electrically conductive. The water/steam mixture would be expanded, as nearly isentropically as possible, in a convergent-divergent nozzle, upstream of which the fluid would pass through an electrically charged grille, emerging as a stream of charged saline water droplets in an uncharged 'wind' of steam. As expansion proceeds in the nozzle, the droplets would explode into a fine atomized spray and would be carried downstream by their released heat energy against an electrostatic pull from a charged ring at the throat. Upstream of the throat the pull would accelerate the droplets: downstream it would retard them. On reaching the discharge grille, the droplets would give up their charge and (ideally) would have expended most of their kinetic energy and would just 'drop out' at the exhaust, leaving the steam to proceed, but at reduced velocity owing to the backward drag of the water droplets. This device could be arranged in consecutive stages so as to limit heat drops per stage and keep velocities to manageable levels, with a condenser at the last stage. Alternatively the device could be used as a topping unit upstream of a turbine, and the retarded water could be flashed off at a lower pressure to provide pass-in steam (Fig. 10.16(b)). There would be formidable difficulties in putting this idea into practice, as EGD is essentially a process involving very high voltages and very low currents (megavolts and milli-amps). Very close spacings – 0.1 mm or less – between the grille plates would be necessary if reasonably high current densities are to be achieved. It is also not absolutely certain that the steam, as well as the water droplets, might not to some extent be ionized; also the mixture might have a low break-down voltage, resulting in flash-backs. Research is needed before a balanced verdict can be pronounced.

10.14.8 *The total flow impulse turbine* [136, 141, 142].   This is simply a proposal for an axial flow impulse turbine to which water/steam mixtures are admitted through expansion nozzles for impingement onto impulse blading. Research was to have been conducted in the Salton Sea area by the Lawrence Livermore Laboratory but the project has now been abandoned. The idea is not new, having already been tried at Pauzhetka in the USSR [141], where it was shown that the power developed from the water/steam mixture from a test bore was virtually the same as that produced from the steam phase alone from the same well, but that the optimal rotational speed was nearly twice as high with steam as with the wet mixture.

10.14.9 *The Biphase turbine.*   The latest development of the 'total flow' principle is the Biphase turbine (Fig. 10.17) in which the pressurized brine or water/steam mixture impinges tangentially upon a free-turning 'momentum wheel' or rotary separator, which is set spinning by frictional drag. By centrifugal action the water forms a rotating liquid ring at the inner periphery of the momentum wheel. Flashing occurs as the fluid passes through the impingement nozzle, and the faster moving steam further accelerates the momentum wheel and the liquid ring, and then passes to a conventional steam turbine (not shown in Fig. 10.17). When the rotating ring of water attains a certain thickness it passes through ports and impinges upon impulse tubes fixed to the arms of a liquid turbine which rotates at about half the speed of the momentum wheel. Both turbines are directly coupled to the same alternator. By means of a diffuser some pressure recovery is effected in the water after leaving the liquid turbine.

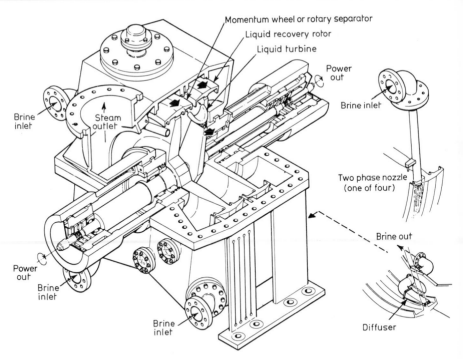

Figure 10.17 The biphase turbine. (By courtesy of Biphase Energy Systems of Santa Monica, California, USA)

10.15 The total flow concept: general

All of the nine devices described in Section 10.14 may be described as 'total flow' devices, in that they seek to handle both the water and the steam from wet geothermal bores. Even the gravimetric loop machine, in common with other binary systems, can in a sense be regarded as a total-flow device since the whole of the natural geothermal fluid can be used in the heat-exchanger to transfer its energy to the binary cycle. However, the remarks that follow apply neither to the harnessing of the total flow by thermal conduction in heat-exchangers, nor to positive displacement machines, but only to those devices in which the energy of the total bore fluid is first converted into kinetic energy before being harnessed – i.e. to the bladeless turbine, the Armstead–Hero turbine, EGD, the total-flow impulse turbine and the Biphase turbine. A fundamental difficulty arises from the fact that the isentropic heat drop, and therefore the ideal jet velocity, of steam will always greatly exceed that of boiling water over the same temperature range. Hence if both fluids are expanded in the same nozzle there will be a drag – the steam accelerating the water and the water retarding the steam. This drag, being frictional, though improving the dryness fraction, *must* reduce the nozzle efficiency. For example, let it be assumed that a 4:1 water/steam mixture is expanded from 149°C to 101°C. The heat drop would tend to produce an ideal steam velocity of 710 m/s and an ideal water velocity of 159 m/s, so that the initial drag would be 710–159, or 551 m/s. If the water were finely atomized, the velocities of the two fluids would tend to become equalized at 269 m/s (the combined momenta divided by the total mass), i.e. $[(4 \times 159) + (1 \times 710)]/5$ m/s.

The total kinetic energy of the two fluids expanding *separately* would be proportional to $(\frac{1}{2}mv^2)$ $[(4/2 \times 159^2) + (1/2 \times 710^2)]/5 = 60\,522$. But if they expanded together at the equalized velocity of 269 m/s, the total kinetic energy would be proportional to $\frac{1}{2} \times 269^2 = 36\,180$. Hence, if drag were eliminated by the equalization of the steam and water velocities, the nozzle efficiency could not exceed $36\,180/60\,522$, or about 60% only. Other expansion ranges would produce other efficiencies.

If, on the other hand, each fluid maintained its own isentropic velocity regardless of drag, the trouble would then be transferred to the discs of the bladeless turbine and to the blades of the total flow impulse turbine which would be struck by two fluids at different velocities, so that the discs and blades could not move at an ideal speed to suit *both* fluids. If a reaction turbine were used instead of an impulse machine the trouble would remain; for a blade speed suited to the steam velocity would be too high for the water velocity, and the water would have to be accelerated and would act as a brake on the driving force on the blades. Alger [142] quotes laboratory tests, as measured by thrust forces, to show that two-phase fluids can expand supersonically in a single nozzle at coefficients of 0.92 to 0.94, equivalent to efficiencies (proportional to the square of the coefficients) of about 85 to 88%. This seems to suggest that in the throat of the nozzles the two fluids are still moving at very different velocities and that sufficient time has not elapsed for drag to have had much equalizing effect. But between the throat of the nozzle and the blades of the total flow impulse turbine (or the discs of the bladeless turbine) the velocity differential of the two fluids will have become reduced, with consequent loss of kinetic energy. This plus the fact that residual drag would cause the two fluids to strike the blades (or discs) at different speeds would suggest low efficiency prime movers even if the nozzles were efficient. The experience at Pauzhetka (see Section 10.14.8) supports the notion that the expansion of two-phase fluids through the same nozzle is unsound practice, particularly if both fluids thereafter strike the same moving part of a prime mover. EGD would also suffer from nozzle drag.

With the Armstead–Hero turbine the two fluids are separated before expansion, so this trouble would not arise, though the water jet would be inefficient (as explained in Section 10.14.5).

In the case of the Biphase turbine the nozzle could not be immune to drag, but the fact that the two fluids impinge onto different turbines after the nozzle means that the machine could make far better use of the kinetic energy than either the bladeless turbine or the total flow impulse turbine.

10.16 The contribution of geothermal power to a composite system

It is elementary knowledge that the cost of producing *anything* is made up of two components: a *fixed* component incurred in providing and maintaining the means of production, regardless of the quantity of end-product, and a *variable* component that is more or less directly proportional to the quantity of end-product.

With conventional thermal power production the fixed costs are mainly the capital charges on the plant, buildings and equipment required for generation, together with the cost of employing the supervising and operating staff: the variable costs mainly consist of fuel, though they may also include some share of the costs of repairs and maintenance. It is for this reason that electricity is normally the subject of two-part costing with a fixed cost related to the kilowatts and a variable cost per kilowatt-hour. it is also well known that a composite power system can be supplied most economically by a combination of two main types of plant:

(a) *Base load plant,* the characteristics of which are high fixed and low variable costs.
(b) *Peak load plant,* the characteristics of which are low fixed and high variable cost.

(Admittedly some plants will carry intermediate loads between the base and the peak, but this does not invalidate the broad division of plants into these two principal types).

With a geothermal power plant, for which there are no fuel costs, the usual practice is to regard *all* the production costs as fixed, with zero variable costs. (This view will be challenged on a point of detail in Chapter 14, but is broadly acceptable.) The generally agreed justification for this attitude is that once geothermal steam has been made available by means of capital expended on exploration, drilling and pipework, it may be regarded as 'free' (much as water may be regarded as free in a hydro-power installation once it has been made available at a suitable head by damming). Geothermal power plants would therefore seem to be ideally suited to serve as contributors of *base load* to composite power systems. All the major power plants in existence are, quite rightly, used to supply base load. Moreover, they can often achieve annual plant factors of 90% or more – higher than obtainable from any other type of thermal or nuclear plant. Nevertheless, a case can sometimes be made for using geothermal power for non-base load purposes [127, 143] in the following circumstances:

(i) *As a sole supplier of small systems* [127]. The advantages of 'scale effect' may sometimes offset the disadvantages of operating at reduced plant factor, so that a geothermal power plant of sufficient capacity to meet the system peak may actually produce cheaper kilowatt-hours, despite the lower plant factor, than could a smaller geothermal plant designed to carry the base load only. In the course of time, as the system grows, a point is likely to be reached when it will pay to install peaking plants of some other type: this would permit the base load to grow until it ultimately matches the geothermal plant capacity, when the production costs of that plant will fall to a minimum. Thereafter, the geothermal plant will take its orthodox place in the system as a supplier of base load, while other types of plant will carry the peaks. The objection to this method of operating a geothermal plant – initially at moderate, and later at high plant factor – is that in the early years, when on non-base load duty, it may be necessary to waste steam that cannot be used at off-peak hours. This is because, owing to the bore characteristics, load reduction can only effect a diminution of steam supply if pressures are allowed to rise, and there is a limit to the permissible pressure variation beyond which wet bores may become unstable and, with both wet and dry bores, the design pressures of the pipework, etc. may be exceeded. Of course, it might be possible to condense unwanted steam at the highest practicable temperature and to reinject the condensate, thus conserving much of the heat within the field; but this will not always be convenient (see [257]). Hence the blowing of steam to waste may be unavoidable during off-peak hours. If the field is large by comparison with the heat consumption of a small geothermal plant, this may perhaps be condoned for a few years until the base load has had time to catch up with the plant capacity; but the wastage of even a small part of a finite source of heat is not generally to be commended.

(ii) *If cheap fuel justifies the superheating of natural steam during peak hours.* This point has already been touched upon in Section 10.9; but the availability of *cheap* heat would be essential.

(iii) *If adequate thermal storage is available* [127]. In the unlikely event of a natural storage cavern of suitable capacity and depth being available, it would be possible to store heat during off-peak hours in the form of boiling water, so that a geothermal power plant could operate economically at variable load. A suggestion has been put forward by Charropin *et al.* [144] and by Meyer and Hausz [145] (see Section 12.4) that natural aquifers could sometimes be used for large scale thermal storage. This raises the possibility of reinjection of bore fluid by pumping (so as to recompress the steam) into the aquifer during off-peak hours so that fluid need not be wasted, but conserved for peak load duty at the cost of pumping energy. The economics could be doubtful, but the proposal would seem to be worth investigation. Now that submersible pumps are becoming a well-established practical proposition (see Section 9.11) it should only be necessary to reduce their output at off-peak times and to circulate sufficient fluid through a rejection well to ensure that the bores are kept hot.

(iv) *For supplying small quantities of secondary peak loads.* (By 'secondary peak loads' are meant small flat-topped strips of load in the duration curve below the extreme peak – see Fig. 10.18.) Non-condensing turbines are cheap in capital but extravagant in heat consumption: they therefore have some of the characteristics of peak load plants if heat be regarded as analogous to fuel. The suitability of using such turbines for supplying small quantities of secondary peak loads has been analysed [143] and shown to be sometimes an economic proposition, particularly where fuel is very expensive. The reason why only small quantities of secondary peak loads can normally be supplied lies in the difficulty of disposing of unwanted steam that cannot be absorbed during off-peak hours, as explained in sub-section (i) above. Of course, if some industrial or other off-peak use for heat can be found, possibly with the help of thermal storage, then there would be no reason why a dual-purpose plant should not be devised for supplying peak load power with little, if any, base load component.

(v) *Emergency or standby plant* [143]. Non-condensing geothermal turbines can usefully serve as emergency or standby plants, in much the same way as diesels are sometimes used in conjunction with conventional power plants. Their high steam consumption for the short time required would be immaterial: their cheapness in capital cost would be their great asset. Moreover, in times of temporary shut-down there will be an abundance of steam available, so that no separate bore(s) need be set aside for supplying such power units.

(vi) *Hot rock circulating systems.* If, and when, power can be extracted from artificially fractured rock by circulating water through the fissures (see Chapter 18), it should be possible to match the heat supply to the pattern of demand, at any load factor, without introducing problems of bore instability or of fluid wastage at off-peak times. This could be done by adjusting the rate of circulation, by the use of a bypass or by means of heat storage where available.

10.17 By-product hydro-power

Circumstances can sometimes arise where geothermal waters can be used 'non-geothermally', as it were [143]. In wet geothermal fields large quantities of hot water are discharged with the steam from the bores. If this water contains toxic substances that could endanger downstream drinking water or irrigation supplies it may not be permissible to discharge it into

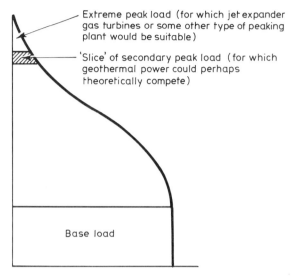

Figure 10.18 Annual system load duration curve, showing 'extreme' and 'secondary' peak loads and also base load.

water courses, in which case reinjection (see Sections 10.4 and 16.4) may be mandatory. Reinjection might also be advisable to avoid ground subsidence. Sometimes, however, huge quantities of waste water will be discarded. Where economically feasible, it is clearly advisable first to extract as much heat as possible from the water before throwing it away, either for power generation or for industrial or other purposes. Regardless of whether or not this is done, the discarded water may possess considerable potential energy if the site of the geothermal field is high above sea-level; and if the terrain is suitable much of this potential energy may be recoverable by passing the water through hydro-turbines. Indeed, at Wairakei, the bore water has for many years been discharged (regrettably, hot) into the Waikato River and finds its way to the sea through a 'staircase' of hydropower plants arranged in series along the river bed. In this way about 2.5 MW of continuous base load is being earned as a by-product of the geothermal installation – a bonus of about 2% on the geothermal energy output, equivalent to an annual saving of about 5000 tons of oil fuel (or its coal equivalent). Elsewhere, according to the configuration of the land, discharged bore waters could perhaps be fed to a separate hydro-plant, even though it might not be practicable to use the full drop to sea-level. The hydraulic energy thus generated could be used either for base load or for peak load purposes. As with all hydro projects, the potentialities would depend largely upon the terrain. Some geothermal fields are situated in mountainous regions – e.g. E. Tatio, in the high Chilean Andes – and substantial hydraulic dividends could perhaps be won in such cases.

There are three ways in which by-product hydro-power could be used:

(i) *For base loads.* The discharge of the bore water is continuous and at a fairly steady rate. Without providing any storage beyond some small forebay pondage it would be possible to generate continuous base load (Fig. 10.19(a)).

(ii) *For peak loads.* By providing storage at the level of the geothermal field for, say 20 h flow, and by letting the water down through the turbines for 4 h during the peak load time, six times as many kilowatts could be generated, though without any change in the number of kilowatt-hours (Fig. 10.19(b)).

(iii) *For augmented peak loads.* If the terrain is such that the construction of a high level reservoir is possible, the principle of pumped storage could be used to give still more peak kilowatts at the cost of some base load kilowatt-hours (Fig. 10.19(c)).

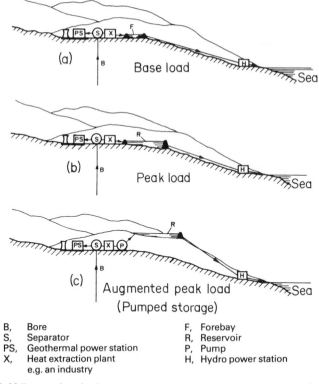

B,	Bore	F,	Forebay
S,	Separator	R,	Reservoir
PS,	Geothermal power station	P,	Pump
X,	Heat extraction plant e.g. an industry	H,	Hydro power station

Figure 10.19 By-product hydro-power plant from wet geothermal field. (After Fig.4 of [143])

10.18 What is a viable geothermal power field?

The question is often asked, especially during the exploration stage, as to what is a commercially viable geothermal field for power generation. Clearly there can be no precise answer, as so much will depend upon the costs of alternative power sources and on other local factors. Nevertheless, as a rule-of-thumb, one well known firm of consulting engineers lays down the following broad conditions as an approximate guide (as in 1976):

(i) The fluid temperature at the bottom of the bores should be at least 180°C.
(ii) This, or greater temperatures, must be found at depths not exceeding 3 km.
(iii) The yield from a 24½ cm (9⅝ in) bore should be at least 20 tons/h of *steam*.

10.19 Photographs and descriptions of existing geothermal power plants

Some idea of existing geothermal power plants can be obtained from the various plates and bibliography references quoted below. The megawatt

Plate 21 The Geysers field, California. Interior view of building housing turbine units Nos 9 and 10 (2×55 MWe). (By courtesy of the Pacific Gas and Electric Company)

Plate 22 The Geysers field, California. Power station and cooling towers serving units Nos 5 and 6 (2×55 MWe). Note the difficult terrain of incompetent rock at the construction site. (By courtesy of the Pacific Gas and Electric Company)

figures given in brackets relate to the gross installed capacities as at the end of 1980.

Ahuachapan, El Salvador (110 MWe). Plate 28.
Cerro Prieto, Mexico (150 MWe) [41, 146]. Plate 29.
Geysers, California (960 MWe) [147, 148]. Plates 21, 22, 23, 24.
Hatchobaru, Japan (50 MWe) [111]. Plates 30, 31.
Krafla, Namafjall and Svartsengi, Iceland (35 MWe total) [149].
Larderello, Italy (169 MWe) [150, 151]. Plates 32, 33.
Makaling Banahaw, Philippines (220 MWe). Plate 34.
Matsukawa, Japan (22 MWe). [152].
Otake, Japan (12.5 MWe) [153]. Plate 35.
Paratunka, USSR (0.75 MWe) [117].
Pauzhetka, USSR (11 MWe) [141].
Wairakei, New Zealand (192 MWe) [40, 107, 154]. Plates 25, 26, 27.

Plate 23 The Geysers field, California. Power station and cooling towers serving units Nos 9 and 10 (2×55 MWe). Photographed in dry weather. Note that the picture was taken before the enforcement of the anti-pollution laws. Note also the ejector discharge stack above the turbine house roof. (By courtesy of the Pacific Gas and Electric Company)

Plate 24 The same Plate 23, but photographed in humid weather. Note the greatly increased quantity of vapour by comparison with Plate 23. Note also the 'rigid' expansion loops in the steam pipes. (By courtesy of the Pacific Gas and Electric Company)

Plate 25 The A and B geothermal power station buildings, Wairakei, New Zealand. Note the clean lines and absence of the unsightly features associated with fuel-fired power stations, such as chimney stacks, smoke, fuel and ash-handling equipment, and fuel storage areas. (By courtesy of the New Zealand Ministry of Works and Development)

Plate 26 Interior view of the A station turbine house, Wairakei, New Zealand [154]. Note the high-pressure and intermediate pressure turbo sets to the left: these six units are all back-pressure sets. In the far background a glimpse can be caught of the low pressure condensing units. In the middle distance are the intermediate pressure steam manifold and inter-stage separators. (By courtesy of the New Zealand Ministry of Works and Development)

Plate 27 30 MWe, 50/0.5 psig pass-in turbo-alternator unit in the B station at Wairakei, New Zealand [154]. (By courtesy of the New Zealand Ministry of Works and Development)

Plate 28 Ahuachapan geothermal power station, El Salvador (110 MWe). Note the 'snaked' steam pipes from the bores to the power plant. (By courtesy of the Japan Geothermal Energy Association)

Plate 29 Cerro Prieto geothermal power station, Mexico. One of the 37.5 MWe turbo-units. (By courtesy of Mr Eric Jeffs, Associate Publisher, *Energy International*)

Plate 30 Hatchobaru geothermal power station, Japan (50 MWe). General view of field and power plant. (By courtesy of the Japan Geothermal Energy Association)

Plate 31 Hatchobaru geothermal power station, Japan (50 MWe). Cooling towers, turbine house and separators. (By courtesy of the Japan Geothermal Energy Association)

Plate 32 Cooling towers and a substation at Larderello, Italy. (By courtesy of Ente Nazionale per l'Energia Elettrica – 'ENEL')

Plate 33 Larderello No. 3 geothermal power station (111 MWe), Italy, showing the turbo generators in the foreground and the gas extractors in the background. (By courtesy of Ente Nazionale per l'Energia Elettrica – 'ENEL')

Plate 34 Makaling Banahaw geothermal power station, Philippines (220 MWe). (By courtesy of the Japan Geothermal Energy Association)

Plate 35 Otake geothermal power station, Japan (12.5 MWe). (By courtesy of the Japan Geothermal Energy Association)

11 Direct applications of Earth heat

11.1 General

It has been explained in Chapter 2 how the obvious commercial attractions of using geothermal energy for power generation should not be allowed to obscure the fact that it can be used far more efficiently in the form of directly applied heat for a wide variety of purposes. The constraint of the Carnot efficiency is removed if the heat can be used *as such* without first converting it into mechanical energy in a prime mover. Further savings can be affected by avoiding the losses incurred in converting the mechanical energy into electricity, in transmitting and distributing that electricity to the point of use, and perhaps in final reconversion of the electricity into the form required by the user. It has also been pointed out that about two-thirds of the world's primary energy consumption is now used for non-electric purposes, and that there are grounds for believing that there may be far more heat recoverable with existing technology from hydrothermal fields of moderate temperature as from fields at the higher temperatures required for acceptably efficient power generation. Despite the greater importance of location in determining the economic value of lower temperature fields (owing to the high cost of heat transmission) the attractions of direct applications of earth heat are rapidly becoming more widely recognized.

This question of location deserves closer attention, as it has had great influence on the history of geothermal development. The facts that Italy pioneered geothermal power generation and that Iceland pioneered geothermal space-heating, both on a large scale, were entirely determined by topography. The Tuscan field was fairly remote from any large energy market; so the ease of transmitting electricity over fairly long distances suggested power generation as the rational use to which that field could best be put. In Iceland, however, the presence of a field very close to the capital city – even within its boundaries – together with the experience of long harsh winters made space-heating the obvious choice.

With the exception of Iceland, Hungary and the USSR the rest of the

Table 11.1 1980 estimate of the approximate total world use of geothermal heat excluding bathing and medicinal applications.

Use	Country	Approximate installed capacity		Approximate maximum rate of heat extraction (MWt)[3]	Estimated or assumed annual heat load factor (%)	Energy extracted from the earth in a year	
		MWe	MWt[1]			GWht	TJ[10]
Electric power generation	World	2586	—	33 000[2]	95[4]	274 630	988 580
Direct applications for purposes other than bathing and medicinal uses (reckoned above 15°C)	Iceland	—	923	1086	50[5]	4 757	17 124
	Hungary	—	619	728	25[6]	1 594	5 738
	USSR	—	555	653	50[7]	2 860	10 295
	China (PR)	—	144	169	45[9]	666	2 397
	USA	—	ca.111	131	40[8]	459	1 652
	Japan	—	81	95	45[9]	374	1 346
	Italy	—	ca. 73	86	45[9]	339	1 220
	France	—	56	66	45[9]	260	936
	Romania	—	36	42	45[9]	166	598
	Czechoslovakia	—	35	41	45[9]	162	583
	Austria	—	2	2	45[9]	8	29
		2586	ca. 2635	33 000 + 3099 = 36 099		274 630 + 11 645 = 286 275	988 580 + 41 918 = 1 030 498

[1] Figures taken from Table 3, Orkustofnun World Survey (OWS), 1981 [156].

[2] Allowance made for under-loading, generation efficiency, type of field, auxiliary consumption, transmission losses, vented steam, bore testing, type of plant, etc.

[3] All figures in this column except the first (33 000) assume 85% extraction efficiency. (See p. 8, OWS, line 23.)

[4] This is NOT the load factor on the generators but on the bores (assumed).

[5] OWS, p. 69, bottom line. 92% of the total is for district heating.

[6] OWS, Table 1, p. 50, for total mixed system.

[7] OWS, deduced from Tables 1 and 2, p. 44 – also p. 40, line 11.

[8] OWS, p. 145, para 2, line 12.

[9] Where the proportions are not known for space-heating, agricultural uses and industry assume a mean annual factor of 45%.

[10] 1 TJ = 277 800 kWht.

N.B. It should be noted that there are considerable discrepancies between some of the figures in this table and others that have been made available to the author. To be consistent, the OWS has here been treated as 'official', but the table should be treated with reserve.

world has been rather slow to respond on a large scale to the opportunities offered by direct heat applications. The broad situation as in 1975 regarding directly applied geothermal heat has been well set out by Howard [155]. By 1980 the extent of world geothermal development was approximately as shown in Table 11.1 *excluding* heat used for bathing and medicinal purposes. The installed electric power capacity has been taken from Table 10.1. The installed heat capacities for direct applications are taken from Table 3, p. 20, of the World Survey of low-temperature geothermal energy utilization prepared by the Geothermal Division of Orkustofnun, the Icelandic National Energy Authority, published for the United Nations in April 1981 and compiled by J.S. Gudmundsson and G. Palmason (Document OS81005/ JHD02)* [156] and are reckoned above 15° C – a figure generally accepted as an approximate mean worldwide value of the earth's surface ambient temperature. The compilers of the OWS point out that the *local* mean surface ambient temperatures should more logically be used, but owing to lack of adequate data the figure of 15° C is considered to be near enough for approximate estimation. The eleven countries appearing in the table exclude several countries that use natural hot waters for bathing and medicinal purposes only, and no heat used for these purposes is included in the figures of Table 11.1. The footnotes to Table 11.1 are amplified, as follows:

[2] Krafla and Wairakei do not have sufficient steam supply to enable these plants to produce their full rated capacities of 30 and 192.2 MWe respectively. This underloading has been allowed for. In reckoning generation efficiencies, the values deduced from Figs 10.3 and 10.4 have been reduced by reasonable factors to allow for the type of field (wet or dry), for auxiliary consumption, for fluid transmission losses, for steam vented to waste for control purposes (see Chapter 13) and for sundry field losses arising from the testing and by-passing of bores, etc.

[4] Although the mean annual load factor on geothermal power plants has been taken at 80.8% for deducing the electrical energy generated (see Section 10.1) it is unlikely that the heat extraction follows the load variations on a geothermal power station. At times of underloading it is more probable that additional steam is vented or by-passed at the wellheads. An arbitrary load factor of 95% has been assumed for the steam supply as such.

[9] The assumed load factor of 45% is deduced by making an arbitrary split to 50% of the total used heat to space-heating, 30% for agriculture and 20% to industry. As the argument applies to only 17% of the total estimated energy extracted for direct applications, the effect of errors in these assumptions upon what is intended to be no more than an approximate

* Hereafter abbreviated to OWS (Orkustofnun World Survey).

evaluation should not be great. On p. 145 of the OWS, para. 2, line 9, it is suggested at 50% and 20% are typical annual load factors for space-heating and agriculture respectively, though of course there will be variations as from one country to another. The load factor for industry could be anything, but as most industries are not seasonal an arbitrary value of 70% has been assumed. The composite load factor may therefore be reckoned so:

Space-heating	50% of the demand	50% load factor	25%
Agriculture	30% of the demand	20% load factor	6%
Industry	20% of the demand	70% load factor	14%
	100%	*Mean load factor*	45%

It will be noted that the estimated world capacity for direct applications (other than baths) amounts to about 2635 MWt, while the installed capacity of geothermal power plants is 2586 MWe. A straight comparison between these two figures, which at first sight suggests that direct applications have advanced further than geothermal electric power, would be most misleading because of the lower efficiencies at which electricity can be generated geothermally. The estimated rates of heat extraction show that direct applications account for only about 8.6% of the total (3099/36099): this serves more to show the inefficiency of geothermal power generation rather than the low level of development of earth heat for other uses. When it comes to the estimated totals of extracted *energy* it will be noted that direct applications account for a still lower proportion of the total–only 4.07% (11 645/286 275) owing to the lower load factors for non-electric uses. (It is here assumed that for direct applications low grade heat is extracted only as and when it is needed). Two factors would tend to credit direct applications with larger shares of the total geothermal uses than these figures would suggest. In the first place the compilers of the OWS state that the figure of 619 MWt (Table 11.1) for Hungary is an under-estimate and that some direct heat applications, being derived from high temperature fields, have been excluded from their findings. Secondly there is a time lag in the data presented: for whereas the electrical demands have been assessed as at the end of 1980, the direct heat demands (though published in 1981) represent in some cases the conditions prevailing earlier, owing to the delays in collecting data.

It is emphasized that the figures in Table 11.1 exclude the use of earth heat for all balneological purposes. The reason for this is to stress the fact that modern direct applications have hitherto been rather neglected, as the figures show. The ancient use of heat for baths needs no advanced technology; and one of the purposes of this book is to stress the economic importance of applying modern technology to direct uses of earth heat. It may come as a surprise to some readers that the heat consumed by bathing establishments is just about *double* that now used for space heating, agriculture, industry and other non-power purposes. If hot baths are taken into account, the total installed thermal capacity in the eleven countries included in Table 11.1 rises from 2635 to 7994 MWt (reckoned above 15°C – an increase of 5359 MWt, or just over twice the lower figure. On this basis, and allowing for 85% heat extraction efficiency as before, direct applications – baths included – extract heat from the earth at about 22% of the total rate for all purposes.

With direct heat applications the base reference temperature assumes great importance, particularly where very low grade heat is used. It is logical to assess the extracted heat in terms of the mean surface ambient temperature – here taken at 15°C – because it is only by reference to this temperature that earth heat has any real meaning. On the other hand, since geothermal fluids are often discarded after use to waste at perhaps 40° C or so, the *useful* heat is less than the *extracted* heat shown by the figures in Table 11.1. If natural hot water is used at, say 90° C, and discarded at 40° C, then the *useful* heat will be $(90 - 40)/(90 - 15)$, or two-thirds of the *extracted* heat as reckoned above 15°C. Even allowing for the assumed 85% extraction efficiency, the overall efficiency would be $0.85 \times 2/3$, or 56.7%, which is still far higher than could ever be attained by generating electricity.

It is worth mentioning that the figure of about 111 MWt of installed direct heat capacity in the USA excludes a very large quantity of geothermal water used in the state of Wyoming for enhanced petroleum recovery. This process consumes something like 10^{13} Btu p.a., which is equivalent to 2930 GWht. The reason for its exclusion is that the oil producers assert that it is the *water* they need – not the *heat*. The fact that the water used comes from artesian thermal springs is alleged to be unimportant. This argument would seem to be debatable, as it seems unlikely that cold injected water would be as effective as hot for the purpose required.

As was shown in Figs 2.1 and 2.2, very many direct applications of geothermal heat can be made at moderate, or even low, temperatures; and low grade fluids are kown to be far more plentiful than the much hotter fluids associated with hyperthermal fields. As semithermal fields and low-grade aquifers are more likely to escape detection without sophisticated exploration methods, they may even be quite common in the many parts of the world that have not yet been subjected to proper reconnaissance. It has been alleged [38] that 50 to 60% of the vast area of the Soviet Union is underlain with economically exploitable hot waters; though absence of local markets matched to the temperatures available and/or unacceptable chemistry could throw some doubt upon the aptitude of the word 'economically'. What is more positively known, as a result of thorough exploration, is that about 40% of the area of Hungary covers semithermal fields [157]. It was estimated in 1970 that the recoverable heat from these fields alone was about half as much as that contained in all the world's then known reserves of oil fuel [158], disregarding grade. The opportunities for exploiting the world's heat resources of moderate and low temperature could undoubtedly be very great, and the savings in fuels that could result from this could be most valuable in helping the balance of payments problems of certain countries. The 11 645 GWht p.a. of extracted heat used for direct applications other than bathing (Table 11.1) represent a saving of about 7¾ million barrels of oil per year. This is a mere 1/3000th part of the present world rate of oil production, which shows how neglected has been the use of earthheat for direct applications other than bathing; but the new awareness of the potentialities will undoubtedly soon raise the oil savings to a significant proportion. Hungary is said to have exploited only one-fifth of the hot water resources that have already been opened up in that country; so rapid developments may confidently be expected there. Growing exploitation of low and medium grade geothermal heat is taking place in the USSR, the People's Republic of China, the USA, Japan and France: other countries too are becoming increasingly active.

Perhaps the most significant development in recent years relating to direct applications earth heat has been the exploitation in France of low-grade aquifers in non-thermal areas. It was in the course of unsuccessful oil exploration that these aquifers were discovered, and they are now increasingly being put to good use. As it had formerly been assumed that only hyperthermal and semithermal fields could be of practical use; and as France, lacking these resources, was regarded as geothermally worthless, this new source of low-grade heat has opened up a new horizon. Several other countries are now actively seeking exploitable low-grade aquifers in non-thermal areas. In the United Kingdom a bore was sunk near Southampton in 1980 and yielded about 6 MWt of low-grade heat at 67° C: it is probable that there will soon be significant developments in southern England. Elsewhere nothing very positive has yet been achieved in the exploitation of aquifers in non-thermal areas, but there are good prospects in Denmark and perhaps in the Federal Republic of Germany, the Netherlands and Sweden.

There are also several other countries having hot spring activities where the heat has not yet been put to use, except for bathing purposes if at all, but where interest has now been aroused and developments may reasonably be expected in the 1980s. Some of these countries, however, are reluctant to develop their low-grade heat resources for purposes other than 'spas', for fear of affecting their tourist trade adversely.

Direct applications of earth heat may conveniently be classified into the following broad categories, to each of which a section will here be devoted:

(a) Space-heating, domestic hot water supplies and air-conditioning.

(b) Farming, including agriculture, horticulture, aquaculture, animal husbandry, etc.
(c) Industry, especially heat-intensive processes.
(d) Miscellaneous applications that do not fall within the other three categories.

11.2 Geothermal space-heating, domestic hot water supplies and air-conditioning

11.2.1 *Approximate situation in 1980.* This is not easily assessed quantitatively. The OWS [156] gave subdivisions into specific uses for only seven of the eleven countries listed in Table 11.1; and the figures for space heating for these seven countries and also for the USA are set out in Table 11.2. They should be treated with reserve. Some of the figures, particularly those for Hungary, are known to be under-estimated; and it is by no means certain that the French figures do not include some heat used in greenhouses, which should properly fall within Section 11.3. Furthermore, the assumption of a 50% mean annual load factor in every case except Hungary – the only country to have given figures of energy consumption as well as of demand – introduces uncertainty in the energy figures shown in the table. Countries having short sharp winters would tend to have lower annual load factors than those having long, chilly autumns and springs.

Table 11.2 Approximate quantitative data for district heating, etc. as in 1980. (All figures reckoned above 15° C).

Country	Estimate of installed capacity (MWt)	Estimated energy consumed in one year (GWht)	(TJ)	Mean annual load factor
Iceland	850.1	3723.4	13 403	50% assumed
USSR	430	1883.4	6780	50% assumed
Italy	84	368	1324	50% assumed
Hungary	58.4	262.8	946	51.4%
France	55.6	243.5	877	50% assumed
Japan	49	214.6	772	50% assumed
Romania	27	118.3	426	50% assumed
USA	20.2	88.5	319	50% assumed

All the figures quoted in Table 11.2 are extracted or deduced from the OWS [156] except in the case of the USA, where they have been indirectly deduced from the US Department of Energy Geothermal Progress Monitor No. 5 of June, 1981 [159].

The seismic belt passes through five of the eight listed countries: earthquakes are rare and of weak intensity in Hungary, France and

Romania. This illustrates a point brought out in Fig. 5.1 – that semithermal fields and low-grade aquifers are not confined to highly seismic zones. Of the three other countries listed in Table 11.1, but not in Table 11.2, one (China) contains seismic zones and two (Austria and Czechoslovakia) do not.

Table 11.2 contains some surprises – notably that the USA features so low in the scale of development. It would seem that that country has pursued a similar policy with direct geothermal heat applications as with geothermal power generation, namely, one of delayed initiation followed by intense development. This is understandable as it is only in recent years that the USA has lacked plentiful energy supplies.

As might have been expected, Iceland, despite its small population, still holds the lead quantitatively, as that country has had about forty years' start over others in large-scale geothermal district heating. Nor is it surprising that the USSR comes second, with its vast area and population and with the seismic belt passing through Kamchakta and the Caucasus.

The countries that are now making the most rapid developments in geothermal pace heating are the People's Republic of China, France, Hungary and the USA, all of which are relative newcomers to this activity.

11.2.2 *Iceland.*   About 98.5% of the population of Reykjavík and its environs were served in 1980 with natural hot water for space heating and tap water: this implies virtual market saturation. Chimneys are nowhere to be seen in this area except on houses of more than about 50 years of age. Also in 1980 about 70% of the entire population of the country were similarly served, and it is expected that the proportion will reach 81.5% by about 1984 or 1985. By the end of the century it is hoped that most of the dwellings in the country, except for a few very isolated villages and farms, will enjoy the benefits of geothermal heat.

In Iceland, thermal waters for domestic heating are mostly drawn from below ground at temperatures ranging from 80 to 120° C from semithermal or from slightly hyperthermal fields such as are conveniently encountered close to, and even within, the capital city of Reykjavík. In one case a field has been operating for over 30 years at an effluent temperature of only 56° C, though this is mainly used for mixing with hotter waters. In general, those fields of low or moderate temperature produce waters of fairly low mineral and gas content, and it is of interest to note that although new thermal fields have been brought into service to cater for the expanding demand growth, the Icelandic fields originally exploited are still providing heat at undiminished output after half a century of service [8]. In some of the newer contemplated Icelandic heating projects hyperthermal fields are to be tapped, containing some rather aggressive saline fluids at 167°C or thereabouts: in such cases dual purpose power/heating projects are being planned, for which heat exchanger/mixers will have to be used so that reasonably pure water may be circulated for district heating purposes [160].

11.2.3 *USSR*.   Geothermal space-heating has been developed extensively in the USSR [38, 161, 162] in three main regions – the Caucasus, Western Siberia and Kamchatka. In 1980, 110 000 people were served with geothermal heat in the Caucasus region alone. There are expectations of further space-heating developments in the region of the Baikal–Amur railway. Most of the hot waters encountered in the Union average about 65 to 70°C, though temperatures of up to 180°C have been known. Heat-exchangers often have to be used in the USSR on account of the chemical aggressiveness of the natural thermal fluids [161] while the use of heat pumps, electricity and supplementary fuel has been proposed in certain instances [163] for peaking purposes or for boosting the temperature of the recirculation water in two-pipe systems.

11.2.4 *Hungary* [157, 164].   In proportion to its population, Hungary has developed its geothermal resources to a greater degree than any other country except Iceland. Exploitation began in 1962 and has developed rapidly. Although only about 10% of the extracted heat is used for space-heating, this is nevertheless a growing activity. The Hungarian Government forces the pace of development simply by withholding oil and gas supplies from their new state-built dwellings in the thermal areas [48]. The country derives its heat from semithermal fields which extend beyond its frontiers into Czechoslovakia, Austria, Romania, Yugoslavia and Switzerland. Temperature gradients are frequently about 50 to 60° C/km and heat flows range from about 2.0 to 3.4 μcal/cm^2 [48] – both sets of figures being about twice the world averages.

Hot water issues from the Hungarian fields at 60 to 100° C [48] from depths of about 1 km, though the aquifer extends downwards to about 3 km in depth. Below these semithermal fields some geopressurized fields are known to exist in places, with tempertures up to 200 to 300° C, and although these could perhaps be exploited for various purposes including power generation, they have not yet been developed. The semithermal fields of Hungary, for the most part, contain waters of low salinity [48]. At first, and in most cases, the water rises to the surface spontaneously, either by thermosyphon or artesian action; but after 4 to 8 years of exploitation the reservoir pressure falls somewhat, so that pumping becomes necessary at times of peak demand. The pumping requirements are small – only about 100 kWe for a well yielding about 5000 kWt of heat.

Updated information is lacking about the demographic extent of the geothermal space-heating services in Hungary, but in 1975 3500 dwelling apartments, having a total space of some 710 000 m^3 were then being heated geothermally. By 1980 these figures would have risen considerably.

Hungary possesses some oil and gas reserves, which now, thanks to geothermal heat, can to some extent be diverted to other uses.

11.2.5 *France.*   France is at present unique in having geothermal space-heating services that derive their heat from low-grade aquifers in non-thermal areas: though other countries may soon be following her example. The first French development of this kind was at Melun, to the south-east of Paris, round about 1970, when hot waters were accidentally discovered when searching for petroleum. After the 1973 oil crisis this was followed by three further such enterprises, also in the Paris Basin; and still later by two in the Aquitaine Basin. In the Paris region the temperature gradients are 30–35°C/km, only slightly above the world average, and water at 55–70°C is drawn from the Dogger aquifer at depths of 1.5–1.8 km [165]. After use it is reinjected into the aquifer, thus maintaining a circulating flow. It is understood that in 1980 about 3500 dwellings were being heated in this way at Melun, and a further 25000 dwellings (approximately) together with about 50000 m² of floor area in offices and public buildings in other places in the Paris Basin. Ambitious plans are afoot for establishing similar heating systems in other parts of France ultimately to serve about 400000 or 500000 dwellings, as deep aquifers abound in that country.

The Dogger waters range in salinity from 8 to 30 grams per litre, and contain traces of H_2S. Heat-exchangers are therefore used so that clean secondary water may be circulated through the public heating systems [166]. Two of the heating systems in the Paris region are of particular interest, as they embody some unusual features. At Creil, heat pumps are being used – not to raise the feeding temperature but to augment the hot water supply. This is achieved by recycling part of the used water, which leaves the radiators at about 40°C, and pumping into it the heat content of the remainder so as to raise the temperature of the recycled water to 70°C and lower the temperature of the rejected water to 7°C. An arrangement of three heat pumps in 'cascade' enables a performance ratio of 5:1 to be achieved. At Villeneuve-la-Garenne the production and reinjection bores are placed closely side by side and are splayed out in the manner shown in Fig. 7.4 so as to strike the aquifer at points separated from one another by about 1 km. In this way a reasonably long life of the system is assured and great economy of surface pipework effected: also, since the wells, heat pumps and other ancillary equipment can be accommodated within an area of about the size of a tennis court, land use is minimized and opportunities are offered for aesthetic concealment behind shrubs. These Paris projects are expected to provide at least 70% of the heating energy requirements; the balance being supplied at peak times by old existing fuel-fired boilers.

11.2.6 *USA.*   Geothermal space-heating has been practised in the USA since about 1890 on a very small scale; but it is only recently that it has been taken up in the form of organized public services after the manner of Iceland. The two most active states in this respect are Oregon and Idaho at present, but other states are now beginning to develop large scale geothermal space-heating. The Geo-Heat Centre at the Oregon Institute of Technology (OIT) has done much to stimulate interest in this activity. Both in Idaho and in Oregon, hot water is found close to the surface. At Boisé, Idaho, 77°C water is encountered at 130 m depth [10] while at Klamath Falls, Oregon, temperatures ranging from 38°C to 115°C are reached at depths of 27 to 500 m, depending on the height of the individual wellhead above sea-level. The water at Boisé is chemically fairly pure, but at Klamath Falls it is so aggressive that heat-exchangers must be used. For small domestic services downhole 'hairpin' bends are used for this purpose: by 1980 almost 500 of these were in use in Klamath Falls alone. They are inefficient, but cheap in capital outlay: they rely on the cross flow of natural underground water currents. The secondary clean water circuits of these small installations usually rely on thermosyphon action for circulation. In 1980 about 25% of the homes in Klamath Falls were heated in this simple way – usually with two homes sharing a single well; but in that year a more sophisticated municipal supply system was established, designed to heat 14 government buildings in 1981, 11 blocks of apartments and shops in 1982 and 54 further blocks in 1983. This system will be supplied from wells located over a fault just outside the city, and will be fed through about 4000 ft of insulated pipeline. Central tube-and-shell heat-exchangers will be used, and the spent water will not be reinjected. Some years previously the OIT complex of buildings had already been heated by a modern system of similar type: the water being pumped from a single well and delivered through a 10-in fibreglass pipe insulated with urethane foam covered by a PVC jacket. This too was a once-through system without reinjection. The supply water for the OIT and the municipal systems is at 192°F: the pH value is 8.3. Though a reserve boiler was supplied for the OIT system it has not been used for 16 years. The Klamath Falls heat demands in 1980 are expected to be doubled when the new municipal system is completed.

11.2.7 *Other countries.*   There is little of special interest to report about space-heating adopted in the remaining six countries that appeared in Tables 11.1 and 11.2, except that China is rapidly extending the use of geothermal space-heating for some of her larger cities, and that Japan uses various types of pipe laminated with synthetic materials [167] in preference to steel, so as to avoid corrosion troubles.

As to countries that do not appear in Tables 11.1 or 11.2, developments on the French model are likely soon to appear in the United Kingdom, particularly as the geology in central southern England is rather similar to that of the Paris Basin. The single experimental bore at Southampton, already referred to, tends to confirm this expectation. There are other areas in the United Kingdom that hold promise for future space-heating activities

[52]. In Australia and Canada there are some modest geothermal space-heating activities of too small scale to warrant their inclusion in Tables 11.1 or 11.2. Even in the sparsely populated regions of the Indian Himalayas [168] some small developments are planned. At Rotorua, New Zealand [169] hyperthermal gradients are exploited to depths of 15 to 366 m, yielding temperatures of 49 to 177° C – the latter at a pressure of about 13 ata, which suggests artesian conditions, as this pressure exceeds the saturation pressure at that temperature. Here again, heat exchangers are necessary owing to the aggressive nature of the fluids.

11.2.8 *Temperatures.*   From Fig. 2.1 (Chapter 2) it will be seen that 80° C is a recommended convenient temperature for domestic heating supplies, though lower temperatures can and are being used. For example, Einarsson mentions temperatures of only 50 to 55° C and upwards [170]; and in Beijing, People's Republic of China, 50 000 m² of floor space were heated in 1980 by means of water at 53–59°C. This was enough to maintain room temperatures of 15–20° C even in winter. Much depends upon the climate and the quality of the building insulation, and there is also the possibility of boosting the heat in exceptionally severe spells of cold weather by means of supplementary fuel or electricity, or even by heat pumps. Allowing for a reasonable temperature drop in transmission, wellhead temperatures should be higher than that of the delivered heat; but it is generally true to say that good *semithermal* fields or low-grade aquifers can be used for space-heating. Hyperthermal fields too can of course be used, and *are* used in some instances; but as this involves some degradation of heat, it is better if a more suitable market can be found adapted to the higher temperatures.

11.2.9 *Duration curve.*   Fig. 11.1 is a typical annual duration curve which shows the statistical distribution of daily temperatures in one particular climate. The optimum room temperature in a dwelling is represented by t_r – say 20–22° C – while the design room temperature will be t_d, a figure less than t_r by a small amount Δt known as 'free heat' – i.e. heat lost by the occupants, by electric lighting and appliances, sunshine, etc. The area *OKLB* will then be a measure of the 'imported' heat necessary to ensure comfort, while the area *LFE* represents (in degree-days) either the amount of discomfort felt in hot weather or alternatively the amount of heat to be removed by artificial cooling if comfort is to be maintained throughout the year. For the present, heating only is under consideration. To effect the required heating wholly by geothermal means would necessitate the provision of a system of capacity proportional to *OB* (integrated for the entire number of buildings in the heated district, and allowing for transmission losses) operating at an annual load factor of *OKLB/OHFB* – or about 30% as drawn. Alternatively a smaller geothermal system, with fewer wells, propor-

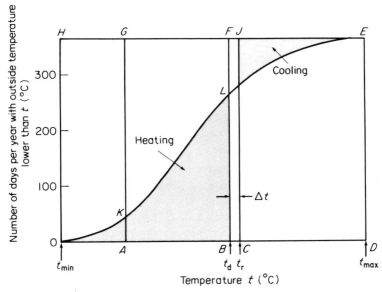

Figure 11.1 Space-heating and cooling annual duration curve. (After Fig. 2 of [170])

tional to AB, and operating at an annual load factor of $AKLB/AGFB$ – or about 42% as drawn – could be installed. The higher load factor would clearly introduce great economies, but the smaller system would leave the problem unsolved of how to deal with the area of extreme cold OKA if people are not to shiver in the depths of winter.

11.2.10 *Peak heating.*   One way of dealing with this area OKA is to provide supplementary heating by means of fuel. The actual amount of heating to be added is small – OKA is only about 7% of the total required amount of artificial heating required, namely the area $OKLB$, as drawn, so that even if fuel is expensive per tonne the total fuel bill would not be intolerably high. The important point is that the capital cost of boiler plant in terms of its peak thermal output is much lower than that of the additional wells, pumping equipment, etc., that would be required if the equivalent maximum heat demand were to be met from geothermal heat. Hence a combination of geothermal base load heating to supply the area $AKLB$ with fuel heating to supply the area OKA for peak demands can achieve an optimum economic arrangement (as with base load and peaking power plants). Supplementary electric heating or even heat pumps could sometimes provide alternative means of meeting the peak heating load, but much would depend upon local energy prices. Fuel will usually prove to be the best solution.

The amount of supplementary heating can be reduced by making use of a characteristic of the aquifer, which permits over-pumping for short very cold periods to draw excess heat from the underground reservoir by lowering the level of the water table, and allowing it to recover later when the heat demand is lower. This, however, as well as the use of large storage tanks for containing hot water, is more of a short-term provision for covering *diurnal* variations of heat demand rather than seasonal variations. In Reykjavík, supplementary fuel, storage tanks and over-pumping are all used to achieve an economic optimum for the entire heat supply system.

11.2.11 *Domestic central heating.*   In Iceland the temperature of the water entering a dwelling is usually maintained at about 80° C by mixing the waters from the hotter and the less hot bores (and with recirculated waters – see below) in suitable proportions. Some of the lower temperature waters flow by gravity from hot springs into storage tanks, while the hotter well waters are raised by means of submersible pumps so as to prevent flashing (with consequent chemical precipitation problems) until after mixing has been achieved and the temperature reduced. In earlier times in Reykjavík a once-through single pipe system was used, in which the cooled water after fulfilling its task in the heated buildings, was discharged into the sewers. More recently a two-pipe system is being used, in which the cooled (but still not cold) water is recirculated and mixed with very hot well water for re-use. The heat requirements of buildings will of course depend upon their purpose, the local climatic conditions, the quality of thermal insulation of buildings and the proportions of window to masonry and roof areas, the thermal insulation of foundations, whether a building is detached, semi-detached, terraced or a flat in an apartment block. Within a dwelling the living room will usually require more heating than bedrooms, passages, kitchen, etc. Einarsson [170] quotes an overall figure for Reykjavík of 19 kcal/h m³ with an external temperature of $-10°$ C and an average internal temperature of 20°C. This is equivalent to about 22 W/m³. In a more temperate climate the requirement would be less.

11.2.12 *Domestic hot water.*   Tap water for baths and washing is of course supplied by the once-through single pipe system, and the used water is discharged to waste.

11.2.13 *Heat transmission.*   Although this subject should perhaps have been covered in Chapter 9 it is thought to be rather more appropriate to this chapter. It is necessary to recognize that an important factor in the direct application of geothermal heat is the somewhat limited transportability of heat by comparison with that of electricity or of piped oil or gas. Hot water can be transported economically over far greater distances than steam; and as so many direct applications of geothermal heat – particularly district heating, the subject of this section – involve the use of hot water, this is perhaps fortunate. To illustrate this point it may be noted that for a given pipe, 2.37 times as much heat can be transported at 180°C (by way of example) in the form of water as in the form of dry saturated steam at the same temperature; for although the permissible velocity of steam is about 20 times that of hot water, and although the enthalpy of steam at 180° C is 3.642 times that of hot water at that temperature, the water has a density 172.6 times that of the steam. Moreover, the proportional heat loss from piped hot water is far less than from piped steam at the same temperature and the same pipe diameter; added to which, the loss of pressure, which can be avoided or limited quite cheaply by pumping when transporting water, can result in a serious loss of energy when steam is piped over long distances. Probably one of the longest distances over which steam is transported is at Wairakei, New Zealand, namely 3.3 km at about 14 ata.

The economic distance over which it is economic to pipe bulk supplies of hot water cannot be closely defined: it must of course depend upon the sending temperature, the required receiving temperature, the quantity of heat to be transmitted, the accessibility and ruggedness of the terrain, the local climatic conditions, the chemical ingredients in the transported water and the value attached to the delivered heat. Scale-forming chemical ingredients in thermal waters can sometimes impose a fairly stringent limit upon the transportable distance. In Japan, water at 70°C is piped over distances of up to 12 km [171]. At Akureyri in northern Iceland water at 96°C is sent over the same distance with only 2 or 3°C drop. The pipeline from Reykir to Reykjavík is 18 km long, with 87°C at the receiving end. Bodvarsson [172] once talked of 50 to 100 km transmission distances as being 'not unrealistic' for space-heating by hot water, though he suggested 10 km as a probable maximum economic distance for transmitting steam. The longest distance yet to be contemplated relates to a project in western Iceland that will involve a 60 km pipeline from Dieldartunga to two other towns, with 25° C drop from 100° C to 75° C. None of these quoted examples is complete with all relevant information, but they give a rough idea of what is practicable. It is possible that in future we may have access to far hotter fluids and that we may need them in much greater quantities than at present, in which case the economic transportable distance will increase; but it could never compete with electricity or piped fuels in this respect. Even now this need not be regarded as a serious obstacle if the heat is sufficiently cheap, and it will become totally unimportant if we ever succeed in winning deep-seated heat in non-thermal areas wherever we will: the real hope for geothermal heat lies in its extreme *cheapness* (see Chapter 14).

As to local heat reticulation, all system pipework in Iceland is laid underground. Pipes are of welded black mild steel laid in concrete channels and insulated with rock wool or aerated concrete if the pipe diameter is 75 mm or more. For smaller mains and house service connections the pipes are insulated with polyurethane foam in the annular space between the steel pipe and an outside projective jacket of high density polyethylene pipe [170].

11.2.14 *Metering.*   Domestic heat consumption can be metered either simply by volume of water entering the premises, assuming a constant entry temperature, or alternatively by means of a heat meter which integrates the product of temperature and flow, so as to give a more accurate measurement of heat.

11.2.15 *Environmental advantages of geothermal heating.*   One feature common to all geothermal public heating systems is the virtual freedom from pollutive problems. Reykjavík, for example, is a smokeless city enjoying clean air. The aesthetic possibilities of locating production and reinjection wells side by side have already been mentioned in connection with the Paris suburban developments described earlier.

11.2.16 *Air-conditioning.*   As a corollary to space-heating, certain hot climates need air-conditioning of buildings for the maintenance of comfort. In former times in such countries as India, the human discomfort suffered in the very hot season was slightly mitigated by the use of *punkhas* or electric fans which did no more than stimulate evaporation from the human skin, or of *khaskhas tatties* – curtains of a specially fragrant grass suspended over windows and doorways and sprayed with water from time to time. Such devices could only help to bring the human environment near to the wet-bulb, rather than the dry-bulb temperature, and thus act as a slight palliative. In very dry climates the effects of a fan could give rise to greater discomfort by drying the eyeballs than could be offset by the very slight relief it brought to the temperature. From the 1920s mechanical air conditioners have largely replaced these crude amenities in the larger tropical cities and more prosperous homes in outlying places, and have undoubtedly done much to mitigate the trials of life in intensely hot climates. Most of these air-conditioners work on the mechanical compression system, requiring external work usually supplied by electricity; but some work on the absorption principle which requires a *heat* supply. Geothermal heat can provide that requirement.

Fig. 11.1 can apply equally to cooling as to heating, except that in a hot climate the ordinate CJ (or t_r) will be shifted to the left: in fact in some tropical places such as Bombay, Colombo, Singapore, etc., where artificial heating is never, or almost never needed, the ordinate CJ will not be far from the left-hand edge of the diagram, *OH*, and the cooling work area *FLE* will be very much greater than in Fig. 11.1 – the heating work area *OBL* more or less vanishing altogether. In variable climates of extreme winters and summers, such as New York, the ordinate *CJ* will occupy a fairly central position in the diagram, so that both heating *and* cooling will be needed for comfort.

Geothermal air-conditioning has hitherto been effected by means of small, or fairly small, units for cooling a single building, apartment or perhaps a small complex of buildings. It has been proposed by Einarsson [8] that whole tropical cities could be served by a public cooling network much in the same way, but 'in reverse' so to speak, as the Reykjavík district heating system. When Managua, Nicaragua, was almost totally destroyed by the devastating earthquake of 1972, Einarsson suggested that such a system might form a feature of the reconstructed city. Unfortunately this proposal was not adopted.

11.2.17 *Combined heating and cooling systems.*   It is reported [38] that lithium bromide absorption machines with a capacity of 2.5 million kcal/h of 'cold' – rather less than 3000 kW thermal – using geothermal heat are being mass-produced in the USSR, capable of two-way operation so as to give heating in winter and cooling in summer. In New Zealand, the International Hotel at Rotorua is served by a plant that provides the whole building with all its heating and cooling requirements from a single geothermal bore – hot tap water for baths and washing, central heating in winter and air condition-ing in summer [7]. The climatic range to be contended with is from $-4°C$ to $+30°C$. The maximum heating load is 0.5 Gcal/h (about 585 kW thermal) while the lithium bromide absorption cooling unit requires an input of 0.575 Gcal/h (about 668 kW thermal), requiring 1.47 heat units per single heat unit of cooling. The bore serving the whole of this complex produces water at $150°C$ and about 7 ata. The heat is transferred in a heat-exchanger to fresh water closed circuits which are heated to $120°C$ for the heating system and for the absorption unit. In Klamath Falls, Oregon, there is a motel in which each bedroom is provided with its own individual heat pump, reversible for cooling in the summer.

11.2.18 *Sundry points relating to space-heating and cooling.*   The technical descriptions set out in this section have of course been over-simplified so that the main principles may be demonstrated. Such secondary considerations as stored heat within the fabric of a building acting as a 'smoother-out' of demand, the load density within the area of supply, the automatic achievement of temperature control by varying the mixture proportions of new thermal waters of different origins and of recirculated water according to demand and climatic variations, the variations of 'free heat' from day to day according to the hours and intensity of sunshine, the effects of high winds upon building heat losses, the decreased efficiency of transmission and of radiators with increased load if fed with water at constant temperature; all these factors, which cannot here be discussed for lack of space, must of course be taken into account when designing a public heating system. These and other technicalities have been admirably expounded by Einarsson [8, 170].

11.2.19 *Conclusion.*   Geothermal space-heating and cooling is a development that is still in its infancy on a world scale. The climates of the world are such that many countries could greatly benefit from artificial heating or cooling, and the 'benison of hot water' is widely appreciated in all lands. Doubtless many thermal fields and deep aquifers, known and as yet to be discovered, will be exploited for this purpose in the coming decades; and if 'heat mining' should become a practical reality by the creation of artificial fields (see Chapter 18) then the future of this activity could indeed be immense. Although in this chapter the broad term 'domestic' heating has often been referred to, the heating of commercial, administrative and industrial buildings is of course also included by intent. If it should one day be possible to construct geothermal heating projects on a huge scale it may become possible to banish permafrost from vast areas of the world and to make those regions habitable and self-supporting. (In parts of the USSR the ground is permanently frozen to depths of about 1500 m [38]). At the same time it could open up many mineral-rich regions to mining operations that are now prohibited by the intense cold.

Apart from thermal fields and deep aquifers exclusively exploited for the extraction of thermal waters, it may one day be possible to endow abandoned oil fields with a second lease of life by putting to good use those same permeable strata that have once yielded petroleum or gas and could be persuaded thereafter to yield their residual heat.

11.3 Farming, including agriculture, horticulture, aquaculture, animal husbandry, etc.

11.3.1 *General comments.*   The boundaries between the four classifications of direct uses of earth heat as defined at the end of Section 11.1 are by no means clearly marked. Much of the heat needed for farming, in the broadest meaning of the word, is simply a form of space-heating; but as the requirements of animals and plants differ widely from those of humans it is convenient to treat them separately. There is also ambiguity as to the classification of refrigeration and of food processing: both are closely related to farming but can equally well be described as industries – as indeed they will be classified in this chapter.

All the heat requirements of farming are of moderate to low grade – from 100°C down to 20°C. As the judicious use of heat can greatly modify, or even remove, the constraints imposed by the seasons upon food production and horticulture, it is evident that geothermal heat could play a very important rôle in raising the organic productivity of the earth. It could render habitable large areas of the world that have access to fresh water but are too cold to support farming activities unaided, and could thus bring relief to the ever-growing pressure of the population explosion. There is undoubtedly a great future for geothermal farming.

11.3.2   *Geothermal greenhouse heating* [9, 38, 156, 168, 173–178].   This is practised in France, Hungary, Iceland, Italy, Japan, New Zealand, Romania, USA, the USSR and even in the Himalayan recesses of India and probably elsewhere. With the aid of earth heat, fruit, vegetables, flowers, potted plants and even cacti can be raised under glass or plastic, which admits sunlight and at the same time retains much of the solar warmth as well as the artificially introduced geothermal space-heating. Double glazing (or double plastic covering) would doubtless improve the efficiency of the process but it has yet to be shown whether this would be economically worthwhile. In Iceland, not only have such produce as tomatoes, cucumbers, lettuces and mushrooms (commonly grown under natural conditions in temperate climates) been raised, but also tropical and subtropical fruits such as bananas (Plate 5), papayas, pineapples, melons and grapes (Plate 4); also exotic flowers like orchids. Table 11.3 shows some figures taken from the OWS [156].

Table 11.3 Some particulars of geothermal greenhouse heating as in 1980 [156].

Country	Area under glass or plastic (m²)	Watts/m² reckoned above 15°C	Heat demands (MWt)	(TJ p.a.)	Remarks
France	17 000				At Melleray only
Hungary	1 900 000		535	3311	Total agricultural needs in MWt and TJ, including soil heating. Implied annual load factor – 19.6%
Iceland	145 000	233–291	51.6		Commercial greenhouse area only shown, excluding unknown area of privately owned greenhouses
Italy	20 000	175	3.5		Galzignano only
Japan	65 000	258	16.8		
Romania	30 000	300	9.0		
USSR	420 000	262	110		

The maximum heat demand in MWt, the annual heat consumption in GWht or TJ, and the mean annual load factor will of course depend upon the crop to be raised, the quality of the greenhouse construction and on the local climate. The only figures that enable a mean annual load factor to be deduced from Table 11.3 are those for Hungary, but as these figures are not exclusively for greenhouse heating it would be unwise to infer too much from them. However, as the total agricultural load is equivalent to 282 W/m² of greenhouse area, which is of the same order as in the other countries quoted except Italy, it would seem likely that soil heating accounts only for a small share of the total agricultural heat load. The relatively low figure in W/m² for Italy suggests either a milder climate than in the other quoted countries or the cultivation of less ambitious crops.

As with space-heating, fuel-fired peaking heaters will usually give better economic results than catering for the very worst conditions by reliance upon geothermal heating alone. Since it is possible, by means of geothermal heat, to create virtually any climate desired – even sunlight can be simulated by means of electric lamps capable of emitting light of the required wavelength – seasonal limitations that apply in nature can be overcome, provided that adequate flows and temperatures are available, so that almost any crop can be raised at any time of year; although naturally it will be more economic to raise crops at times when the local climate can offer the greatest possible encouragement. At present, agriculture consumes over ten times as much geothermal power and four times as much heat energy as domestic space-heating in Hungary, which emphasizes the importance attached by the Hungarian Government to geothermal farm aids.

Gutman [173] has described how a small hydroponic installation has been established in north-eastern California at a height of 1250 m above sea-level, where the winter temperature sometimes falls to $-33°C$, aided by a natural artesian thermal spring yielding hot water just below atmospheric boiling point. Cultivation is effected beneath corrugated plastic giant barrelled cloches placed parallel with one another at 13 ft (about 4 m) spacing so that they do not shade one another from the sunlight. Humidifiers are necessary in summer.

Some idea of the economic importance of geothermal greenhouse heating may be gathered from the fact that in 1967 Iceland produced about 64% of her vegetables and 98% of her cut flowers and seedlings with the aid of earth heat. In a country where the potato is about the only vegetable that can be grown naturally, this application of geothermal heat is of immense importance. The same could doubtless apply to many other places in the world having inhospitable climates.

11.3.3   *Soil heating and cooling* [174]. Artificial soil heating can be effected either by means of submerged hot pipes or by the use of warmed irrigation water, or both. The example of Agamemnon has been cited in Section 1.1.6, but more closely controlled use of warm thermal waters for irrigation is practised in several countries. Soil heating can of course be combined with greenhouse heating. It is not known whether soil *cooling* has ever been attempted by means of geothermally activated cooling plants, but the author suggests that this might well prove to be an attractive proposition for growing fruits and vegetables (naturally raised in temperate climates) in hot tropical countries in partial shade – e.g. strawberries.

11.3.4 *Nursery gardens and botanical gardens.*   By the joint use of greenhouse heating and soil heating (and/or cooling) the growth of seedlings, saplings, flowers, shrubs, etc. can be encouraged by means of geothermal heat in climates that would naturally be hostile to their requirements. Several botanical gardens have been established in Japan [175] in this way. The raising of forest seedlings has been greatly aided and accelerated in the USA and other countries.

11.3.5 *Soil sterilization.*   The heating of soil to sufficiently high temperature, by means of geothermal heat, can effectively sterilize it against insect and bacterial pests before the soil is used for cultivating seedlings. This is practised in Japan [175].

11.3.6 *Crop drying* [9, 179].   Grass, seaweed, onions and other crops can be effectively dried by means of geothermal heat. This is practised in Iceland, the USA, the USSR and elsewhere.

11.3.7 *Animal husbandry* [9, 175, 176, 180].   Farm buildings, such as cattle stalls, pigsties, stables, dairies and hen coops may be geothermally heated in winter for the greater comfort and health of farm workers and animals. This is particularly effective in poultry raising. Komogata *et al.* [176] claim that the yield of poultry meat can be raised 2½ times by weight in terms of fodder consumed, by means of geothermal heating. Pig food may be cooked from various wastes, milk can be pasteurized, chicks can be incubated, biodegradation of organic wastes may be effected, wool may be washed and dried – all by means of geothermal heat.

11.3.8 *Fish breeding.*   Hatcheries for trout, eels and other fish may be served by means of only slightly heated waters – down to about 20°C. Cooling water effluent from geothermal power plants could be used for this purpose unless they are chemically contaminated. The writer even knows of a quite successful fish hatchery in Scotland where a very small infeed of warm water from some deep source ensures that the pool never freezes nor even drops in temperature to below 7 or 8°C, which is apparently sufficient for the purpose even though rather warmer temperatures are preferred.

11.4 Industry

11.4.1 *General comments.*   The criterion on which the suitability of earth heat for an industry can be judged is its degree of heat intensiveness, which may be expressed as the number of kilogram-calories absorbed in producing finished product to the value of $ US 1, having due regard to the *grade* of heat required. Líndal [9] draws a parallel between the great hydroelectric developments of the early twentieth century which attracted electricity-intensive industries from far and wide to places where very cheap power was available (e.g. the aluminium industry) to the probable appearance within the near future of large geothermal developments which will produce such cheap and abundant heat that it will pay to establish heat-intensive industries close to the available heat more or less regardless of location. This argument which is expanded in Section 14.14.1 may indeed be sufficiently powerful to overcome the one disadvantage from which geothermal energy suffers when compared with hydroelectricity, namely, the fact that it is seldom economic to transport heat over more than quite moderate distances – at least in the quantities of commercial interest in present times. Electricity, on the other hand, can be economically transported over hundreds of kilometres (in sufficiently large blocks). Thus, whereas the siting of an electricity-intensive industry can compromise to some extent with a source of cheap power, a heat-intensive industry must perforce have to be sited fairly close to a geothermal field, wherever it may be. This fact could perhaps be turned to social advantage in that it could lead to the habilitation of unpopulated waste lands that are now regarded as uninhabitable, and thus help to solve to a small extent the problem of the world's over-population. In the past, there has been a regrettable tendency to write off remote geothermal fields as worthless owing to the difficulty of heat transportation; but if heat is sufficiently abundant and sufficiently cheap, thermal fields in the most unlikely and remote places could become economic for certain very heat-intensive industries. If this is true, it follows that less heat-intensive industries could be economically justified if the source of heat is close to the required raw materials and labour.

Líndal [9] lists 35 random industrial processes, mostly chemical, which show a ratio of about 45:1 in their heat-intensiveness (which he expresses in pounds of steam required to produce product to the value of $ US 1). Although his actual figures will doubtless now have been overtaken by inflation, their relative proportions are probably still more or less valid. Nineteenth in his list, in descending order of heat-intensiveness, was the Kraft process for producing paper pulp; and yet this process is being used in the world's largest existing geothermal industrial installation – the mills of the Tasman Pulp and Paper Company at Kawerau, New Zealand. This is an illustration of a less heat-intensive industry being economic where the raw materials (forests in this case) and labour are available close to a thermal field. Many chemical industries are several times as heat-intensive as the Kraft pulp process – e.g. the production of caustic soda, ethyl alcohol, acetic acid, lactose and heavy water.

If we should ever succeed in winning very high temperature geothermal steam from magma pockets (Section 18.5) or from great depths in natural or artificial fields, the scope of application will be further extended despite the inefficiency of power generation as such, because certain industries which need either high-grade heat or electricity cannot be served by the relatively low grades of heat now available.

The number of industries that could probably be served with geothermal heat is immense, and is likely to increase with time. To record them all would in itself involve a major work of research. The list which follows therefore makes no claim to being complete – in fact many of those mentioned by Líndal [9] do not even appear in it – but is indicative of the very great scope offered. Many of them are already practised: others have only been considered.

11.4.2 *The chemical industry.* Many chemical industries are heat-intensive without requiring very high grade heat. Some of them are as follows:

(i) *Extraction of sodium and magnesium salts from sea-water* [176, 181–186].
(ii) *Extraction of valuable minerals from geothermal fluids* – e.g. lithium [185], bromine and the chlorides of potassium and calcium [9, 184] from 'normal' geothermal waters. A wide variety of other useful substances is believed to be commercially extractable from the highly concentrated hot brines found beneath the Imperial Valley, California [186]. Geothermal gases may also contain useful constituents such as H_2S, CO_2, CH_4, B, NH_3 and other vapours [187, 188].
(iii) *Recovery of elemental sulphur deposited in volcanic craters, fumaroles and hot springs* [176]. See also Section 1.1.3.

   Note: The value of substances recovered from sea-water, geothermal fluids and geothermal surface manifestations is often of an intermediate nature. Some of them, of little intrinsic value in themselves, can be used for the production of much more valuable materials such as metallic sodium, magnesium and titanium, also sulphuric acid and chlorine (which can be used for a host of other chemical processes [186] when further treated with chemical additives, electricity and additional heat. Multiple-effect evaporation is often used for items (i) and (ii): very hot geothermal fluids are self-evaporative.

(iv) *Heavy water production.* This was once proposed for Wairakei as part of a dual-purpose project, using a counter-flow water distillation process for producing the heavy water. There are other methods of production, of which the hydrogen sulphide–water isotope exchange process is favoured as the most economical by Valfells [189]. The potentialities of this application will largely depend upon 'fashions' in the nuclear power industry.

(v) *Sulphuric acid from elemental sulphur and H_2S occurring in geothermal discharges.*

(vi) *Fermentation of molasses to produce ethyl alcohol, butanol, acetone, citric acid and other products* [9].

(vii) *Production of proteins, vitamins and ammonia* [38].

(viii) *Ethanol production* [190]. This is a very promising process that is already in operation in four geothermally activated plants in the USA. It is of particular interest as an example of a working partnership between two 'alternative' energy sources – solar and geothermal – with the aim of producing a combustible fuel. The solar energy provides *biomass* in the form of wood wastes, root crops, grain or farm wastes such as straw. The four American plants referred to are already producing 5 million gallons of ethanol per year. In doing so they consume about 3×10^{11} Btu, or 88 GWht annually, equivalent to about 51 720 barrels of oil p.a. The product is a 180 proof spirit suitable for mixing with nine parts of gasoline to form 'gasohol', which can be burnt in motor cars. By 1984, if all the 53 geothermally activated ethanol factories now under construction, proposed, and the subject of feasibility studies come into operation, no less than 286 million gallons of ethanol will be produced annually in the USA alone. The great attraction of producing artificial fuels from geothermal energy is that the product is endowed with *mobility*. Regardless of the efficiency of the process, energy from a local source can be transported in the form of fuel to a remote point of use.

11.4.3 *Mining and upgrading of minerals*

(i) *The production and refinement of diatomaceous earths.* This is being practised as a highly successful industry at Lake Mývatn, near Námafjall, Iceland [9], Plate 36.

(ii) *Production of boric acid by applying steam heat to ores.* This method is being practised at Larderello superseding the older, less economical method of recovering the boric acid from the fumarole vapours [9].

(iii) *The facilitation of mining operations in permafrost areas* – e.g. in Siberia, where the climate is so severe that, without the introduction of extraneous heat, mining operations would be impossible [38].

(iv) *The drying of peat.*
(v) *The production of alumina from bauxite by the Bayers process* [9].

Plate 36 Námafjall diatomite plant, near Lake Mývatn, Iceland. An example of a successful industrial application of geothermal energy. The plant uses about 50 tonnes per hour of geothermal steam at 10 atü/183° C. (By courtesy of Baldur Líndal, Consulting Engineer)

11.4.4 *Food processing*

(i) *Cane sugar production,* thus releasing the bagasse, formerly used as a fuel, but which is now a valuable raw material used for other manufacturing purposes.
(ii) *Sugar beet manufacture,* requiring 3300 to 5500 kcal/kg of sugar [191].
(iii) *Production of powdered coffee.*
(iv) *Production of powdered milk.*
(v) *Crop drying and dehydration of foods.*
(vi) *Fruit and juice canning.*
(vii) *Production of cattle meal from Bermuda grass.*
(viii) *Rice parboiling.*
(ix) *Fish drying and fish meal production,* as practised in Iceland.
(x) *Freeze drying of foodstuffs* [192].

11.4.5 *Various industries*

(i) *Rayon manufacture.*
(ii) *Other textiles.*

(iii) *Pulp and paper manufacture* [9, 191], as practised in New Zealand and Australia.
(iv) *Timber seasoning* [9, 193] as practised in New Zealand.
(v) *Veneer manufacture* [193], as practised in New Zealand.
(vi) *Production of cotton seed oil.*
(vii) *Manufacture of aggregate cement building slabs* [9], as practised in Iceland.
(viii) *Brewing,* as practised in Japan [9].
(ix) *Synthetic rubber manufacture* [38].
(x) *Refrigeration and gas liquefaction* [177].
(xi) *Dye industry,* as practised in China.
(xii) *Lurgi process for total gasification of coal,* if sufficiently high pressure steam is available.
(xiii) *Sinter extraction* (alum) [175].
(xiv) *Plastics industry.*
(xv) *Process steam* for many types of industry.
(xvi) *Wool drying.*

Very few national statistics of industrial heat quantities are available, but some meagre information is contained in Table 11.4.

Table 11.4 Some limited information about industrial applications of earth heat

Country	No. of industries using geothermal heat	Maximum demand (above 15°C) (MWt)	Annual heat consumption (above 15°C)		Mean annual load factor	Remarks
			(GWht)	(TJ)		
Hungary	21	24.5	131.4	473	61.2%	
Iceland		18.3				
New Zealand		140	442	1590	(36%)	Load factor is for paper factory alone.
USA	18	817				Excluding enhanced oil production

The nature of many of the industries that are already using geothermal heat is not known; but the following illustrate some of them:

Australia Paper manufacture
China Textiles, printing, dyeing, cement production, chemical extraction, glue, tanning and paper manufacture
Iceland Seaweed drying, recovery and upgrading of diatomaceous earths, fishmeal production, aggregate cement block manufacture, salt production, sea chemicals extraction (sugar from beet under consideration)
Italy Chemical industries

Japan Brewing, salt production, sulphuric acid production, elemental sulphur recovery

New Zealand pulp and paper manufacture, timber seasoning and veneer fabrication, alfalfa drying

USA Crop drying (alfalfa, onions, grain), food processing, ethanol production, wood pellet production from sawmill waste for fuel, vaporization of LNG, enhanced petroleum recovery, laundering, refrigeration

The main stimulus to the development of geothermally activated industries is now evident in the USA. If enhanced oil recovery be disregarded (see the introductory section of this chapter) the industrial projects in the USA that in 1980 were under construction, proposed, or the subject of feasibility studies – all planned to use geothermal heat – totalled to a demand of $19\,751 \times 10^9$ Btu p.a. as compared with a 1980 demand of 817×10^9 Btu p.a. (excluding enhanced oil recovery) – a 24-fold increase, to be achieved perhaps by the mid-1980s. This is more indicative of the new awareness of the importance of direct applications of earth heat than any other available statistic. It is not suggested that this ambitious goal will necessarily be fully achieved in the USA or that other countries will accelerate their development to the same spectacular extent; but trends of this sort are apt to be 'infectious', and a few successful examples will doubtless be followed elsewhere.

11.5 Miscellaneous

Several other useful functions that can or could be served by geothermal heat but which do not fall properly within the scope of Sections 11.2, 11.3 or 11.4 are mentioned below. Doubtless other uses will be thought of in time.

11.5.1 *Distillation* [194, 195]. With the increasing rates of population growth and of water consumption per head, due to rising living standards and increased industrialization, the supply of adequate quantities of fresh water has long been becoming a matter of growing urgency in many parts of the world. Suitable sites for rainwater catchment reservoirs are becoming more difficult to spare in highly populated countries like the United Kingdom. Other densely populated countries like California can at present rely to a limited extent on long distance transportation of water from remote catchments. Newly found prosperity from oil wealth has led to great population growth in almost totally arid areas in the Middle East. Recourse is becoming increasingly necessary to the desalination of sea and brackish waters. Desalination is a heat-intensive process that can be most effectively achieved by means of a multiple-effect or flash evaporators. Since a high performance ratio plant costs much in capital but little in heat, and *vice versa*, it follows that where heat is expensive a very high performance plant can be justified, but where heat is cheap a lower performance ratio is acceptable economically. Geothermal heat is generally very cheap, so that distillation plants of low capital cost could be used in conjunction with it. Desalination requires heat of moderate grade – say 120° C – so that hyperthermal fields would be more or less necessary for efficient geothermal desalination. The chances of the coexistence of a hyperthermal field and an arid, but highly populated, area are not great, but such places could well be

Plate 37 Alligator-breeding in the Atagawa tropical gardens, Japan, by means of geothermal heat [197]. (By courtesy of the Japan Geothermal Energy Association)

found in California, Mexico, North Africa, the Canary Islands, Greece and perhaps in parts of the Middle East. Moreover, if and when we ever succeed in creating artificial thermal fields in semithermal or even non-thermal areas (see Chapter 18) the opportunities for geothermal desalination will become very great indeed. It is even conceivable that we shall one day succeed in distilling sea or brackish water economically, not merely for drinking purposes but also for *irrigation* – a feat that has not yet been achieved.

11.5.2 *Balneology and crenotherapy.* These applications have been touched upon in Section 1.1.1. Chiostri [196] states that in 1971 in Italy alone 15 million people were treated at more than 200 thermal clinics for various ailments, while in Russia more than 10 million are medically treated annually with thermal waters. The Japanese [175] have an even more impressive record, claiming that about 100 million visitors annually frequent the hotels located near the 1500 (approximately) hot spring resorts in their country. To what extent crenotherapy, or the healing of certain ailments by the external or internal application of thermal mineral waters, owes its popularity to faith cannot be determined, nor is it altogether relevant; but in general the medical profession has given its blessing to the practice. At any rate it is an undisputed fact that thermal balneology and crenotherapy are responsible for a gigantic tourist trade of immense commercial value to those countries that are fortunate enough to possess the necessary facilities. The fact that many thermal areas occur in scenically attractive places may account for part of this trade. Geothermally heated swimming pools can also be a good source of revenue to well situated communities and hotels.

11.5.3 *Zoology.* In Japan (Plate 37), the breeding of alligators and tropical fish in geothermally heated waters is proving to be a great attraction to tourists in Peppu, Atagawa and Nagashima. In combination with geo-thermally heated greenhouses displaying tropical flora, these alligator and fish farms are becoming a great sightseeing attraction.

11.5.4 *Bottled mineral waters.* Many hot springs in widely scattered parts of the world contain minerals with alleged medicinal properties. A large international trade in these waters, after bottling, has been practised for more than a century and is still growing.

11.5.5 *Cooking.* In parts of Japan, and in the island of Lanzarote (Canaries) cooking is performed on a small scale by means of geothermal steam and vapours. Cooke [198] described a modern adaptation of the ancient Maori *hangi* – a steam cooking device for use with natural steam. These are extensively used in New Zealand for cooking food both for humans and for animals.

11.5.6 *The prevention of freezing of public water supplies and of fire-fighting water, and the melting of snow on roads in winter.*   Both these adaptations of geothermal heat are practised in Japan [175, 199]. In Canada, in the towns of Whitehorse and Mayo in Yukon Territory the freezing of municipal water supply pipes is prevented by the application of geothermal heat.

11.5.7 *Scenic attractions.*   Many natural geothermal phenomena possess both scenic beauty and great curiosity value. Tourists are thus attracted from far and wide, bringing revenue to the district – e.g. Yellowstone Park, Wyoming, USA; the geysers of Iceland; Rotorua, New Zealand; the thermal areas of Japan and elsewhere.

12 Dual and multi-purpose projects

12.1 Scope for dual or multi-purpose projects

The potentialities of dual and multi-purpose projects for the efficient exploitation of earth heat were briefly mentioned in Chapter 2. It is of course the very wide spectrum of usable temperatures and also the frequent availability of two-phase fluids that provide the great theoretical scope for such projects, as shown up to some extent in Figs 2.1 and 2.2 (Chapter 2), and the range of possible combinations of different uses of earth heat could be great. More multi-purpose plants have hitherto been talked about than have actually been realized, though a few are now in the process of development. It is nevertheless only fair to recognize that despite their obvious ideal advantages in principle there are often considerable difficulties in establishing them in practice. The fact that the two or more end products must be produced in the same locality can raise difficulties of relative marketability, and perhaps of raw material availability in the case of industries. Disparity of demand fluctuations can also introduce problems of storage or of inefficient heat usage.

A distinction should be made between *parallel* and *series* uses. Obviously if a supply of hot fluid can be split into two or more streams flowing in parallel with one another, each of which is used for different purposes, then the result could indeed be described as a multi-purpose complex. For example, a flow of hot water could be divided in various proportions to supply, say, a district heating system, a group of greenhouses and perhaps some heat-intensive industry requiring low-grade heat – e.g. crop drying, beet sugar extraction, a synthetic rubber factory. All these applications would be operating in parallel with one another, and all would require a heat intake at more or less the same temperature. In so far as an available supply of hot fluid may exceed the demands of any one of these component applications by an adequate margin, this would certainly be conducive to the efficient use of the available heat; and if there should be a diversity in the times of incidence of the component maximum demands, the overall composite utilization factor of the heat could perhaps be better than that obtainable by any one of the component uses individually. Where it is possible to put to separate uses the steam and hot water of wet mixtures in water-dominated fields, this too would be an example of parallel uses.

Of greater economic appeal, but more difficult to achieve in practice, is a group of different applications that use *the same fluid* over different bands of temperature, and thus operate in *series* with one another. Since every application uses and rejects heat over a certain temperature range, a supply of hot fluid could be put to the greatest possible use if the rejection temperature of one process were the same as, or very slightly higher than, the intake temperature of the next 'downstream' process.

Parallel processes can be regarded as not true multi-purpose applications, but rather as *separate* entities. It may be that a field or a pipeline serves in a multi-purpose capacity, but the processes themselves are individual. With series operation the processes are truly inter-dependent and can be regarded indisputably as 'multi-purpose'.

12.2 Present examples

These are rather few. The borax production and power generation at Larderello, Italy, was formerly a true series multi-purpose complex. Now, separate steam is used for the two processes, and it has become a parallel arrangement. The diatomite factory at Námafjall, Iceland, and the associated power generation is also a parallel arrangement, as a common steam supply feeds both processes separately. A few examples of series operation are listed below.

(i) *Kawerau, New Zealand* [9]. The exhaust steam from a 10 MW back-pressure turbine is used as process heat for the pulp and paper factory.

(ii) *Svartsengi, Iceland* [149, 200]. Bore fluid is flashed in two stages. The higher pressure flash steam is used for power generation and the lower pressure flash steam is used for district heating. This, however, is an over-simplified description of a more complex system, which is illustrated in greater detail in Fig. 12.1. As the quantity of high pressure flash is greater than the requirements of the turbine, the balance is used to supplement the district heating process. The originality of the Svartsengi plant is the use of flash steam for district heating, because the geothermal fluid contains too high a concentration of dissolved solids for it to be piped directly to the heat consumers without serious scaling troubles. The lower pressure flash steam is mixed with cold fresh water and then de-gassed at atmospheric pressure after being heated by the turbine exhaust. The process is not efficient, but by comparison with the alternative of oil heating it is very attractive commercially. Condensate is discharged at different temperatures to suit the needs and the distribution systems of the various communities served with district heating.

(iii) *Horticulture and zoology.* In Japan, at the tropical gardens of Atagawa, Peppu and Nagashima, hot geothermal water is first used for greenhouse heating and then passed on to the alligator pools at a lower temperature.

(iv) *Space-heating and aquaculture.* At Klamath Falls, Oregon, USA, the discharged water from the heating system of the Oregon Institute of

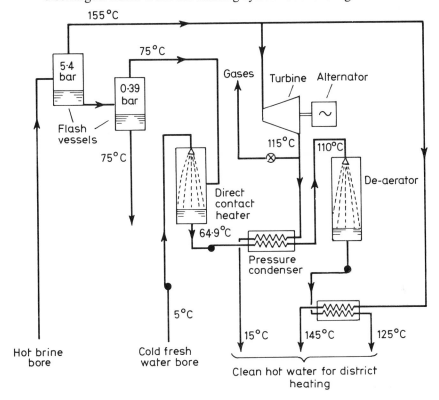

Figure 12.1 Combined heat/power installation at Svartsengi, Iceland. (After Fig. 7 of [149], by courtesy of Orkustofnun, the National Energy Authority of Iceland)

Notes:

(i) The fresh cold water is drawn from a 40 m thick 'lens' that floats over a deeper reservoir of hot brine.

(ii) The temperature in the first stage flash vessel is high enough to retain silica in solution, thus avoiding precipitation.

(iii) The second stage flash vessel is gas-free.

(iv) Development: Stage I 2 MWe + 50 MWt
            1981   6 MWe + 75 MWt added
                  ─────────────────────
                    8 MWe + 125 MWt

Technology buildings is used for warming a pool in which a species of large prawn is bred. Several similar schemes are in operation elsewhere in the USA for other kinds of fish hatcheries – sometimes with greenhouse heating interposed between the space-heating and the aquaculture.

(v) *Multiple domestic uses.* In Iceland it is not uncommon for householders to use geothermal water consecutively for space-heating in the home, greenhouse heating, and finally in a swimming pool.

(vi) *Space-heating and de-icing.* In Japan and in the USA the use of discharged space-heating waters is applied to road de-icing in winter.

(vii) *In Kamchatka, USSR* the rejected heat from the 11 MWe geothermal power plant at Pauzhetka is used to heat 80 000 m² of greenhouse area which yields from 2000 to 2500 tonnes p.a. of vegetable produce, and also to provide space-heating [38].

There are other multi-purpose geothermal installations in operation. Even though some of them are of a modest scale, they apply a sound principle.

12.3 Planned and discussed examples

In contrast with the few working examples of dual or multi-purpose plants many other projects have been planned, or at any rate considered: some of them are understood to be under construction. A few are briefly described, as follows:

(i) *Heavy water/power project originally planned for Wairakei.* One of the first dual purpose geothermal projects to be planned, and which would have been impressive had it materialized, was the original proposal for a joint chemical/power installation for Wairakei, New Zealand, in the middle 1950s. Geothermal steam at 180 psig was to have been expanded in 13 MWe of back-pressure turbines exhausting at about 50 psig. The vapour discharged from these turbines was to have activated a twin tower distillation plant of the 'bubble plate' type to produce a 50% solution of heavy water that would have been concentrated further by electrolysis. The distillation plant would have discharged vapour at slightly above atmospheric pressure, and this was to have been passed through condensing turbines of 33.5 MWe capacity. Thus the total power production was to have been 46.5 MWe. Unfortunately, owing to a slump in the heavy water market, the distillation plant was cancelled after the power plant has been ordered and partly manufactured, thus leaving an unfilled 'pressure gap' between the exhaust from the high-pressure turbines and the inlet to the condensing turbines. This gap was then filled by installing 22.3 MWe of additional back-pressure turbines so that the project became a 'power only' enterprise. As further steam had meanwhile been discovered, the capacity of the whole installation was also raised to 192 MWe. The rather complicated arrangement of turbines at Wairakei was entirely due to this unintended historical development.

(ii) *Desalination combined with other processes.* Several proposals have been made for using geothermal energy for desalination in combination with power generation and/or mineral extraction from hot brines [186, 194, 201].

(iii) *Off-shore power and mariculture.* Palmer *et al.* [202] have advanced an ambitious proposal for generating electricity on the sea-bed over thermal areas in the continental shelf, and using the waste heat to encourage the breeding of fish on a very large scale.

(iv) *Raft River Valley development proposal.* Kunze *et al.* [203] proposed in 1975 an elaborate and ingenious geothermal complex intended to combine power generation with several farming activities (poultry, dairy cattle, greenhouse produce, cheese and sugar production) and with food processing, canning, refrigeration and fertilizer production.

(v) *Miscellaneous.* Many other projects have been mooted for various combinations of two or more productive activities – electric power, farming, district heating, ethanol production and various industries [204, 205]. Many of these will undoubtedly come into operation in the 1980s.

12.4 The problem of demand balancing

If the patterns of demand for the various series processes in a dual or multi-purpose project should differ from one another it will be necessary to provide either storage or bypass facilities, or both. Bypassing is wasteful of power potential and involves the degradation, but not necessarily the *loss*, of heat. Storage can take the form either of *heat* storage or of *product* storage. Heat storage is difficult to provide, but product storage requires only warehouse or tank capacity. An industrial process that produces goods for which there is a seasonal demand may be operated more or less continuously at full capacity provided that sufficient space is available to absorb the imbalance between production and distribution. An ethanol factory, for example, may operate steadily at or near full capacity, but there must be enough tank storage to cater for intermittent shipments from the factory to the market.

Thought has recently been given to the notion of using an aquifer for heat storage [145], the principle of which is illustrated in Fig. 12.2 in its simplest terms. Two wells, forming a 'doublet' and each capable of serving either as a production well or as a reinjection well, enable water to be circulated through an aquifer in either direction. At times of surplus available heat, water would be extracted from the 'warm' zone of the aquifer and passed through a heat-exchanger so as to absorb much of the surplus heat, and then

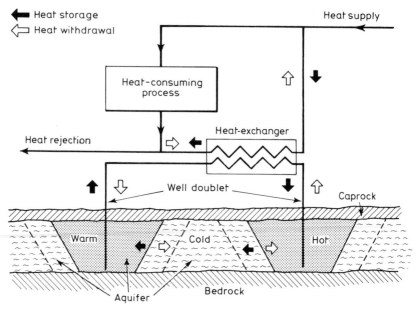

Figure 12.2 Diagram to show how an aquifer could be used for heat storage. (Based on Fig. 2 of [145])

injected back into the aquifer at the 'hot' zone. At times of heat storage the flow cycle would be reversed so as to enable heat to be yielded up from the hot zone of the aquifer and returned, in large measure, to the hot water transmission system. This method could be used for conventional co-generation, or 'total energy' projects that do not rely in any way on geothermal energy as a *source* of heat. It could also be adapted to serve as a means of adjusting the imbalance of demands between two or more processes of a geothermal multi-purpose complex.

13 The control and safety of geothermal installations

13.1 Control: general

As shown in Chapter 8, every geothermal bore will have its own individual characteristic relating mass flow and pressure: no two bores will behave in exactly the same manner. Where several bores simultaneously supply thermal fluids to a utilization plant, whether for power generation or for other purposes, it is necessary to devise a system of control whereby the bores, despite their different characteristics and allowing for possible changes in these characteristics, will collectively provide the required quantities of fluids at the required pressures and with a minimum of waste, and whereby any unwanted fluids are automatically disposed of. In a dry field there may also be variations in the degree of superheat, which can generally be accepted as a dependent variable of little importance requiring no special means of control. In a wet field, the dryness fraction of the bore (yet another variable) is normally dealt with by wellhead separators from which the two component phases are separately controlled. Thus the control of steam and the control of hot water may here be treated separately. Armstead and Shaw [101] have shown in some detail how both fluids may effectively be controlled so as to enable a utilization plant to function smoothly and correctly. If the use of submersible pumps or the adoption of two-phase fluid transmission (see Sections 9.11 and 9.12) should become regular practice, the problems of control should become greatly simplified. The description which follows therefore deals with the most arduous conditions that need be considered.

13.2 Steam control

With a *conventional* steam power plant the main task of the steam control system is to ensure that the turbine receives whatever quantities of steam it may need for meeting the load with minimum variation of temperature and pressure at the turbine entry. This is achieved by varying the rate of fuel feed and the quantity of combustion air accordingly. In contrast with a boiler, a geothermal steam bore cannot be controlled in this manner. Not only is there no fuel or combustion air to be regulated but it is impossible to vary the rate of steam flow from a bore without at the same time altering the wellhead pressure and, as already mentioned, each contributing bore will have its own individual pressure/flow characteristics. How then is it possible with a number of bores working in a parallel, each having different (and perhaps changing) characteristics, to ensure that the plant receives the amount of steam it needs at the required pressure? Within this primary problem three subsidiary problems may be discerned, namely:

(i) That of primarily ensuring that the bores deliver the required quantity of steam at adequate pressure.
(ii) That of primarily ensuring that a more or less constant pressure is maintained at certain points.
(iii) That of ensuring an economic balance of steam flows.

13.2.1 *Steam quantity primarily assured.* The upper part of Fig. 13.1 shows a hypothetical steam transmission system in which four bores A, B, C, and D are connected to a pipeline feeding a plant X, which may be a power plant or some other steam-consuming process. Let it be assumed that for its full performance the plant requires a quantity of steam Q at a pressure P, and that the characteristics of the four bores are such that they can just provide these requirements – no more and no less. Under these conditions the pressures at the plant, at the bores and along the main and branch pipelines would be somewhat as shown by the full line (1) in the lower part of

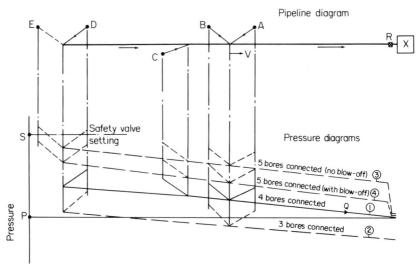

Figure 13.1 Steam supply control; steam quantity primarily assured. (From Fig. 1 of [101])

Fig. 13.1. Now let it be assumed that one of the four bores, say C, is disconnected. This would have the effect of lowering the pressures at all points of the system somewhat as shown by the broken line (2) in Fig. 13.1. Although the reduced pressures at the bores would cause some increase in the steam yields from each of the remaining three connected bores, the total steam delivered at the plant would necessarily be less than when all four bores had been connected. Thus the supply of steam to the plant would be deficient both in quantity and in pressure.

If C be now restored and a fifth bore, E, connected to the system, the pressures at all points would rise. In order to limit both the admission pressure and the quantity of steam at the plant to the required values it would be necessary to throttle the steam by means of some pressure-reducing device R. As this device would have some inherent 'regulation' some small rise in the admission pressure above P would have to be tolerated. The pressures throughout the system would now be somewhat as shown by the broken line (3) of Fig. 13.1. It could well happen that the bringing in of the fifth bore would raise the pressures at some of the wells above the setting pressure S of the wellhead safety valves, thus causing these valves to lift. This would be undesirable for the reasons given in Section 9.6, and should be prevented except in the last emergency by blowing off a controlled quantity of steam through a suitably designed vent valve V placed at some convenient point on the pipeline (Plate 13). By thus blowing off a sufficient amount of steam the pressures throughout the system could be reduced to such levels that no wellhead safety valve need lift at all (see the broken line (4) of Fig. 13.1). The more the blow-off, the lower will be the pressures at all points. Obviously it would be highly improbable in practice for a precise number of bores to give exactly the required amount of steam at the required pressure (as in line (1) of Fig. 13.1). In order to ensure that the plant may receive what it needs it will always therefore be necessary to connect rather more bores than are theoretically needed to achieve this, particularly if allowance is to be made for the changing of bore characteristics with time and for the possible loss of a bore through blockage, its removal for testing or maintenance or for any other reason. Hence it will nearly always be necessary to throttle the steam at entry to the plant, to blow off some steam to waste, or both. Of these unavoidable procedures, throttling is the less wasteful in energy. The provision of vent valves automatically controlled by the pressure in the steam pipeline close to their point of installation can also prevent excessive pressure rises if the utilization plant steam requirements should suddenly fall – e.g. if a turbine in a power plant should trip off load.

The throttling device R must be designed so as to restrict the steam flow as soon as the downstream pressure rises above a predetermined (and adjustable) level. With a turbine this can quite simply be incorporated with the governing gear: it would have the effect of limiting the steam flow virtually to the normal full load consumption regardless of any excess upstream pressure, and thus preventing the overloading of the alternator. The vent valve should preferably be sited close to the steam main but fairly far from the utilization plant, because if the whole plant should suddenly shut down through some mishap the vent valve would have to blow off the full load plant steam requirements into the atmosphere, and this could be tolerable only at some fairly remote point. Silencers should be provided to muffle the noise from vent valve discharges.

The purpose of the vent valve is solely to prevent the lifting of wellhead safety valves. Its setting should be such that the following requirements are fulfilled:

(i) When the plant is at full performance, either no steam at all, or a bare minimum of steam, should be discharged to waste. Whether it is possible to avoid altogether the discharge of steam will depend upon the number and characteristics of the connected wells and on the pressure drops along the various parts of the transmission system.
(ii) When the plant is shut down, though with the full quota of connected wells as for full performance, the vent valve must be capable of discharging the whole of the bore steam without the pressure at any point in the system rising to a level where a safety valve would lift.

Further details of the setting and peformance of vent valves have been given by Armstead and Shaw [101].

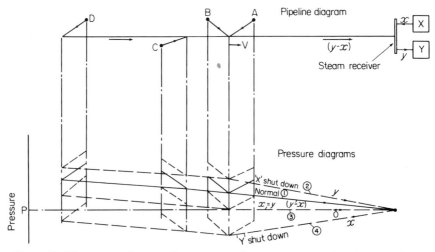

Figure 13.2 Steam supply control; constancy of steam pressure primarily assured.
(Fig. 2 of [101])

13.2.2 *Constancy of steam pressure primarily assured.* Such a require-ment could apply to a steam receiver (Fig. 13.2) *into* which is fed exhaust steam *x* from some plant unit X, and *from* which is drawn a steam feed *y* to some other plant unit Y. An example of such a receiver is to be found at Wairakei, where high pressure turbines exhaust into a manifold from which lower pressure turbines take their supply. Constancy of pressure in the steam receiver at the required level P can only be assured if the bores supply *y−x* of steam directly to the receiver. Under steady conditions let it be assumed that this constancy of pressure and balance of flows is achieved by feeding steam from four bores as shown in Fig. 13.2: the pressures along the pipework and at the bores would then be as shown in the lower part of that figure by the full line (1). Now let the balance be disturbed by a sudden cessation of *x*. The only way of preventing a fall of pressure in the receiver would be to raise the steam flow from the field to the plant to a value of *y*, thus restoring the balance of flows; and the only way of bringing about this higher flow would be to raise the pressures in the bore field so as to impel more steam along the pipeline. However, higher field pressures would result in *lower* steam yields from the bores, so the additional steam would have to be found from elsewhere. This could be arranged by connecting a vent valve to some convenient point along the pipeline. If, under normal conditions (line (1) of Fig. 13.2) this vent valve were discharging to waste a quantity of steam somewhat in excess of *x*, and if this valve were to close as soon as the inflow from X is cut off, pressures in the field would rise, as shown by line (2) of Fig. 13.1, and steam would be diverted from waste to the plant. Similarly, if Y were suddenly shut down the pressure in the receiver would rise unless the whole of the quantity *x* could be diverted elsewhere. This could be done by opening the vent valve sufficiently wide to enable the field pressures to fall, as shown by line (4) of Fig. 13.2, thus bringing about a reversal of flow along the pipeline from the plant to the field. If at any time *x* and *y* became equal to one another, no balancing flow would be needed to or from the field and the vent valve would have to discharge enough steam to ensure that the pressure adjacent to it was equal to P, as shown by line (3) of Fig. 13.2.

Although only the extreme conditions of the total shut-down of X and Y have here been considered, a satisfactory control system would have to cater for any variation of either *x* or *y*, however small. Control may be effected by using the pressure at the steam receiver to actuate the position of the vent valve so that a fall or rise in pressure may cause the vent valve to move towards the closed or open position, as may be required. Under transient conditions some momentary rise or fall of receiver pressure would be inevitable, but the characteristics of the control system must be such that the pressure deviation is kept within acceptable limits. If the vent valve were sited at the plant (and the only objection to doing this would be the nuisance of large volumes of escaping steam), simple proportional control with a very sensitive setting would be satisfactory, but if it were sited remotely from the plant this type of control would be unstable owing to the time lag inherent in a long stretch of pipeline. A 'zero error' two-term type of control with a small proportional and a large integral term can give satisfactory control. The addition of a derivative term could give rise to hunting. The temporary pressure deviations at the receiver will depend upon the magnitude of the disturbance to the balance of flows.

The hypothetical example here chosen would involve an unacceptable waste of steam under normal conditions (something more than x), but it would seldom be necessary to cater for such a drastic condition as the total shutdown of X, or at least it might be possible under such a condition to curtail the value of y. It would then be possible to arrange for a satisfactory control with only a moderate 'normal' discharge through the vent valve. If so, the control would give a greater range of operation in one direction than in the other, for if the valve is normally only slightly open it cannot move very far in the closed direction. If a disturbance exceeded the capability of the vent valve to deal with it, automatic action would have to be supplemented by manual adjustments. For further details of this system, see [101].

Vent valves may consist of single or multiple units arranged to operate in series or in parallel with one another. This applies both to those required for the assurance of steam quantity and of constant steam pressure. At Wairakei, where both systems are adopted, each is provided with two valves operating in parallel. This allows for limited control if one valve should be taken out of service for maintenance, after suitably adjusting the setting of the other valve remaining in service.

13.2.3 *Economic flow balance assured.*   This question has already been considered to some extent in Section 12.4, and has been described at greater length by Armstead and Shaw [101]. 'Economic' flow balance implies that as far as possible the removal or curtailment of any one plant unit in a multiple assembly of plant units shall not affect the performance of any other plant unit. The required balance of flows can be obtained by manual manipulation of reducing valves and dump condensers, but more speedy operation can sometimes be achieved to some extent by automation. For example, the opening of reducing valves could be affected by proportional control from the downstream pressure of a parallel plant unit. Back-pressure governing could also prevent dangerous pressure rises occurring in low pressure pipework, at the same time helping to avoid the lifting of safety valves at the plant. There will generally be a limit to what can be achieved automatically: the bringing in of a dump condenser for example, would require too many operations for automatic actuation. A combination of automatic and manual operation control will usually be needed.

13.3 Hot water control [101, 106]

The problems of hot water control are of two kinds. First there is the automatic disposal of unwanted hot waters collected at the wellheads in wet geothermal fields, and this has been covered by Section 9.4. The more difficult problem of controlling piped hot water at near-boiling temperatures will now be discussed.

In Section 9.9 the dangerously explosive nature of superheated water was emphasized. This alone would be sufficient to justify very careful control when transmitting such fluids, but further controls are necessary to ensure that supply and demand of the hot water are continuously reconciled. As mentioned in Section 13.1, the advent of a successful submersible pump which would enable superheated water to be transmitted at pressures far exceeding the saturation pressure corresponding with the temperature would greatly simplify the problem of control: so too would the adoption of two-phase fluid transmission (Section 9.12). The only known case where slightly pressurized water at 200° C or more has been transmitted successfully over 1.5 km or so was the experimental pilot hot water scheme described in Section 9.9, where the needs for pumping, attemperation and controlled valve movement speed were emphasized. The method of controlling that transmission system is thought to be of sufficient interest for it to be briefly described here, as it is always possible that a similar scheme may be introduced elsewhere in the future, should hopes of submersible pumps and of two-phase transmission not be entirely fulfilled.

The Wairakei experimental scheme is illustrated in Fig. 13.3 which shows one (of several) h.p. and one (of several) i.p. water collection drums, with the steam spaces connected to their appropriate steam transmission systems. A low lift pump, X_1 raises the h.p. water into the 'floating' head tank HT, (mentioned in Section 9.9) which serves not only as a means of pressurizing the water transmission pipe to the power plant and of taking up short term fluctuations of hot water yields, but also of providing means of actuating the control of the flow rate of hot water into the first-stage flash vessel at the power plant. Hot water from an i.p. collection tank is injected into the hot water transmission line by a pump, X_2 having higher delivery head than X_1 (equal to the sum of the static pressure from the head tank (Plate 17) plus the pressure difference between the two steam systems). Both pumps are of the self-regulating cavitating type, like the extraction pumps normally provided in conventional power plants for removing the condensate from the condensers, which will deliver into the hot water pipe whatever quantities of water are supplied to the collection tank, up to the maximum pump rating. Should this capacity be exceeded, the water level in the collection tanks would rise and spill the excess water to waste by means of a float-actuated relief valve HLR. At the plant end of the transmission main is an automatic control valve, CV, with a flow meter, F, just upstream of it. Under steady

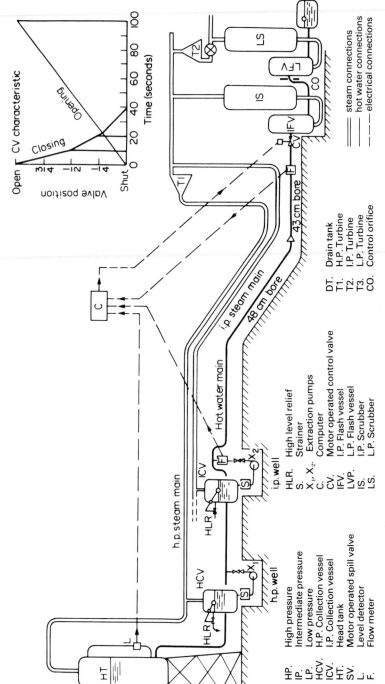

Figure 13.3 Experimental hot water transmission scheme tried out at Wairakei for power generation from two stages of flashing, showing method of control. (From Fig. 4 of [106])

Note: For relationship of collection vessels to wellhead gear, see Fig. 9.1.

HP.	High pressure
IP.	Intermediate pressure
LP.	Low pressure
HCV.	H.P. Collection vessel
ICV.	I.P. Collection vessel
HT.	Head tank
SV.	Motor operated spill valve
L.	Level detector
F.	Flow meter

HLR.	High level relief
S.	Strainer
X_1, X_2.	Extraction pumps
C.	Computer
CV.	Motor operated control valve
IFV.	I.P. Flash vessel
LVP.	L.P. Flash vessel
IS.	I.P. Scrubber
LS.	L.P. Scrubber

DT.	Drain tank
T1.	H.P. Turbine
T2.	I.P. Turbine
T3.	L.P. Turbine
CO.	Control orifice

conditions, this control valve will take up a position that allows the passage of exactly the total quantity of hot water, h.p. and i.p., being pumped into the system: the level in the head tank will then remain steady. As soon as a difference arises between the hot water *supply* from the pumps and the hot water *demand* as determined by the position of the control valve, the water level in the head tank will rise or fall accordingly. This change of level is detected by a level-sensitive device L, which sends a signal to a computer C, which automatically adjusts the control valve position in such a manner as to restore the balance between inflow and outflow. This is effected by calibrating the head tank water level so that under steady conditions its height is proportional to the flow through the control valve. A rise or fall of head tank water level will therefore immediately cause the computer to detect a discrepancy between the *apparent* flow, as registered by the head tank level, and the *true* flow as registered by the flow meter. Any such discrepancy causes impulses to be sent to the control valve causing it to change its position until the discrepancy disappears and a new steady state is achieved with a different head tank level. The relationship between the detected discrepancy and the corrective control valve movement is a two-term one with a proportional and an integral element. (As with the constant pressure stem control system described in Section 13.2.2, the addition of a derivative term could give rise to 'hunting'.) A special mechanical drive ensures a characteristic valve movement as shown in the inset to Fig. 13.3, so as to limit surge heads induced by the valve movements to acceptable values. The rate of opening must be much slower than the rate of closure because the negative surge heads induced by water acceleration tend to cause boiling in the pipeline. The rate of closure may generally be more rapid, but this too must be restricted as the valve approaches the closed position because the reflected surge heads which follow the moment of closure will depend upon the rate of valve movement just before that moment.

As the transmitting capacity of the pipeline will depend upon the quantity of attemperating water injected into the system, it is necessary to curtail the volume of transmitted water if the amount of attemperation should fall below a certain critical figure. Hence a further signal must be sent to the computer from a flow meter that summates the total flow of i.p. water, and when this signal indicates an insufficiency of attemperating water an appropriate sequence of tripping off h.p. well pumps is initiated so as to ensure adequate curtailment of the total hot water flow according to the degree of attemperation. A hot water transmission system which does not rely on attemperation, but only on pressurizing, could of course dispense with this refinement.

A final measure of control is needed to ensure that the utilization plant can make use of all the hot water delivered to it. On arrival at the plant the hot water is flashed into steam (in the Wairakei experimental scheme) and if the plant is unable to use all of the flash steam it would be necessary to spill some of the hot water to waste. This can be done by means of a remotely controlled spill valve connected to the head tank. The operator can adjust the amount of spill by remote control from the power plant, or alternatively the spilling may be automatic by the actuation of the spill valve from the pressure of the flash steam, which would tend to rise if it were being generated more rapidly than it could be used. The hot water spill valve has a similar function to that of the vent valves in the steam control systems – that of trimming the supply to match the demand.

13.4 Safety precautions: general

In common with all installations that make use of fluids at high temperatures and pressures a geothermal installation must be provided with the necessary means for protecting personnel against injury and plant against damage, both under normal and under emergency conditions. In Section 9.3 the means have been described whereby dust, grit and rock particles ejected with the hot fluids from the bores may be intercepted at the wellheads, while in Section 9.6 the provision of safety devices to prevent the occurrence of excessive pressures at wellheads and in the fluid transmission system has been described, as well as the precautions to be taken in wet fields against the entry of dangerous quantities of hot water into the steam transmission system. These steps, however, do not exhaust all the precautionary measures necessary in a geothermal installation.

13.4.1 *Pressure and vacuum limitation.* Although the bursting discs and safety valves at the wellheads will protect the entire steam collection and transmission system against excess pressures, they can do nothing to safeguard downstream plant and equipment designed to operate at lower pressures. For example, in a power station such as Wairakei, where back-pressure turbines exhaust into lower pressure manifolds, it is necessary to protect these manifolds against excess pressure by means of safety valves in case outflow should be interrupted while inflow continues. Manifolds feeding condensing turbines must also be protected against vacuum conditions; for if the supply of steam into the manifold should be interrupted, the condensing turbines would continue to 'suck'. Low pressure pipework is of necessity of large diameter owing to the high specific volume of steam, and would normally have thin walls: it would thus be liable to collapse under external atmospheric pressure. The provision of stiffening rings and internal struts could give protection against a moderate fall of internal pressure, but it may also sometimes be necessary to provide vacuum-breaking relief valves to admit air if the pressure drop should exceed a safe limit. The operation of such relief valves should be avoided except in the last extremity, as it could lead to corrosion of troubles.

13.4.2 *Water entrainment.*   Despite the ball check valves (Fig. 9.1, Section 9.1) it is always possible that one of them may be defective and that large gulps of water may enter the steam pipework system. At all costs this water must not be allowed to reach the exploitation plant (turbines being particularly vulnerable). If the entry of water is only moderate, the normal pipeline drain traps (Section 9.10) should be adequate to remove it during its passage from the wells towards the exploitation plant; but if excessive, these drain traps may lack the necessary removal capacity. A back-up defence system of stategically placed water-detectors should therefore be provided at intervals along the main steam pipelines. The presence of water at the detector most remote from the utilization plant would simply send an alarm signal to the plant operators to alert them of a potential hazard. If the drain traps can deal with the water present, the trouble may pass and normal conditions restored; but if water should reach the next downstream detector the resulting signal should be used to disconnect *part* of the exploitation plant (e.g. trip off one or two turbines in a power station). This would slow down the velocity of steam in the mains and would give the drain traps a better change of removing the remaining water in the mains. Should the water still reach subsequent downstream detectors, further plant components would have to be disconnected consecutively. Finally, if water should reach a point fairly close to the exploitation plant entry, *all* plant should be tripped off so as to halt the steam flow altogether. A plant shutdown, though very inconvenient, would be preferable to a disaster. A system of this sort takes progressive and discriminate action, avoiding 'panic' action until the last possible moment. The presence of water detectors at the junctions of branch steam lines with the steam mains would facilitate the location of the source of trouble and thus shorten the time taken to deal with the offending well. Water carry-over into the mains could be due to too small a waste discharge orifice coupled with a sticking ball check valve, or by the failure of the high level spill device in the head tank of a hot water transmission system.

13.4.3 *Gas detection.*   Nearly all geothermal steam contains hydrogen sulphide – a lethal gas when present in quite modest proportions. This gas is rendered more dangerous by the fact that it paralyses the olfactory nerves when fairly strongly concentrated. Thus its characteristic and rather unpleasant smell, which is noticeable when it is present in harmlessly low concentrations, may fail to give warning just when such warning is most needed. As the density of H_2S is 18% greater than that of air at the same temperature, it is apt to accumulate in low-lying pockets such as building basements. Methods of limiting the escape of H_2S into the atmosphere will be dealt with in Chapter 16, but as a precaution against possible imperfections of those methods it is essential that all possible 'gas traps' be adequately ventilated. It may also be advisable periodically to take gas samples at all potential danger spots, and perhaps to install gas detectors coupled to alarms; and also to limit the length of time permitted to personnel to stay in danger areas.

13.4.4 *Overspeeding.*   The steam turbines of a geothermal plant will normally be designed to run safely at an overspeed of about 10% above synchronous speed (as in conventional steam stations). With a mixed system, however, they may have to operate in parallel with hydro-turbines designed to run safely at speeds of up to 40% above synchronous speed. In the event of a system disturbance in which a large proportion of the load is lost, there may be a danger of a synchronized combination of hydro and geothermal plants running away to speeds which, though safe for the water turbines would be dangerous to the steam units. To protect the latter from such a contingency it may be necessary to provide 'overspeed protection' which will trip off the geothermal plant if the hydro-plant tries to run away with it. The setting at which this tripping occurs must be below the safe overspeed limit of 10% but high enough to prevent unwanted tripping under minor system disturbances. About 7.5% overspeed is of the right order. In systems containing mostly steam plant, with only a small element of hydro, this precaution may not be necessary; but in New Zealand, where hydro-plant greatly predominates – or *did* predominate when Wairakei was built – this overspeed precaution was very necessary.

13.4.5 *Standby corrosion.* Protection against this hazard will be discussed in Section 15.8.

13.4.6 *Load limiting in cold weather.*   If a condensing turbine is fully loaded in the daytime it may become overloaded at night when the temperature of the cooling water falls. A simple load-sensing device can give the necessary protection [206].

13.5 Monitoring

At some central control point it is necessary to monitor certain essential measurements so that overall operating control may be exercised over a geothermal installation, or group of installations, by means of local and/or remote control. Typical of the basic indications required for this purpose are:
  (i) Steam pressures in various pipe manifolds and at key points in the steam transmission pipework, such as at vent-valve locations.
 (ii) Steam flows into the plant, summated where several pipes operate in parallel.
(iii) Total kilowatt output, if the utilization plant is a power station.

(iv) Position indicators for reducing valves and certain other important steam valves.
(v) Head tank level and control valve position, and perhaps water temperatures at certain key points, for a hot water transmission system controlled in the manner described in Section 13.3.

In addition to indications, facilities will be required to enable the operators at the plant to take action by remote control, where necessary, for such operations as the following:

(i) Varying the lifting pressure and operating range of the vent-valves for a steam control system as decribed in Section 13.2.1.
(ii) Varying the zero point and the proportional and integral settings of the vent-valves for a steam control system as described in Section 13.2.2.
(iii) Varying the opening pressure and degree of sensitivity of automatic reducing valves.
(iv) Varying the proportional and integral settings of a hot water control system such as that described in Section 13.3.

Alarms related to safety precautions must also be brought to the control centre to indicate, for example, entrained water in the steam pipes, high pressure in steam manifolds, high water level in the head tank of a hot water transmission system, etc.

As to the bores themselves, the cost of monitoring conditions such as valve positions, steam and water flows, operation of water interceptors, the tripping of hot water pumps, etc., can seldom be justified. Experience of geothermal bores suggests that so long as steady conditions are maintained at the utilization plant the behaviour of individual bores generally remains remarkably constant and the bores can be left unattended for long periods. Operators soon acquire the 'feel' of a geothermal installation and can infer from pressures and steam flows whether, for example, the bursting disc of a bore had ruptured, in which case the field maintenance staff can be instructed to locate and remedy the fault. Routine inspection in the field, with regular logging of pressures and flows, is cheaper then remote instrumentation and usually just as effective. However, good communications by telephone or radio between the control point and the field are imperative, so that any abnormality sensed at the control point can be located and rectified by the field attendants, and so that any abnormalities detected by the field staff can be reported immediately to the control centre and appropriate instructions may be issued. A mimic diagram at the control centre can be useful, so that operators can see at a glance which bores are in service or under maintenance, which valves are open or closed, and what plant units are operating. Such a diagram need not, however, be automatic: hand setting will usually suffice.

13.6 Maintenance

Although briefly mentioned in Section 10.11, it is important here to repeat that with a geothermal installation – especially with an integrated *system* of installations as at Larderello or in the Geysers area – it is most advisable to have a carefully planned scheme of preventive maintenance. It also pays to keep an adequate stock, at convenient places, of spare parts – particularly complete turbine blade assemblies, both fixed and rotating – so that plant outages may be kept to a minimum. Ricci and Viviani [129] and Matthew [130] have described the arrangements adopted at the Larderello and Geysers fields. Limited workshop facilities are available on site at all major geothermal power plants in the world, while for major overhauls it may sometimes be necessary to send faulty plant to the nearest city having the necessary facilities, though such steps seldom have to be taken. At the Geysers there are facilities for turbine rotor balancing and blade work in the field area, and a workshop for medium sized equipment repairs at the Administration Centre close to the bore field; but a full-scale workshop capable of handling plant repairs of all kinds, mechanical and electrical, is to be constructed at the Administration Centre, with a machine shop for medium-sized equipment repairs at a Satellite Centre close to unit 13 (the largest unit in the field).

Corrosion hazards, a serious cause of troubles in the early days of geothermal development, have become almost negligible by the adoption of suitable resistant materials for the duties required (see Chapter 15). At the Geysers, turbine units are overhauled normally at 2-year intervals (by comparison with 6 years for conventional steam plants), but the outage time averages only 4–6 weeks (by comparison with 7–10 weeks for conventional steam plants).

For the maintenance of bores it is constantly necessary to log the fluid outputs so that the pattern of declining yields may closely be observed. Timely reaming of bores liable to show calcitic blockage can sometimes restore a bore yield to much higher outputs, and a study should be made to determine the economic period for such reaming. A maintenance rig should always be held available to cover the contingency of an unforeseen blockage or casing failure, while the sinking of replacement bores should be carried out at an adequate rate to cover the ultimate decline of older bores to uneconomic yields. Sometimes the life of a bore may be extended by perforating the casing at well-chosen depths (see Section 7.11).

13.7 Manpower, automation and decentralization

At Wairakei, when the station was first operated at its full potential output, the staff employed for supervising, operating and maintaining the field equipment and 192 MWe of plant (in 13 units) was as follows:

Power station   51 or 0.266 men/MWe installed
Field equipment 31 or 0.161 men/MWe installed
    Total   82 or 0.427 men/MWe installed

These figures included allowances for absences on leave and sickness. There was a degree of flexible give-and-take between the station and field staffs, which were not rigidly separated into 'water-tight' groups.

Later, when it was found that frequent modifications were needed to adapt Wairakei to the changing field conditions, the manpower had to be increased. By 1980 it was as follows, including clerical staff:

Power station   87 or 0.453 men/MWe installed
Field equipment 66 or 0.344 men/MWe installed
    Total   153 or 0.797 men/MWe installed

(These figures are only approximations, as they involve certain arbitrary apportionments.)

In 1980 at the Geysers a staff of 222 (including supervising, technical, storekeeping and clerical staff) were operating and maintaining 960.4 MWe of power plant – but *not* the steam supply system. The men/MWe installed were thus 0.231; or rather more than half the Wairakei figure of 0.453, which seems reasonably comparable for an installation almost five times as great (though in 15 units). No less than 45 of the 222 men were accounted for by the H_2S abatement plant (see Chapter 16). But for these, the manpower would have been only 0.184 men/MWe installed, or about 41% of the Wairakei figure.

Reduction in manpower per MW has been achieved at the Geysers by the high degree of automation necessitated by the decentralization of the power plant over several scattered buildings. The primary purpose of this decentralization was to economize in field pipework so that the average distance between bores and power plant could be minimized. If each plant were to be fully manned on the scale of Wairakei, where all the 192 MW of plant is concentrated at one site (albeit in two adjacent buildings), much of this economy would have been nullified. A high degree of automation and remote control thus became an economic necessity. In the Tuscan fields 418 MW of plant are now spread over 15 sites, so there too decentralization and automation have become the natural consequence of expansion. (At Wairakei there has of course been no expansion since the original installation.)

At the Geysers all plants have been designed for unattended operation with 'shut safe' devices to warn the Control Centre (at units 5 and 6) when they act. The operating staff works round the clock on roving duty, aiming to make two visits per shift per plant. All other staff normally works on single shift duty (40 hours/week) except when urgent repairs require longer hours from the maintenance men.

The Control Centre is computer-based and is manned continuously. Analogue computers are used for instrumentation: digital computers for 'on/off' monitoring. The system despatcher takes appropriate action, in collaboration the the operating staff where necessary, for starting up, shutting down and effecting load and voltage adjustments, based on transmitted signals and alarms. If a unit trips off load, the connected wells are allowed to blow to waste through silencers until it has been ascertained that the unit cannot soon be restarted.

At Wairakei there were two good reasons for the concentration of the whole power plant at a single point. In the first place there was an abundant supply of cool river water and a convenient flat riverside site available. Secondly the original intention (see Section 12.3) was to have a dual-purpose plant which would have necessitated the placing of the power plant close to the distillation plant.

13.8 Annual plant factor

The reliability of controls, the keeping of an adequate stock of spares, the enforcement of a strict preventive maintenance programme, together with the inherent reliability of thermal fluid flows should collectively ensure that geothermal power plants attain higher annual plant factors than any other form of generating plant. A similarly high degree of high availability for non-power applications of earth heat can also be expected. Matthew [130] mentioned in 1975 that during 56 turbine-years at the Geysers there had only been 12 enforced outages averaging 3 weeks each – i.e. 1.24% of the time – though there had admittedly been unintended load *curtailments* that had raised the total effective time of enforced outages to 6.2% which is greater than the corresponding figure of 2.5% for conventional reheat turbo units. On the other hand the availability of geothermal steam supplies is considerably greater than that of boilers, so that the overall availability of complete geothermal installations would appear to be generally better than that of conventional steam power plants.

Annual plant factors for individual geothermal power plants exceeding 90% are not uncommon. In Japan 85–90% is regularly achieved and at Cerro Prieto 91% has been achieved.

At Wairakei where 41 MW of the 192 MW installed are provided as spare capacity – unmatched by a correspondingly adequate steam supply system, thus leaving only 151 MW available for normal service – an annual plant factor of 98% was achieved in 1975/76 if reckoned on the 151 MW figure. This however, is misleading as it is probable that some of the 41 MW of spare plant was sometimes used to consume some surplus steam capacity over and above the amount required for the 151 MW which it is normally intended to serve; moreover the availability of so much reserve plant would preclude direct comparison of this achievement with that of other geothermal power installations. The annual plant factor based on the total installed capacity of 192 MW was 77% for the same year, but this is an unfairly *low* figure, as the steam supply system would have been incapable of producing 192 MW.

In expanding systems such as those in Tuscany and in California, annual plant factors of only 76 or 77% have often been recorded, but these figures considerably underrate the potential utilization of geothermal power plants because they are distorted by new plants coming into service and operating only spasmodically for part of their first year of operation. 'Steady state' plant factors for such expanding systems should be around 85 to 90% (annual); but at the Geysers there was an unfortunate decline in the late 1990s owing to the disruption to smooth operation caused by the retro-fitting of gas abatement plant (as will be described in Chapter 16): in the three years 1978/79/80 the average annual plant factor for all the Geysers plants collectively was only about 70%. When the gas abatement programme has been completed, a return to much higher plant factors may confidently be expected.

14 Some economic considerations

14.1 General

There is of course more to economics than mere costing, although costs must always be important. Such factors as national self-sufficiency, environmental impact and energy conservation may now often overshadow considerations of cost. Nevertheless, costs can never be ignored. It is also important to recognize at the outset that the days of very cheap energy have probably gone – if not for ever, then at least for a very long time. Those days ended in October 1973, when politically motivated swingeing increases in oil prices rocked the economies of many nations. The alarming rates of inflation that subsequently afflicted so many countries, frequent movements in the currency exchange markets, and great instability in the rates of interest on borrowed money have all contrived to make it increasingly difficult for rational comparisons to be made between geothermal energy costs in different countries (or even between the costs of different forms of energy in the same country when the foreign currency component differs greatly between one form and another). Moreover, when making energy cost comparisons between one country and another it is important to bear in mind the national average income *per capita* as a yardstick of acceptability.

Difficult though the task of advance costing of geothermal energy may be, it is one that cannot be shirked; for decisions of whether or not to proceed with geothermal development will always have to be made at some stage or other against a background of many uncertainties. Unfortunately the author forgets who it was that once aptly likened the task of electricity cost accountancy to that of a merchant who buys by the cubic metre and sells by the ton a commodity whose specific gravity is both difficult to determine and constantly changing. The task of geothermal cost estimation in advance of actual development is just about as forbidding.

14.2 Unpredictability of geothermal costs

Even in stable times, so many uncertain factors can arise between the moment of intent to develop an energy source and the moment when the energy starts to flow, that the precise forecasting of costs for *any* form of energy has always been a precarious business. However, with geothermal energy, the task of predicting costs has always been rendered far more difficult by a number of other uncertainties peculiar to this form of energy. The instabilities of modern times have aggravated the difficulties. Exploration costs can seldom be foreseen with any degree of accuracy. A sum of money may be devoted to exploration, based on a rational plan of action, but until quite a lot of data have been collected and analysed it is virtually impossible to predict how much money should be spent on each of the various exploration techniques available. The costs of exploratory drilling are particularly hard to foresee, as the success ratio can vary so greatly. Also it is hard to foretell the depths to which the bores must be sunk to reach the productive horizons (and, as will be shown in Section 14.10 drilling cost are extremely sensitive to depth); nor can the quality of rock to be penetrated by the drill be known with any certainty beforehand. Again, we can never know in advance either the bore characteristics, or the practical bore density per square kilometre, which will greatly influence the fluid transmission costs. There are other uncertainties such as the foundation conditions for a plant, the siting of which cannot be chosen until the pattern of productive bores is at least approximately known; or the ability or inability to discharge waste fluids into water courses without risk of pollution. All these uncertainties, however, should not be used as an argument against geothermal development, for somewhat paradoxically the very monetary troubles from which so much of the world is now suffering are tending to *improve* the economic prospects of geothermal energy and to make the taking of financial risks more worthwhile. Geothermal exploration always involves some measure of risk, but experience has shown that where the risks have been taken on the basis of reasonably promising evidence, it has nearly always paid handsome dividends. Moreover, exploration techniques are continually improving and the risks are steadily being reduced. A useful analysis of the problems of geothermal costing was made by Leardini in 1970 [207].

14.3 Geothermal cost experience until 1973

When the rate of cost inflation was lower, it was possible to make a reasonable attempt to rationalize geothermal costs, both in terms of heat and of electrical energy. Such an attempt was made in about 1970 [208]: the results were reasonably valid for the 1960s, although they have since become hopelessly outdated by inflation. Nevertheless, it was the *comparative* costs between geothermal and alternative energy sources that were of greater significance than the absolute figures. In those days of cheap fossil fuels it would have been approximately true to say that geothermal heat from exploited fields could be expected to be of the order of 12 US cents/Gcal at the wellhead. After allowing for the cost of collecting and transmitting the thermal fluids from the bore field and of separating and (usually) rejecting the hot water at the wellheads, the cost of geothermal heat from a wet field piped to a single delivery point for use in some exploitation plant might have been approximately quadrupled to, say, 50 US cents/Gcal. In those same years, the cost of fuel heat ranged from about 100 to 250 US cents/Gcal. If this heat was required in the form of steam or hot water, the price (after allowing for the capital charges, operation, repairs and maintenance incurred by the necessary boilers or heat-exchangers) was raised to anything from about 150 to 400 US cents/Gcal according to the location and scale of the enterprise. A fair average figure would have been about 275 US cents/ Gcal – *about* 5½ *times* the cost of geothermal heat. This, of course, is a very broad statement, and individual cases would have differed widely, but it may be taken as a statement of a general trend. In the case of small industries or micro-power plants that could be served by a single bore without costly collection and transmission pipework, the cost advantage of geothermal heat could have been even greater. These approximate cost comparisons were based upon oil fuel as the alternative source of heat. Heat from coal or lignite could sometimes have been considerably cheaper, but not nearly cheap enough to cover this greatly advantageous factor of about 5½. In Reykjavík, Iceland, even after adding the extensity city heat reticulation system, the cost in 1961 of geothermal heat delivered to domestic consumers was only about 58% of the cost then prevailing of heat supplied from oil burned in small domestic boilers [209]. According to Einarsson [8] this percentage had fallen to 25–30% by 1974: by 1980 it had fallen to 12%.

In the case of electric *power,* the advantage in the 1960s in favour of geothermal energy, though still positive, was less striking. This was because conventional power plants could operate at very much better thermal efficiency and could make use of power plant costing much less per kilowatt than geothermal power plants owing to the higher temperatures and pressures and larger capacities that could be used. Nevertheless, it was probably broadly true to say that geothermal power in those days cost only from 50–60% of conventionally generated thermal power at the same annual plant factor.

14.4 Effects of inflation

From about 1970 world price inflation started to accelerate, and in late 1973 oil prices rocketed, with other costs rising sharply, though less markedly than oil. By 1980 the price of oil fuel had risen by a factor of about 13 since the end of the 1960s and other fuel prices had risen in sympathy, though not so steeply. However, the fact that fuel prices have risen more rapidly than material and labour costs has rendered geothermal energy still more attractive *relatively* than conventional thermal energy, despite the higher interest rates on investment. This can be demonstrated by means of the following arbitrary, though not unrealistic, example.

Let it be assumed that in 1969 a geothermal power plant could generate electricity at 55% of the cost of that generated by a conventional thermal power plant of the same capacity and serving the same duty, and that the proportional composition of the production costs were as follows:

Conventional		Geothermal	
Fuel	61%		
Operation, repairs and maintenance }	12%	Labour and materials	20%
		Capital charges	80%
Capital charges	27%		
	100%		**100%**

Let it further be assumed that by 1980 fuel prices had risen 13-fold, that labour and material costs had trebled, and that capital charges had risen from 9% to 14%. The relative production costs from new plants would now be:

Conventional

Fuel	$61 \times 13 = 793$
Operation, repairs and maintenance	$12 \times 3 = 36$
Capital charges	$27 \times \frac{14}{9} \times 3 = 126$
	Total $\qquad$ 955%

Geothermal

Labour and materials	$20 \times 3 = 60$
Capital charges	$80 \times \frac{14}{9} \times 3 = 373$
	$\overline{433\%}$ of 55%
	$= 238\%$
Relative production costs	$\dfrac{\text{geothermal}}{\text{conventional}} = \dfrac{238}{955} = 25\%$

Thus, whereas geothermal energy was costing about 55% of conventional thermal energy in the late 1960s, it would now be costing only about 25% This is probably a cautious deduction, for exploration costs have become relatively more moderate with improved methods, though they too have of course inflated.

This comparison, while making no claim to accuracy, clearly illustrates the relatively improved economic position of geothermal energy resulting from the fact that fuel costs have inflated more rapidly than other cost components. In so far as industrial heat is concerned, the improved position of geothermal energy should be considerably greater owing to the higher fuel content of the cost of conventional heat. The reduction, from 1961 to 1980, in the Reykjavík domestic geothermal heating costs from 58% to 12% of the oil-fired boiler alternative, supports this belief. The trend, favouring geothermal heat economically, could well continue.

14.5 The nature of geothermal costs

It has earlier been stated that geothermal costs are virtually fixed. Traditionally, it has been the custom to regard them *all* as fixed, but in the author's opinion it would be logical to assign a small *variable* component arising from the exploration costs. This is because the fruits of exploration, if successful, are a finite quantity of energy, which may be expressed in MW-years (see Fig. 5.7; Section 5.17) or in teracalories. If the resources of a geothermal field were infinite, there would be no justification in regarding the exploration costs as anything but fixed, but it is generally believed that each field has a finite extractable heat content, though that content may be difficult to assess with any degree of accuracy. This finite quantity of heat may be squandered quickly in a large plant, or it may be eked out slowly for a longer time in a smaller installation. It may be used wastefully in an inefficient plant, or economically in an efficient plant. It may be used 'now' or, like fossil fuels, it may remain stored in the ground for future use. The general belief is that the infeed of magmatic and/of radioactive heat is small by comparison with the rate of draw-off in all exploited fields, so that the commercial development of a thermal field may be likened to the mining of a finite seam of fossil fuel. When we exploit a geothermal field, we are tapping a capital store of energy that has been gradually built up over thousands of years. The energy, except for a small fraction, is not strictly 'renewable' unlike hydro-power which is seasonally, or at least periodically, replenished (see Section 4.6).

If a geothermal installation is operated at a fairly constant annual plant factor, the assessment of a variable cost to cover exploration expenditure would be of little significance, once we have decided upon the method of exploitation. However, when we come to consider different ways of exploiting a field, the existence or non-existence of a small variable cost component assumes some importance. For example, it becomes significant when we consider the relative merits of using condensing or non-condensing plants and whether the latter could be used for carrying peak loads.

The actual evaluation of the variable cost component defies precision because of the inevitable uncertainty as to the thermal capacity of a field, and also because 'exploration' is a process that may well be continued spasmodically for some time after a field is first exploited. Still, in theory the variable cost could be assessed by dividing the total costs incurred over the years of financing the exploration work (including interest and amortization of borrowed money) by the estimated total exploitable energy capacity of the field. Both numerator and denominator – especially the latter – are difficult to assess. In terms of generated electrical energy, where appropriate, the variable cost assessed in this manner is likely to be small – probably well under 1 US mill/kWh* in 1980 for condensing plant and rather more than 1 US mill/kWh for non-condensing plants (which would consume more steam per kilowatt-hour than condensing plants). The corresponding variable cost component of *heat* at the wellhead might be of the order of 5 US cents/Gcal.

When it comes to considering the use of non-condensing plants for peaking purposes, about 10% should be added to the variable cost component to allow for scavenging steam to exclude air from the turbines when not in service.

These rough figures are intended to be no more than illustrative for 1980 conditions. By comparison with the variable cost component for conventional thermal power plants, the geothermal variable costs would be extremely small.

14.6 The relative economics of using condensing and non-condensing turbines

The normal method of exploiting a thermal field for power generation is to pipe the steam from a number of bores to a central power station, where electricity is generated by means of condensing turbines. This practice is followed because of the relatively greater efficiency of condensing plants compared with non-condensing plants. An alternative method of exploiting the field would be to install a group of relatively small non-condensing plants close to the wellheads and to interlink them electrically. Such plants would require minimal pipework, the simplest of weather protection and foundations, no condensers and no cooling water system. Moreover, by siting the turbines close to the bores, the heat losses inherent in long steam mains can be reduced and higher turbine admission pressures can be adopted or more well steam could be extracted by using the same turbine admission pressures. It is true that their higher steam consumption would mean sinking about 70% more bores in order to achieve the same output, but this can be more than offset by the economies effected. It would not in fact be unreasonable to expect that the capital cost per kilowatt of the non-condensing plant arrangement might perhaps be from 12 to 14% lower than for the central condensing plant alternative, and that the production costs per kilowatt-hour (despite the higher variable cost, as suggested in Section 14.5) could be about 10% lower.

---

* 1 US mill = US 10^{-3} or $\frac{1}{10}$th of a US cent.

This would seem to pose a paradox: if power can be generated more cheaply by means of small non-condensing plants scattered around the bore field than by means of a central piped-up condensing plant, how is it that all major existing geothermal power plants make use of condensing turbines? The answer is that production costs per kilowatt-hour, though important, are not the only criteria on which the economic choice of the type of a geothermal power plant should be based. Another criterion of even greater importance is *energy conservation*. This illustrates the truth of the opening sentence of Section 14.1.

Let it be assumed that the assessed thermal capacity of a field, based on the use of condensing turbines, be estimated at 3400 MW-years. Since non-condensing plants would consume about 70% more steam, the capacity of the field based on the use of non-condensing plants would be only about 2000 MW-years. In other words, by using condensing plants the life of the field would be about 70% longer for a given installed capacity. Would it be right to squander about 1400 MW-years for the sake of saving perhaps 10% in production cost per kilowatt-hour? There is a difference between the *cost* and the *value* of a kilowatt-hour. One way of looking at the problem is to compare the geothermal production costs with those of the most nearly competitive *alternative* source of base load energy. To take a hypothetical example, let it be assumed that the production costs of the alternative source are 90 US mills/kWh while those of condensing and non-condensing geothermal plants are 30 US mills and 27 US mills/kWh respectively. Over the life of the field, the savings effected by using geothermal energy rather than the alternative source would be:

For a central condensing plant $\qquad 3400 \times 1000 \times 8760 \times \dfrac{90-30}{1000}$

$$= \text{US \$1787 million}$$

For non-condensing plants $\qquad 2000 \times 1000 \times 8760 \times \dfrac{90-27}{1000}$

$$= \text{US \$1104 million}$$

Although the latter would save 63 US mills/kWh, as against 60 US mills for the former, by comparison with the alternative source of base load, the larger saving could be applied to a much smaller number of kilowatt-hours, so that the *long-term profitability* of the costlier condensing plant would be *US $683 million* greater than if the cheaper non-condensing plants were used. This, of course, is an over-simplification, for it assumes a constant money value over the life of the field. In an inflationary market the savings in favour of the condensing plant would probably be far greater if the alternative energy source were fuel-consuming. The example serves to show, however, that the normal considerations of 'merit order' generally applied to thermal plants in terms of their variable cost component per kilowatt-hour would usually have to be overruled if any non-condensing geothermal plants were connected to the system; otherwise such plants would tend to be offered base load. There would, however, be two exceptions to this over-ruling – pilot plants and non-condensing plants using geothermal steam of high gas content.

14.7 The economics of geothermal pilot plants

In the early stages of developing a geothermal field it is customary for the first few successful bores to be blown to waste at high output for a considerable time in order to study the bore behaviour and the field characteristics. A pilot plant could put this steam to good use. Since the steam thus used would otherwise be wasted, it could be of greatest value if consumed by a plant in almost continuous operation, so that it could contribute its mite towards the system base load. Pilot plants are of an experimental nature, installed to foster confidence and to gain experience. They should be cheap, as their function is comparatively ephemeral, so non-condensing units would be suitable for the purpose. They could be quickly installed, and as they would make use of a single bore, or very few bores, they would be of fairly small capacity – say 2 to 5 MW. Since they would only use steam that would otherwise be wasted, they should not be expected to bear any share of the costs of the bores or wellheads, these being primarily provided to serve a later and larger, more permanent plant. Nor, for the same reason, need they be burdened with any variable (exploration) costs (see Section 14.5). However, if a pilot plant is to enjoy these exemptions, it would be unfair to compare its production costs (as computed on the basis of these exemptions) with the *full* production costs of other plants. It would be more logical to compare them only with the direct incremental savings earned by the generated output of the pilot plants. For practical purposes, these savings may be taken as the value of fuel saved in any conventional thermal plants that may be contributing energy to the integrated electricity system. As pointed out in the Introduction, *every kilowatt of pilot plant capacity, if supplying base load, can save about 2 t of oil fuel per annum* (or equivalent for other fuels) that would otherwise have to be burned elsewhere in the system. Even under 1980 conditions of inflated costs it should be possible to operate a non-condensing geothermal pilot plant at an annual cost of perhaps US $100/kW or less. Such a plant would not only pay for itself but would also contribute some modest excess revenue if the cost of fuel oil exceeded US $50/t – a condition sure to be found in non-oil producing countries.

The fact that a pilot plant may have served its original purpose within 2 or 3 years of its installation need not mean that its life is finished after that period. It could be shifted to a new field for further pioneering work, retained *in situ* as an emergency generator, or even used for peak load purposes (see Section 14.8).

14.8 The economics of geothermal power for contributing to peak loads

It is well known to power system planners that cheap kilowatts can sometimes be as valuable as cheap kilowatt-hours, since it is necessary to cater for peak loads as well as for base and intermediate loads. Non-condensing geothermal power plants could sometimes provide relatively cheap kilowatts, but since considerations of pressure regulation and bore stability make it difficulty to switch geothermal turbines on and off as theoretically convenient, the potential usefulness of such plants for peak load purposes is subject to constraints. Ideally, if a use can be found in a dual purpose plant for off-peak steam for industrial or other profitable application, possibly with the help of thermal storage, it could conceivably pay to use the power production element of such a dual purpose installation to contribute to the system peak and intermediate loads more profitably than to the base load. However, the problem is not a simple one. It has been analysed at some length elsewhere [143], where it has been shown that any possible usefulness of geothermal power for contributing to system peaks should preferably be confined to 'secondary peaks' rather than to the extreme peaks. By 'secondary peaks' are implied flat-topped slices of load in the duration curve just below the extreme peaks (see Fig. 10.18).

14.9 Scale effect

With conventional thermal power plants the capital cost per kilowatt installed is sensitive to what is generally known as the 'scale effect'; that is to say, a very large plant will tend to cost less per kilowatt than a small plant of similar type. An approximate formula for this effect, which broadly applies to conventional thermal plants, is:

Total capital cost $\propto$ kilowatt capacity$^{0.815}$.

Thus, if plant A has 10 times the capacity of plant B, its total cost would be about $6\frac{1}{2}$ times that of plant B and the cost per kilowatt of plant A would be about 65% of that of plant B. This advantage in favour of the larger plant is partly due to the spread of overheads over a greater number of kilowatts, partly to other general economies of large scale manufacture, and partly to the fact that the larger plants can use higher pressures and temperatures which are conducive to higher efficiencies, to more compact blading and the accommodation of a larger number of stages on a single shaft.

   With geothermal plants the scale effect is much less pronounced (with the reservation that follows below) for the following reasons:

(i) Drilling costs are more or less directly proportional to the installed plant capacity (except that since an integral number of bores must always be used, the proportion of a fractionally unused bore will tend to be higher for very small installations).

(ii) Turbo-generator costs are less subject to scale effect, partly because there is a more limited choice of the maximum plant unit size and partly because admission pressures and temperatures are usually modest, being influenced by factors other than mere plant size.

The net result is that for geothermal power installations the scale effect formula would be more like:

$$\text{Total capital cost} \propto \text{kilowatts}^{0.875}.$$

Thus, if plant A has 10 times the capacity of plant B its total cost would probably be about 7½ times as great, and its cost per kilowatt would be about 75% that of plant B. The advantage of size is therefore less marked with geothermal than with conventional thermal plants. However, this is approximately true only if the costing theory outlined in Section 14.5 – that of assigning a variable cost component per kilowatt-hour to cover the exploration costs – is accepted. If the exploration costs were treated simply as capital expenditure, the scale effect of a geothermal plant would be *greater* than for a conventional thermal plant, and the formula would become something like:

$$\text{Total capital cost} \propto \text{kilowatts}^{0.7}.$$

That is to say, if plant A were 10 times the size of plant B, its total cost (including exploration) would be about five times as great, and its capital cost per kilowatt would be about 50% of that of plant B. This formula, however, should be treated as indicative only, since exploration costs can vary so widely from project to project.

It is clearly more logical to treat exploration costs as a variable cost per kilowatt-hour so that the scale effect is less marked. The alternative of treating exploration as capital expenditure means that if only a small installation were first adopted, the exploration costs (unless partly allocated to a suspense account to be charged against expected future development – which cannot always be assured with certainty) would have to be fully recovered on that first installation and that a large quantity of residual 'free heat' would remain in the ground for use in future developments which, if realized, would quite illogically be exempted from any share of the exploration costs.

The three approximate formulae quoted above are illustrated in Fig. 14.1.

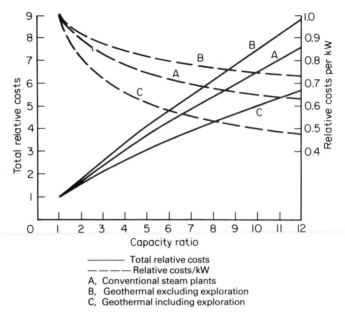

Figure 14.1 Approximate cost scale factor.

14.10 Drilling costs

Quite apart from inflation and variations in international exchange rates, it is virtually impossible to be at all specific about drilling costs for the following reasons:

(i) A drilling contractor must incur substantial 'mobilization costs' for acquiring his equipment, transporting it to where it is needed, preparing his drilling sites, providing access thereto, hiring a team of skilled men and providing them with temporary housing and other social facilities in out-of-the-way locations. He may also have to provide water supplies and construct wellhead cellars, which may, or may not, be included under his contract obligations. Hence, very substantial sums of money are necessarily spent before even a single metre of borehole has been sunk. The essence of reducing the total costs of perforating a field with bores is *time*, so that maximum total penetration may be effected within a given period, with a minimum of 'idle' or non-productive time. Delays due to breakdowns of equipment, moving from site to site, the need to plug zones of lost circulation, held up supplies of casings, mud, bits, drill-stem tubes, etc., can all affect the overall costs of drilling enormously.

(ii) Exploration holes may be two or three times as expensive as production holes of the same depth, because of the need to obtain continuous cores and to take frequent down-hole measurements and samples.

(iii) Drilling costs are very sensitive to depth.

(iv) Drilling costs depend greatly upon the nature of the rock to be penetrated. For example, a hole 3000 m deep drilled in granite could perhaps cost about three times as much as if drilled in sedimentary rocks.

(v) Drilling tends to be far cheaper in those countries or regions where there are already extensive drilling activities, and where rigs, skilled crews and contractors abound. At a remote site, unaccustomed to such operations, all the necessary facilities must either be imported or at least moved over great distances.

Regarding point (iii), it is sometimes loosely said that drilling costs vary exponentially with depth. This may be roughly true at depths exceeding about 2 km but it is an over-simplification for shallow depths. The sloping straight line of Fig. 14.2 has been taken from a graph published in 1979 [210] and claims to show typical drilling costs (in 1979 dollars) for geothermal wells of low and moderate temperature. An unfortunate graphical error in the spacings of the vertical scale introduces an element of ambiguity as to the actual costs of the 18 wells plotted; but if the extremities of the sloping line are taken to be correct the implied cost/depth relationship would conform with either of the following equations:

$$C = 95\,000\,(1.153)^{D}$$
$$or\ C = 95\,000\,(1.0014)^{d}$$

where C is the total cost of a bore in US $
D is the depth in hundreds of metres
d is the depth in metres.

Whether 18 bores are sufficient to warrant the deduction of a statistical cost/depth relationship is one matter for debate: what is also here contended is the implication of a simple exponential relationship. The figure of $95\,000 can have only one interpretation: it must represent the mobilization and setting-up costs that are incurred before even a single metre has been penetrated – it could mean nothing else. But there is absolutely no justification for assuming that the actual cost of penetration thereafter should be a function of those setting-up costs – for that is what the equations imply.

The use of logarithmic scales is a well known device for trying to detect a relationship between points that show a wide degree of scatter when plotted to linear coordinate scales; but although the scatter in Fig. 14.2 looks moderate, the upper ringed point lies above the sloping line by a factor of

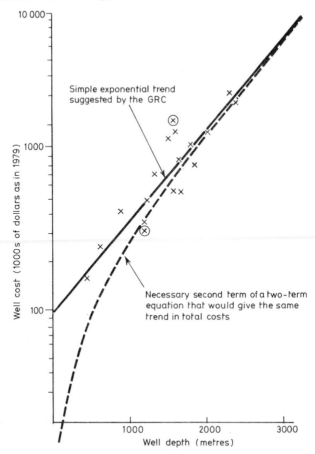

Figure 14.2 Graph of drilling costs in terms of depth for 18 American wells sunk in formation of low to moderate temperature. (From pp. 3–21, Special Report No. 7, Geothermal Resources Council, 1979 [210])

2.77 while the lower ringed point shows a cost of only about two-thirds of what the line suggests.

A more logical expectation of the form of drilling costs would be a two-term equation in which the first term would be the setting-up costs and the second expressing the cost of actual penetration as a function of depth – the two being added. If we could accept the setting-up costs at $95 000 and the sloping line of Fig. 14.2 as the true mean drilling costs, then the second term would have to conform with the curve shown beneath it, being asymptotic to the vertical ordinate and to the sloping line; but this curve by no means respresents a simple exponential relationship except at great depths. In any case it would be too much of a coincidence if the sum of the two terms were to follow a simple exponential trend. Although many investigators have favoured this exponential relationship it would not be fair to deduce this from the 18 points plotted in Fig. 14.2. There is really no reason why the trend, even with a logarithmic vertical scale, should be linear. The fact is that an infinite variety of curves could provide equally good 'fits' as a straight line to suit the meagre data plotted in Fig. 14.2; so it must be concluded that it is not really justifiable to deduce any positive cost/depth relationship from such an inadequate number as 18 points.

As far more drilling cost data are available for oil wells than for geothermal wells it is of interest to note curve A of Fig. 14.3, which shows the trend deduced from a whole year's statistics of a very large number of oil well drilling costs recorded and published by the American Petroleum Institute. Although the vertical scale would doubtless be different for geothermal wells it would not be unreasonable to expect the general *shape* of the trend to be similar. The very great number of wells from which the curve has been deduced ensures that the four erratic influences listed at the beginning of this section, other than (iii), have been smoothed out. The curve clearly shows a quasi-exponential trend similar to that of the arbitrary second term curve of Fig. 14.2. The higher percentage cost increments at shallow depths could be due to the fact that the costs of cement and production casing, as also that of bit wear at shallow depths, are more or less directly proportional to depth. Direct proportionality would mean that the second 100 m penetrated would account for 100% incremental cost; but as these particular cost components

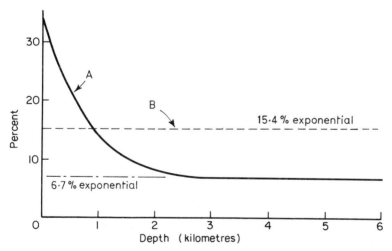

Figure 14.3 Proportional incremental costs of every additional 100 m drilled at various depths, for conventional rotary drilling. A, for oil wells, deduced from API statistics; B, for geothermal wells of low to moderate temperature, as deduced from the sloping line of Fig. 14.2.

form only a moderate proportion of the total costs, the incremental percentage at shallow depths is only of the order of 30% – not 100%. At greater depths the influence of other cost components that are not directly proportional to depth would preponderate. By way of comparison, the implications of the sloping line of Fig. 14.2 have been transferred to Fig. 14.3 as the horizontal line B. As the line B represents only bores of low to moderate temperature the relatively high incremental percentage is probably due to a greater proportion of cost components that are directly proportional to depth.

Fig. 14.4 shows the relative average cost per metre in terms of depth, as derived from curve A of Fig. 14.3: it exhibits a pronounced minimum at a depth of about 500 m, whereas the corresponding minimum deduced from the sloping line of Fig. 14.2 occurs at about 700 m. These minima are of no great significance, as mere depth as such is not the object of drilling; the true goals being high temperatures and fluid yields.

Since in a wet field the temperature rises less rapidly than the depth increases (Fig. 5.6), and even in hot dry rock the temperature rise is probably more or less directly proportional to depth, it follows that the drilling cost per million kgcal of heat won must rise rapidly with depth. This would be so even on the improbable assumption that the fluid yield from a bore remains constant regardless of depth. It is true that the *grade* of heat rises with increasing depth and that its potential range of applicability therefore increases, but the fact remains that conventional drilling is a process that yields diminishing heat returns, per unit of cost, with increasing depth after attaining an optimum. It is for this reason that geothermal drilling can seldom be justified at present for depths exceeding 2000 m or so, except perhaps in the case of geopressurized fields where the prize to be won is not confined to hot water and steam. Thus the recovery of very deep heat is likely to remain commercially unachievable until we can develop new methods of penetration for which the costs are far less sensitive to depth, or unless the quantity and grade of heat at great depth are so attractive as to offset the cost of reaching it. Even now we are *technically* capable of drilling to depths of the order of 15 km by conventional means, but the cost of doing so would be prohibitive. As long ago as 1971 it was estimated that a hole of this depth and about 25 cm diameter, even under ideal conditions, would take about 4¾ years to sink and that it would then have cost about US $20 million [211]. It is probable that by 1980 the cost would have risen to something exceeding US $50 million. To justify such a vast capital outlay a gigantic emission of heat would be necessary.

What are the reasons for this rapid rise in drilling costs with increasing depth? There are several. In the first place a very deep hole needs a larger and heavier rig to handle the great weight of long casing strings and drill stems, and more powerful driving units. Secondly, a very deep hole may

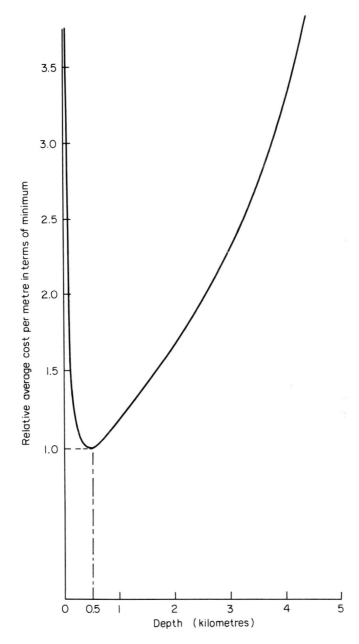

Figure 14.4 Proportional drilling costs per metre (derived from curve A, Fig. 14.3 for oil wells).

need more than the conventional three concentric casings above the slotted liner (Fig. 7.2); and if the bottom diameter of the hole is to be retained at a required practical minimum, the upper casings will have to be larger and heavier to cater for the greater number of stepped enlargements. All this requires more steel, heavier loads more grouting cement, longer times to place and grout the casings in position, and slower penetration rates at the upper levels owing to the larger bits needed. Then, penetration rates, which may be as much as 200 to 300 ft/day in sedimentary rocks at fairly shallow depths will be far less in the harder igneous rocks liable to be encountered at great depths – perhaps as low as 30 ft/day. Yet another factor is that bits have a finite life, depending upon the hardness and temperature of the rock; so that in general the deeper the penetration the shorter the bit life. This not only raises the cost of the bits, but adds enormously to the time wasted in replacing them, for every time a bit is replaced it is necessary to raise and dismantle the entire drill stem, fit the new bit, reassemble the drill stem and lower it into the hole. The time required for this may be about 10 hours at 3000 m depth and about 24 hours at 10000 m depth. As the life of a bit may sometimes be as short as 24 hours, the wasted replacement time virtually halves the effective penetration rate at this latter depth. Deeper holes require larger quantities of drilling mud to be stocked and better facilities for heat dissipation. All these factors contribute to the apparently quasi-exponential cost/depth relationship of drilling.

The outcome of this digression is unfortunately negative. Although it seems quite likely that there may be an underlying trend for the cost/depth relationship to be quasi-exponential – that is, it may at great depths approach a simple exponential form while at shallow depths the proportional (though not absolute) rise in costs with increasing depth may be more rapid – the evidence in support of any such belief is far from convincing. The five random factors listed early in this section are such that there can be no simple way of estimating or for extrapolating geothermal drilling costs. Only approximations can be made by experienced drilling engineers furnished with all available information as to the geology and accessibility of the site and with local economic conditions: even then, estimates are likely to be subject to wide errors, against which contractors must be expected to cover themselves.

The virtual impossibility of deducing a clear cost/depth relationship is well illustrated by Fig. 14.5, which shows the costs in 1979 US $ of 62 bores in the USA. It would be rash to propose from these any straight line or curve that betrays an underlying trend of a cost/depth relationship. The lower sloping line attempts to express this relationship for oil and gas wells in the USA as a simple exponential function: its equation is

$$C = 19660(1.0998)^D$$

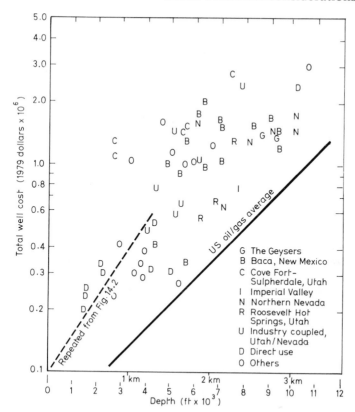

Figure 14.5 Some North American geothermal well costs in terms of depth. (From [212])

where C is the well cost in 1979 US $ and D is the depth in hundreds of metres.

This means that in place of curve A of Fig. 14.3 a mean horizontal line has been chosen at a constant incremental cost of 9.98% per 100 m of additional depth.

Also in Fig. 14.5 has been repeated the sloping line of Fig. 14.2 which accords reasonably well with the 'direct use' points to which it is intended to apply. Its greater slope by comparison with that of the oil/gas well costs confirms what has been noted earlier. Fig. 14.5 relates to one country only, so that the influence of unstable rates of exchange and of different levels of national labour and material costs has been eliminated and that of the fifth factor mentioned at the beginning of this section would be minimal. Nevertheless the scatter of the points is very great.

Other drilling cost data are hard to obtain, and such figures as are available are seldom accompanied by full particulars of geology or bore diameter. Table 14.1 quotes a few figures which serve to illustrate the random nature of drilling costs. Comparison of the later with the earlier figures brings out the effects of inflation over the last 25 to 30 years.

The cost of *directional* drilling, as used in the Paris suburban district heating schemes (Section 11.2.5) has been estimated at 28% greater than that of vertical bores reaching the same depth [165]. This should not be regarded as a universal figure because it is largely influenced by the degree of divergence and hence the additional length of casing. As the object in Paris

Table 14.1 Some random drilling costs.

Year(s)	Field	Depth (m)	Bore diameter (in)	Geology	Cost (US $)
1950s	Larderello, Italy	660 (average)	8½–12¼	Shales, limestones and sandstones	50 000
1955–60	Wairakei, New Zealand	600 (average)	8	Pumice breccia with 120 m of caprock	100 000
1960s	Broadlands, New Zealand	1216 (average)	8	Pumice breccia and rhyolite	200 000
1975	United Kingdom non-thermal areas (general)	3000 4000 8000 3000 4000 8000		Sedimentary Granite	480 000* 665 000* 2 400 000* 1 385 000* 1 910 000* 6 900 000*
1975	Los Alamos, New Mexico, USA	3650	9.625	Granite	500 000*
1977	USA selected thermal sites	500 1000 1000 1000 1500 1500 2000 2000 2000		Hard Hard Medium hard Soft Hard Soft Medium Medium hard Hard	656 000 984 000 615 000 601 000 2 437 000 655 000 (av.) 1 003 000 2 138 000 2 281 000
1979	Iceland: Svartsengi Námafjall Krafla	1603 1923 1830	10½ 10½ 10½	Basaltic Basaltic Basaltic	558 000 703 000 823 000
1980	Japan (general)	1500 (average)			800 000* to 1 200 000*

*Approximate estimates only.

was to gain the greatest possible separation between the bottoms of two adjacent bores, this percentage is probably an approximate maximum. The extra cost is small by comparison with the savings arising from having a single drilling site and pad and from reduced surface pipework.

14.11 Actual geothermal power costs

In view of what has been said earlier in this chapter it will be understood that the author quotes actual recorded geothermal cost figures with reluctance. Apart from the inflationary trend that has inevitably increased both the investment and production costs of recently constructed geothermal instal- lations by comparison with similar ones of older vintage, it is also necessary to recognize the wide range of individuality of developed fields. Differences of geology, aquifer depth, underground permeability, fluid temperatures, fluid corrosiveness, capacity of installation, site accessibility, local wage levels, national statutory constraints and other local factors can result in very wide differences in costs between two or more developments constructed at the same time in different parts of the world. Moreover, production costs are often published without specifying the rates of capital charges borne by the enterprise, and these have differed, and continue to differ, widely from one concern to another. Also, some quoted capital costs cover the entire works and activities involved, including exploration, drilling, fluid delivery, exploitation plant and equipment and perhaps (in the case of power projects) of power transmission lines for delivering the energy to the national or state grid system. This is usually so where a single owner is involved throughout – e.g. the Government of New Zealand or the Municipality of Reykjavík. Elsewhere, as in California, more than one enterprise may be involved in exploration, drilling and supplying the steam, while another enterprise buys the steam in bulk from these suppliers and uses it to generate and transmit electricity to its consumers. In such cases of multiple ownership it is impossible to arrive at the true overall costs of exploitation, since the price paid for steam by the electrical enterprise includes the profits of the steam suppliers. Hence the production costs quoted by the electricity suppliers are a mixture of true costs and of an entrepreneur's profit.

For all these reasons it would be unwise to seek uniformity or pattern in the figures quoted below, while comparisons between one set of figures and another could be invidious. Nevertheless, the quoted costs are not without interest.

14.11.1 *The steam purchase agreements of California.*   When a steam supplier provides steam to one of the power plants belonging to the Pacific Gas and Electric Company, the price is not fixed but is variable according to an agreed formula relating to the costs of alternative sources of electricity derived from fossil fuels. This linking of steam price to the costs of alternative electricity sources is not without its critics, for it is argued that the OPEC countries have only to raise their oil prices and the steam supply concessionnaires of California will immediately collect more revenue for supplying exactly the same quantity of steam without their costs having in any way been affected by the action of OPEC. The argument of the steam supply concessionnaires, however, is that the arrangement offers them a fair reward for the risks they take, that it provides them with funds for further geothermal exploration and that it offers encouragement to other concessionnaires. The purchaser (the Pacific Gas and Electric Company) is also satisfied with the arrangement because of the incentive it provides to the developers of thermal fields. This somewhat surprising acceptance has its explanation in the historical development of the electrical industry in the USA. In the early days of natural gas development, fixed cheap prices were agreed which made natural gas economically attractive by comparison with other fuels, so the whole nation more or less shifted to a gas economy. Although the developers did fairly well out of it for a time, there was no extra reward to encourage further exploration. They therefore lost the incentive to develop more gas only to have it rigidly controlled in price, so they invested in other more rewarding ventures or did nothing at all. Suddenly the country found itself geared to a gas economy and running out of gas. Despite clear warnings, the country had failed to develop its coal industry and other energy sources because of the artificially low gas prices. The Pacific Gas and Electric Company do not want history to repeat itself: they prefer to observe the rules of the market place and to make sure that they will get geothermal steam for as long as it is available at a competitive price rather than no steam at all. The price formula is such that geothermally produced power will always be substantially lower than power from fossil fuels: there is no risk that geothermal steam will price itself out of the market. At one time, when the steam price was low, the developers of the Geysers field showed signs of losing interest in continuing exploration because they found better and more rewarding ways of investing their capital. Furthermore, since some of the steam concessionnaires were also involved in oil prospecting, they diverted all of the then scarce tubular supplies to that more rewarding effort. Another factor that has justified higher revenue for the steam suppliers is the increased costs of drilling. Some of the earlier wells cost only about $50 000 each. More recently, partly because of rising costs and partly because the nature of the field is necessitating much deeper drilling in more difficult formations, the cost of some of the wells is well over the $1 million mark. Yet another factor is increasing bureaucracy which has delayed the necessary sanctions for power station construction. In one case it took 3 years to get a permit to build a unit identical to one already in service, and this delay occurred after the developer had completed his exploration for the necessary steam and had given assurance as to its availability. In this particular case the developer had to wait about 7 years before he received any return on his very substantial investment. Where entirely new projects are concerned it is not unusual in the USA for as much as 11 years to lapse from the beginning of reconnaissance to the time when power comes 'on line'. It is not the manufacturers who are responsible for this – for example, units 7 and 8 (55 MW each) at the Geysers were built in 23 months: unit 12 (110 MW) in 31½ months. The delays are administrative – application and granting of permits, establishment of steam ownership rights, conformity with anti-pollution law, etc. While high rates of interest prevail, and while these delays persist, there is a serious risk that the oil companies who have hitherto been involved in developing geothermal fields will lose interest and divert their efforts into more rewarding enterprises. Hence the steam price formula is not a matter of mere profiteering, as might at first be thought by those who are not familiar with the historical facts.

Another feature of the steam purchase agreements of California is that the steam is not metered. It is considered sufficiently accurate, and is certainly far simpler, to charge for the steam at so-much per kilowatt-hour generated. Thus if a power plant were shut down, no steam would be consumed, no power would be generated and no charge would be made by the steam suppliers despite the fact that their costs are more or less fixed and continuous, regardless of the power station output. This arrangement of indirect metering of steam supplies by means of the kilowatt-hour meter has been accepted by the steam suppliers in view of its simplicity and in the light of experience as to the probable annual production of electricity from each power plant – i.e. on the assumption of an average expected plant factor.

Until recently the Pacific Gas and Electric Company was the only large-scale purchaser of steam in the Geysers field. Now others have appeared – the Sacramento Municipal Utility District (SMUD), the California Department of Water Resources (CDWR) and the Northern California Power Agency (NCPA).

14.11.2 *Californian power costs.*   As explained above, it is possible only to quote the costs incurred by the power producers. Their capital investment excludes the costs of exploration and the winning and supply of steam, but their production costs include the bulk purchase of steam. Table 14.2 shows the *actual* capital costs of the various power plants in the Geysers field that were in service at the end of 1980, and the *estimated* costs of plants then on order or under construction, in 1981 dollars. These costs include all power plant, buildings and civil engineering works from the point of delivery of the steam supplies at the various power stations, and include step-up transformers, substations and electrical transmission to the nearest convenient point in the state grid.

Table 14.2 Actual and estimated capital costs (as in 1981) of geothermal power plants in the Geysers field, California.

Power producer	Unit no.	Year of actual or expected commissioning	Net rating (MWe)	Capital cost (US $)	Incremental cost per net kilowatt (US $)		
	1	1960	11 ⎫	4 010 000	167.1		⎫
	2	1963	13 ⎭				
	3	1967	27 ⎫	7 610 000	140.9		
	4	1968	27 ⎭				
	5	1971	53 ⎫	12 756 000	120.3 ⎫ 114.5		
	6	1971	53 ⎭				
P G and E Co.[1]	7	1972	53 ⎫	11 520 000	108.7 ⎭ average		Actual
	8	1972	53 ⎭				
	9	1973	53 ⎫	13 520 000	127.5		
	10	1973	53 ⎭				
	11	1975	106	19 666 000	185.5		
	12	1979	106	27 580 000	260.2		
	13	1980	129	52 800 000	409.3		
	14	1980	110	27 966 000	254.2		
	15	1979	55	25 530 000	464.2		⎭
	16	1983	110	42 700 000	388.2		⎫
	17	1982	110	41 592 000	378.1		Estimated
	18	1982	110	48 882 000	444.4		
NCPA[2]	2	1982	110	28 000 000	254.5		⎭

[1]Pacific Gas and Electric Company.
[2]Northern California Power Agency.

Some of the fluctuations in the costs per net kilowatt are due to the fact that additional transmission capacity is not required for each power unit added.

The figures are of limited significance, partly because of the exclusion of the steam supply system with exploration and drilling) and partly because of the wide variety of unit capacities and 'repeat' orders. The sudden drop in the cost of units 7 and 8 by comparison with units 5 and 6 is presumably explained by the gains from a repeat order; so it would be wiser to treat these four units together at an average cost per kilowatt of $114.5. It would be difficult to attempt to allow for scale factor because some of the quoted costs are for single units and some are for pairs. The effects of inflation, though masked by other factors, are noticeable; but the higher costs per kilowatt shown in the more recent figures are in no small measure due to the additional interest during construction arising from bureaucratic delays (see Section 14.11.1).

Unfortunately the absence of steam supply capital cost figures makes it impossible to estimate the total geothermal investment in the Geysers enterprises: but it is of interest to note that at Wairakei the heat supply system accounted for approximately half of the total costs excluding the electrical transmission. However, the Wairakei heat supply system costs included the experimental hot water transmission system, and the distance from the bore field to the power plant was much greater than at the Geysers. The *gross* capital costs at the Geysers, therefore, including exploration, drilling and steam supply, would probably be considerably less than double the figures quoted in Table 14.2. No idea of the Geysers steam supply investment can be gleaned from the price charged for steam to the Pacific Gas and Electric Company owing to the flexibility of this price, as explained in Section 14.11.1.

The geothermal production costs of the Pacific Gas and Electric Company's electricity has grown in the following manner since the Geysers field was first exploited:

	Year(s)	Price of steam	Other costs	Total costs	
	1960–63	2.5	2.16	4.66	
	1978	13.0*	10.4	23.4	mills/kWh
	1980	18.63*	12.37	31.0	
Mean annual growth	1960–80	10.56%	9.12%	9.92%	

*Including 0.5 mill for reinjection.

In 1980 the following *comparative* estimates were made by the Pacific Gas and Electric Company of the production costs of electricity from various thermal sources – all for an assumed annual plant factor of 80%:

	mills/kWh	Relative cost
Nuclear	25	0.806
Geothermal	31	1.0
Coal	44	1.419
Oil (distillate fired)	91	2.936

In 1971 [213] it had been estimated that electricity from nuclear fuel would then cost 33% *more* than from geothermal energy. This decline in geothermal competitiveness is largely attributable to the heavy costs incurred in complying with the Californian environmental laws (see Chapter 16). This also explains to some extent the fact that power from geothermal energy costs about one-third of power from oil fuel, whereas in Section 14.4 the apparent advantage was shown at one-quarter – though only on the basis of arbitrary inflation rates.

14.11.3 *New Zealand power costs.*   The Wairakei capital costs (192.2 MWe installed, commissioned 1958–63) were as shown in Table 14.3. Collectively they are now of little more than historic interest, as the costs were mainly incurred in the second half of the 1950s. The proportions of the component items are, however, of some interest. The Wairakei costs should not be regarded as 'typical' for the period when they were incurred, owing to the following special factors; the first three of which tended to inflate, and the last two to deflate, the total costs:

Table 14.3 Capital costs at Wairakei.

Item	£m	% of total	£/kW	US $m	US $/kW
Steam winning	4.392	23.44	22.85	12.298	63.98
Heat transmission	4.975	26.55	25.88	13.930	72.46
Power house and plant	7.441	39.71	38.72	20.835	108.42
Cooling water system	1.262	6.73	6.57	3.534	18.40
Substation	0.669	3.57	3.48	1.873	9.74
Total	18.739	100.00	97.50	52.470	273.00

The dollar conversions have been made at the then exchange rate of £1 = US $2.80.

(i) Owing to a change of design 'in mid-stream' from a multi-purpose (power generation and heavy water production) to a 'power only' project, a complicated arrangement of 13 relatively small units [40] became necessary and undoubtedly cost far more than if a purpose built 'power only' plant had been designed from the start, with a different choice of working pressures and turbine arrangement.

(ii) The adoption of river cooling, and the subsequent discovery that the centre of gravity of the field was further away from the river than had at first been thought, led to very long pipelines.

(iii) The heat transmission costs included the cost of the abortive hot water transmission scheme described in Section 9.9, which would not have been undertaken had it been foreseen that the selected wells would 'dry out' so soon.

(iv) No costs have been charged to the electrical transmission line to interconnect the station with the North Island grid.

(v) The New Zealand Government charged only part of their exploration costs to the project, holding the balance in a 'suspense account' which would have been booked to a subsequent extension then planned but in fact not carried out. It is not known how they have since dealt with the unbooked balance of this account.

The *production costs* at Wairakei were estimated in 1961 at 4.9 US mills/kWh [40], but the interest rate was then only 4% p.a. A life of 20 years was assumed for all the assets except the bores for which a 10 year life was assumed. (It is customary with most geothermal projects to depreciate the bores more rapidly than the other assets, as their output tends to decline with time. Thus there will always be a proportion of relatively new high-yield bores together with some older low-yield bores in simultaneous service. In assessing the production costs allowance should be made for the re-use of wellhead equipment and part of the branch pipelines moved from old to new bore sites, while charges are made for the re-adaptation and partial wastage of such equipment.) By the late 1960s the Wairakei production costs had risen to over 6 US mills/kWh due to labour and materials inflation, and by 1980 they were estimated at about 11 US mills/kWh. The fact that this figure is so much less than for the Californian plants is of course mainly due to the station having been built so long ago, when capital costs were so much lower.

14.11.4 *Mexican power costs.*   For the 75 MW geothermal power plant at Cerro Prieto, first commissioned in 1973, the total investment was US $14 400 000, or $192/kW(gross) at the generator terminals, which would probably work out at slightly over $200/kW(net) after allowing for auxiliary consumption. These figures compare favourably with the Geysers figures as set out in Table 14.2, considering that they include exploration, drilling, steam supply as well as power plant and buildings, etc., and that only two units of 37.5 MW each were installed. No further capital estimates are available since the station capacity was doubled to 150 MW. The only production costs for Cerro Prieto were those *estimated* in 1974 at 12.56 US mills/kWh generated, based on an assumed annual plant factor of 80%, a 30 year plant life and a 10 year bore life.

14.11.5 *Italian power costs.*   In 1964 the *estimated* production costs in the Larderello complex ranged from 2.6 to 2.74 US mills/kWh for non-condensing units and from 2.38 to 2.96 US mills/kWh for condensing plants, based on an assumed interest rate of 6% p.a. [214]. These costs were of course considerably lower than for geothermal power plants in other countries at that time because the Italian plants were constructed earlier and (by comparison with Wairakei) to rather more austere standards. In 1976 the production costs of geothermal power were updated to the figures shown in Table 14.4 by the Italian authorities themselves [215].
The following figures were valid for geothermal power installations at Larderello in 1974 [216]:

2 units of 62 MW (condensing)	US $318/kW	including a share
1 unit of 15 MW (condensing)	US $374/kW	of exploration
		costs, drilling,
1 unit of 15 MW (non-condensing)	US $347/kW	steam supply and
1 unit of 4 MW (non-condensing)	US $420/kW	plant

By 1980 these figures could have about doubled.

Table 14.4 Production costs in Italy (1976).

Item	Cost			
	Condensing plants		Back-pressure plants	
	Lire/kWh	US mills/kWh	Lire/kWh	US mills/kWh
Capital amortization and interest	3.7	4.77	1.9	2.45
Research and drilling	2.2	2.84	4.4	5.67
Exploitation and maintenance	2.1	2.70	1.2	1.55
Total	8.0	10.31	7.5	9.67

No more recent Italian production costs are available.

14.11.6 *Icelandic power costs.*   The only Icelandic geothermal power capital cost available is a statement in 1979 [43] that the cost of the 60 MW Krafla power installation was then expected to be about US $55 million including steam supply, when completed. This is equivalent to US $917/kW. The figure had been inflated by the damage caused by the volcanic disturbances experienced at Krafla since late 1975. As the whole area might well experience further seismic shocks, it is quite impossible to foretell what the final capital cost will be when the station is generating its full designed output: nor will the figure, when known, be very significant as the site is exceptional. No estimates are available for the production costs of Krafla.

In 1970 it was estimated rather vaguely [217] that the production costs for the 2.5 MW non-condensing plant at Namafjall were 'from $4\frac{1}{2}$ to $5\frac{1}{2}$ US mills/kWh, with interest at 8% p.a. and with assumed asset lives of 7 years for the bores and 20 years for the other assets. This figure is not very informative as the turbo-set used was secondhand; but it compares reasonably well with the Italian figure of 9.67 mills/kWh quoted in Table 14.4 for 1976.

14.11.7 *USSR power costs.*   The only published costs discovered by the author relate to the small 5 MW plant at Pauzhetka, in Kamchatka, for which an energy cost of 7.2 US mills/kWh was quoted [38] in 1970. Considering the small size of the unit and its very remote situation, this can be regarded as a low figure even for 1970. It was claimed that this production cost was 30% lower than could have been achieved from any alternative energy source in the area. No capital investment costs for Russian geothermal power installations are known to the author.

14.11.8 *Japanese power costs.*   In March 1976 the capital cost of the Kakkonda 50 MW geothermal power installation was estimated at the figures shown in Table 14.5 [100].

Table 14.5 Capital costs of the Kakkonda 50 MWe geothermal power installation, Japan.

| Item | Capital costs (1976) | | | |
	Yen (millions)	US $ (thousands)	US $/kW	%
1½ year's basic investigations	250	828	16.6	2.08
Roads and temporary housing	300	993	19.9	2.49
Exploration and test boring (1½ years)	1000	3311	66.2	8.29
Production wells (1½ years)	2500	8278	165.5	20.74
Reinjection well	500	1656	33.1	4.15
Power house, steam system, cooling towers etc.	7000	23 179	463.6	58.10
Environmental expenses	500	1656	33.1	4.15
Totals	12 050	39 901	798.0	100.00

Conversion at 302 yen = 1 US $.

The *production* costs for Kakkonda were estimated at the same time at 6¼ to 7 yen/kWh, or 20.7 to 23.2 US mills/kWh, based on a 20-year asset life, 10% p.a. interest and an annual plant factor of 91.3% (8000 hours).

More recently the following general estimates have been made for Japanese geothermal power installations in the 10–300 MW range [171]:

| | Estimates (US $/kW) (1979 dollars) | |
	Low	High
Exploration	58	168
Fluid production	333	667
Separation and gathering system	125	250
Generating plant	500	830
Effluent disposal	167	417
Totals	1183	2332

The corresponding *production* costs have at the same time been estimated at 37.5 to 70.8 US mills/kWh on the basis of the following assumptions:

 (i)   15 years amortization for wells.
 (ii) 30 years amortization for plant.
 (iii) 12% p.a. interest.
 (iv) 1.5% p.a. of capital cost for operation and maintenance (including tax).
 (v)   85% annual plant factor.
 (vi) 50 MW plant capacity.
 (vii) 8 to 10 years from start of exploration to power generation.
(viii) Exchange rate: 240 yen = 1 US $.

All these Japanese geothermal costs appear high in relation to geothermal capital and production costs in other countries. Nevertheless the Japanese claim that by comparison with oil-fired thermal power production geothermal energy is competitive. Even in 1976 it was costing about 20% less than power from oil fuel, and the advantage will have improved since then with the subsequent rapid rises in oil prices. It is also difficult to make direct comparisons between Japanese and other costs, as the yen has appreciated in value very much in the 1970s.

14.12 Actual geothermal heat costs

A very useful analysis was made in 1974 by Einarsson [8], in which he estimated typical Icelandic costs of domestic heat of geothermal origin, in terms of the temperature at source, the transmission distance from the source to the market, and the capacity of the system (or size of the market) as represented by the diameter of the transmission pipe. He showed the effect of these three parameters in graphical form in Fig. 14.6 (using the inner left-hand vertical scale), alongside which he showed the prices at which fuels must be available for them to be competitive if used for domestic heating. The diagram clearly shows that with rising fuel prices it would pay to transmit geothermal heat over longer and longer distances before its delivered cost becomes too great to compete with fuel as the alternative. Einarsson expressed his heat in thermal kilowatt-hours; if kgcals are preferred a simple conversion can be made at 1 kWh = 860.5 kgcal. He wisely gave warning that, although conservative, his figures made no claim to great accuracy in view of shifting markets and different local site conditions. The distribution costs were based on those applicable in 1974 to Reykjavík city, where the space-heating density is about 20 MW/km², and it was assumed that distribution is at 80° C regardless of the temperature at source.

Unfortunately conditions have changed greatly since 1974. By 1980 the costs of labour and materials had about doubled. On this assumption it is still possible to use Einarsson's diagram by using the right-hand vertical scales. It

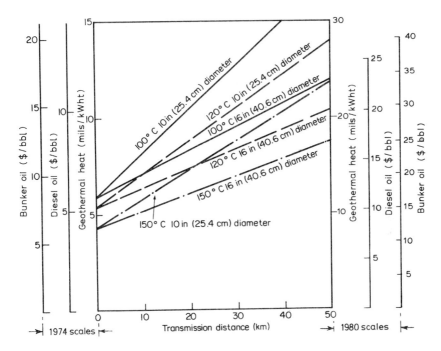

Figure 14.6 Approximate geothermal heat costs for Iceland in terms of transmission distance, pipe diameter and temperature (with hot water as the heating medium). (After Fig. 7 of [8])
Note: The left-hand vertical scales were valid in 1974; the right-hand vertical scales would be valid if heat costs had doubled since then – a not unreasonable assumption for conditions in 1980. The outer scales show fuel prices above which geothermal heat becomes competitive.

will be seen that with bunker oil at US $30/bbl (barrel) heat transmission would apparently be economic over distances of about 50 km for hot water at 120° C or more in 16-in pipes. The diagram should be used in this way only with great caution, simply to arrive at very rough approximations; particularly as conditions in countries other than Iceland could be very different.

Published data of geothermal heat costs in other countries are scarce. From the available information concerning the Creil geothermal heating system near Paris [165] it can be inferred that the cost of heat there in 1976 was about 10 US mills/kWht, with water at 65°C and virtually no heat transmission. Allowing for two years' inflation from 1974 to 1976 this is reasonably in accord with Fig. 14.6. In October 1979 it was stated that the cost of geothermal heat at an ethanol factory in California, at 177° C – again with no appreciable transmission – would be approximately 8 to 9 US mills/kWh. This figure also does not differ greatly from the cost obtained from Fig. 14.6 using the right-hand scale.

Fig. 14.6 is based on the transmission heat in the form of hot water. Much has already been said in Section 11.2.13 about the greater merits of hot water by comparison with those of steam as a heat transmission medium. In 1962 the author prepared the following comparative costs which were then valid for New Zealand. Subsequent inflation should not have greatly affected the proportions.

Saturated steam at 5 atü average pressure	7.88
Saturated steam at 10 atü average pressure	5.86
Saturated steam at 15 atü average pressure	4.87
Hot water at 100° C average temperature	2.46
Hot water at 150° C average temperature	1.46
Hot water at 20° C average temperature	1.00

These figures are only ratios, and are proportional to the reciprocals of the heat transmissible through an arbitrary length of pipe – 1000 ft (about 300 m) × 20-in (about 50 cm) diameter. A pressure drop of 5% of the absolute average pressure (or about 10% of the pressure at the entry end) is assumed in the case of steam, and a fluid velocity of 7 ft/s (2.133 m/s) is assumed for the hot water.

In the late 1970s saturated steam at still lower pressures was transmitted at Wairakei over distances of 2 to 3 km (see Section 10.6). A rough extrapolation of the above figures would suggest that this is costing at least ten times as much as that of transmitting hot water at 200°C (without the terminal equipment associated with the latter); but in this case it was necessary to consider how to make the best possible use of capital plant costs already incurred.

It will be noted that hot water at 150°C was shown by the figures quoted above as being only about 60% as expensive to transmit as hot water at 100°C, whereas according to Fig. 14.6 the proportion is slightly over 70%. The difference could be explained by differences in assumptions and in local conditions; but in both cases the advantages of the higher temperature are clear.

When we consider the relative costs of transmitting hot fluids for power generation purposes, the pattern is entirely different, because of the low power potential of hot water as disclosed in Fig. 10.4 and because of the relatively high cost of the terminal equipment (pumps, head tank, flash vessels and control valves and equipment) which greatly reduce the relative advantages of hot water as a heat transmission medium. Moreover, the position is complicated further by the fact that flashed hot water makes use of low-pressure turbine stages which are intrinsically costlier than the high pressure stages. It may be noted that in 1970 Bruce suggested that in the Geysers field the economic distances for supplying steam to 25 MW and 100 MW power plants were then about 1500 and 2500 ft respectively if tower cooling were adopted [206]: with good river cooling facilities available considerably longer distances would have been justifiable. The formidable rises in the price of fuels that have occurred since then will certainly have had the effect of justifying much greater distances.

The economics of two-phase fluid flow transmission in a single pipe (see Section 9.12) cannot yet be properly evaluated owing to lack of sufficient working experience. It seems probable that if no serious technical difficulties are encountered, it could prove very favourable.

The comparative costs of the French use of low-grade aquifers against traditional oil-fired boilers for the heating of 2500 dwelling apartments in 1980, consuming on average 400 000 MWht annually have been given as follows:

| | Method of heating | | |
	Oil	Geothermal	Remarks
	Francs p.a.	Francs p.a.	
Fuel	5 050 000	1 265 000	Peaking fuel for
Electricity	—	390 000	geothermal
Repairs and maintenance	600 000	1 000 000	
Overheads	200 000	300 000	
Total	5 850 000	2 955 000	
Cost per kWht	Fr. 0.146	Fr. 0.074	
Capital costs:			
Bores		14 000 000	
Other costs		11 000 000	
Total		25 000 000	or *Fr. 10 000 per apartment*

It will be noted that geothermal heating in France (1980) cost just about half as much as oil heating. It was stated in Section 14.3 that in 1980 in Reykjavík this proportion was about 12%. The large difference is due in part to the much lower temperatures used in France and in part to the far smaller scale of the enterprise.

14.13 Scope for cost improvements

It is important that we ask ourselves in what way can we seek to *reduce* the costs of geothermal exploitation. There is possible scope in several directions.

14.13.1 *Exploration costs.*   In the early days of geothermal exploration, when our knowledge of the nature and performance of geothermal fields was rudimentary, the search for exploitable earth heat was largely a 'hit-or-miss' operation. There was much 'wild-cat' drilling; bores were sunk in the vicinity of surface thermal manifestations without any proper understanding of underground processes. Drilling for steam or hot water was largely a gamble that sometimes did, and sometimes did not, pay off. During the last 30 years, however, exploration techniques have become greatly refined and improved, and we have learned much about subterranean processes. The skills of the earth scientists and the drilling engineer have become so coordinated as to enable exploration to become much more systematic than in former years. Methods still continue to improve, and the risk element, though not yet entirely removed, is steadily being reduced. The chances of finding productive drilling sites have been much improved, so that the costly sinking of abortive bores now forms a far smaller part of the total exploration costs than formerly.

14.13.2 *Two-phase fluid transmission and submersible pumps.*   The technical feasibility of using submersible pumps and two-phase fluid transmission has been touched upon in Sections 9.11 and 9.12. If either should become standard practice for wet fields the economic savings could be very great, not only in wellhead equipment and fluid transmission costs, but also from the fact that by bringing the hot water right up to the plant the exploitation of its flash potential would be so greatly simplified. Moreover, the scale effect of supplying and maintaining few very large separators at the plant, by comparison with having to provide very many small separators scattered all over the bore field, would also be considerable.

14.13.3 *Binary fluids.*   As already shown in Section 10.2.6, the use of binary fluids has both advantages and disadvantages. It is too early to draw positive conclusions concerning its economic aspects. It is possible, however, that for exploiting low-grade heat appreciable economic gains might sometimes be achieved. It is important to understand that the use of binary fluids does not imply any improvement in cycle efficiency as such: it simply enables use to be made of heat that would otherwise never enter the turbine cycle at all, being rejected to waste.

14.13.4 *Dual and multi-purpose plants.*   Such plants have been discussed in Chapter 12. Their potentially great economic advantages are self-evident.

14.13.5 *By-product hydro-power from wet fields.*   This question has been discussed in Section 10.17. Clearly by-product hydro-power could, where practicable, bring modest economic advantages, though these would be too individual to particular projects for any generalizations to be made. The question would arise of whether a value (variable cost) should be assigned to the water. This could be quite a logical allocation, as indeed it would be if the heat content of the bore water were first usefully exploited. In any case, any *variable* cost which may be assigned to the steam in accordance with the arguments presented in Section 14.5 should be appropriately reduced, since the charges on the exploration costs should be spread over *both* fluids – logically in proportion to the utilized energy content of each.

14.13.6 *Novel drilling methods.*   Some of these have been mentioned in Section 7.17 and others will be discussed in Chapter 18. It may simply be stated here that these are other possible sources of economy and extension of scope in geothermal development.

14.13.7 *Heat pumps.*   The use of heat pumps is by no means exclusively relevant to geothermal development. Nevertheless, where low-grade waste heat must unavoidably be discarded from a geothermal exploitation plant it could sometimes be economic to upgrade the heat at a fairly high performance ratio so that the natural fluid is rejected at a lower temperature and more heat extracted therefrom, while at the same time providing a reasonably large quantity of higher grade heat that could perhaps be used for some other practical purpose. Where tepid aquifers having a horizontal fluid cross-flow can be found at shallow depths there could be scope for upgrading the natural heat for practical application without risk of freezing the subterranean water.

In the USSR [161] there is a project for combining low-grade heat with a heat pump for space heating, while at Creil in the Paris Basin heat pumps are already being used for boosting the input and improving the efficiency of a district heating scheme (see Section 11.2.5). Too much optimism should not, however, be placed upon the potentialities of the heat pump, which has a seductive appeal arising from the often misunderstood conception of 'performance ratio'. To the unwary, the heat pump is apt to suggest that something can be had for nothing. The heat pump is, of course, only a device for *transferring* heat from one place to another, and the amount of energy required to effect the transfer may perhaps be considerably less than the amount of transferred energy.

If electrically driven heat pumps are used there is no net gain in useful heat unless the performance ratio exceeds the reciprocal of the *delivered* efficiency of the electricity (i.e. of the product of the power station, transmission and distribution efficiencies). For example, if the power station efficiency is 35% and the combined transmission and distribution efficiency is 80%, the performance ratio would have to be greater than $1/(0.35 \times 0.8)$, or 3.57. Hence the 5 : 1 performance ratio achieved at Creil in the Paris Basin heating scheme (Section 11.2.5) is energy-profitable.

The heat pump certainly has its uses, but generally it is serviceable if the vapour pressure of the refrigerant is not too high; i.e. it is profitable only at warm and tepid temperatures. It is also important not to fall into the common error of proposing the use of heat pumps to upgrade the heat from the warm condensing water from a steam-field power plant, for it can never pay to degrade heat to a low level and then pump it 'uphill' again. In such cases it would be better either to use back-pressure or pass-out turbines, the exhaust temperature of which is sufficiently high for direct application to space heating or some other purpose. The only conceivable exception to this dictum would perhaps arise where two ownerships are involved. If a power generating enterprise has failed to use a back-pressure or past-out turbine in this manner, it might conceivably pay *some other party* to apply a heat pump to the warm water discharged from the condensers of the power plant, even though the overall efficiency of the double process would be less than that of the more correct use of the heat.

14.13.8 *Excess geothermal power capacity.*   In Section 10.16(i) mention was made of the possibility of taking advantage of 'scale effect' to operate a geothermal power plant at a reduced plant factor in order to gain in capital costs per kilowatt installed, with consequent improvement in the production costs per kilowatt-hour. This could sometimes bring economic benefit to small systems, where the base load demand is (at present) smaller than the capacity of a reasonably sized geothermal plant, although with very large composite systems in which available geothermal capacity is likely to form only a fraction of the system base demand, this would not apply [127].

14.14 Further economic points

14.14.1 *Importance of location.*   Looking to the future when it may perhaps become possible to win earth heat at any place we choose, we shall of course extract the heat as closely as possible to the markets. At present, however, we are constrained to exploit this form of energy only where it can be found; and sometimes it occurs in most inconvenient places. It is argued by some that if a good field should be found in some remote situation far from populated areas, it is necessarily worthless. This is not inevitably true; for even those who quote this theory are obliged to define the word 'remote', which is a relative term. If the field is hyperthermal and of good quality and capacity, the generation of electricity could perhaps economically justify the transmission of power over one or two hundred kilometres or even more. However, if the field is semithermal, or if there is an exploitable aquifer in a non-thermal area, it is clear that the word 'remote' would, at best, have to be limited to some tens of kilometres (see Fig. 14.6) if a 'local' market is being sought. Nevertheless, as implied in Section 11.4.1, if heat (even of moderately low grade) is sufficiently abundant and cheap and of an adequate grade for some large industry, it could sometimes pay to establish such an industry at the field, to transport the necessary raw materials and labour to the site – even across national frontiers – to set up a community centre there, self-sufficient in food and with adequate amenities and high standards of living comforts, and to ship the end products of the industry to the markets. Hence the known or suspected presence of large thermal fields in very remote locations should not be dismissed out of hand as valueless. If the possible scale of their exploitation is sufficiently large, fields in the most 'off-the-map' places could be of great commercial value.

14.14.2 *Value versus cost.*   The cost of geothermal heat is not the only criterion of its value. The costs of local alternative energy sources of equal adaptability, and the political and environmental advantages of geothermal exploitation, are factors that must also be taken into account. Judgments of these criteria will not necessarily be valid for all time. In one year it could be concluded that geothermal development is uneconomic, whereas later – perhaps as a result of higher fuel prices or of strained relationships with another country from whom energy has hitherto been imported, or even of the discovery of cheaper methods of geothermal exploitation – the situation could be reversed.

14.14.3 *Diminishing returns.*   Until, or unless, universal geothermal energy becomes available, it is only to be expected that the most promising hyperthermal fields will first be exploited and that the economically less attractive fields and low-grade aquifers will follow later, due account being taken of location and the size of the market. This tendency has been historically true of hydro-power development, in which countries have tended first to 'skim the cream' off their hydro-resources and later to develop less attractive projects under changing economic pressures. However, this will apply only for as long as exploitation is confined to present methods. When, as will almost certainly happen one day, large-scale deep earth-heat mining becomes economically practicable (Chapter 18), a new and far greater lease of life should follow.

15 Chemical and metallurgical problems

15.1 Chemical and physical composition of geothermal fluids

Hitherto, apart from various passing references to non-condensible gases, geothermal fluids have been treated in this book simply as H_2O – hot water and/or steam. However, these fluids are invariably accompanied by gases and, in the case of wet fields, by water-soluble substances, some of which can be potentially dangerous to the construction components of a geothermal installation and to the environment. In dry fields the steam is also apt to contain small amounts of dust that has eluded capture in the axial separators in the steam lines. This dust may cause turbine blade erosion, or if it is mixed with oil it can form a deposit on the first row of turbine blades and reduce the power output.

The proportions of *non-condensible gases* vary widely from field to field, ranging from only about 0.1% by weight (in relation to the steam phase of the geothermal fluid) at Hveraderdi in Iceland to figures of 12%, 25% or even more in the Monte Amiata field in Italy. Typical intermediate figures (by weight) are:

Wairakei	from 0.35% to 0.5% (to steam)
The Geysers	from 0.6% to 1.0%
Cerro Prieto	about 1.25%
Larderello	from 4.5% to 5%
Matsukawa	about 1.1%

Of the non-condensible gases CO_2 always forms by far the largest component – from 63 to 97% by weight – with H_2S usually as the next largest ingredient – from 1% to 21%. The remaining gases consist of small quantities of CH_4, N_2, H_2 and sometimes traces of NH_3, H_3BO_3 and the rare gases.

The *water-soluble components* in geothermal waters usually consist of chlorides, silica, perhaps metaboric acid, sulphates and traces of fluorides. Waters are sometimes alkaline, as at Wairakei (pH = 8.6), and sometimes

312

acid, as at Matsukawa (pH about 5). Rare cases have been known – e.g. the Tatun field in Taiwan – of pH values as low as 1.5, where the acidity has destroyed bore casings in a few hours and the field has been unusable. Sometimes traces of valuable minerals are to be found in thermal waters.

In wet fields, almost all the non-condensible gases remain with the steam phase while almost all the water solubles are associated with the water phase, though each phase may carry small traces of the chemicals normally to be found in the other.

15.2 Dangers of chemical action on plant components and on the environment

With the presence of all these chemical ingredients in geothermal fluids, it is only to be expected that corrosion or deposition is liable to occur at various places in a geothermal installation unless special precautions are taken. Likewise, since some or all of the chemicals are likely to escape at some stage into the environment, pollution can be caused unless suitable steps are taken to prevent, or at least to mitigate it. This will form the subject of Chapter 16. For the present only the vulnerability of exploitation equipment and plant will be considered. Strict attention must be paid to the metallurgy of all component parts, taking into consideration the chemical and physical conditions of the fluids where they come into contact with them. Wet fields tend to be rather more troublesome than dry fields owing to the greater variety of impurities to be found in the fluids and also to the catalytic action of the presence of water or vapour. In the early days of geothermal development it was common practice, before manufacturing the plant, to submit for long periods stressed and unstressed test pieces of all materials, proposed to be used in construction to environments of bore steam, air/ steam mixtures and bore water, so that careful measurements could be made of corrosion rates and samples could be subjected to microphoto-graphic examination for sub-surface changes of metallic structure. In this way, vulnerable materials could be avoided and resistant materials used for the construction of the plant in appropriate places. Such tests are still carried out to some extent, though so much is now known of the effects of bore fluids upon construction materials that much less empirical testing is now necessary.

15.3 Chemical deposition

This is apt to occur at the following points:

15.3.1 *Calcite deposits in wet bores.*   Certain waters rich in calcium salts are apt to give rise to the build-up of hard deposits in wet fields at the points where the uprising hot water starts to flash into steam (see Fig. 7.3). Such deposits at first reduce the fluid output owing to the restriction they form, and ultimately they may almost choke the bore altogether. The only practical solution to this trouble is to ream the bore periodically to clear the deposits. If this is necessary at rare intervals, perhaps about once annually, it is a trouble with which one can learn to live; but if calciting builds up too rapidly, it may be necessary to abandon the bore. The frequency of reaming depends largely on the fluid yield when the bore is clear of deposits, as well as the rate of build-up. With bores or very large output quite a large degree of calcite deposition can be tolerated for a bore to become economically unserviceable. The occurrence of calcite blockage of bores is related not only to the calcium but also to the CO_2 content of the bore fluid. Experience in New Zealand fortunately suggests that the rate of deposition tends to decline with time.

Calcite deposition may sometimes also occur in the rock formation if the permeability is not very good, as flashing may take place before the fluid reaches the bore walls. In such cases the fairly early abandonment of the bore may become unavoidable.

15.3.2 *Silica deposits from wet bores.*   At Wairakei and other wet fields silica may be present in the bore water to saturation. As the temperature falls from the original value at depth, there will be a tendency for the silica to come out of solution. There are various allotropic forms of silica, of which (fortunately) the commonest is capable of remaining in solution in a state of super-saturation for long enough to prevent precipitation until the thermal water is clear of the plant. Thus at Wairakei, no silica deposition occurs in the bores, wellhead separators or flash-tanks; but in open discharge channels carrying waste hot water which is rapidly cooling, very thick deposits are built up which have to be periodically and laboriously chipped away. In the course of time, the underground waters feeding the base of the bores through rock fissures may slowly deposit silica if flashing occurs in the formation where fluid velocities are low. This can gradually reduce the fluid yield of a bore and, perhaps after a few years, lead to its abandonment. Thórhallsson *et al.* [218] have described the troubles experienced with silica deposition in Icelandic heating systems, and the remedy of dilution with fresh water that has shown some degree of success.

15.3.3 *Turbine blade deposits.*   If separation of the steam from the water, in wet fields, at the wellheads is imperfect, and particularly when the bores are close to the generating plant (and the scrubbing effect of pipelines – see Section 9.10 – is not therefore very effective), traces of solubles may reach the turbines of a geothermal power plant, and may be deposited from the small residual water fraction onto the blades as the temperature falls with expansion. This results in the gradual build-up of hard chemical deposits which can reduce the output of the turbine and perhaps also upset its dynamic balance. This trouble was at first experienced at Cerro Prieto in Mexico, where the power station is fairly close to the bores, and effectively reduced the station's annual plant factor to a figure of only about 70% (as against more than 90% sometimes achieved in other geothermal power plants) during the first two or three years of operation. Blade deposits there caused a loss of power of 4 or 5 MW (out of 75 MW nominal) and turbines had to be taken out of service from time to time for the deposits to be removed by sandblasting. After a few years, however, this trouble was cured quite simply by raising the height of the steam take-off pipes in the wellhead separators as described in Section 9.3. The result was improved separation and the virtual elimination of turbine blade deposits.

In dry steam fields too – both at the Geysers and at Larderello – deposits on the turbine blades have been caused by silica, carried by the steam, though not to a serious extent. Steam-washing with oil has been found to be effective in Italy [129] in reducing the quantity of blade deposits and in facilitating their removal. The risk of blade deposits is a factor that favours the adoption of fairly low working pressures, which are conducive to larger steam passages (see also Section 8.8).

15.3.4 *Iron sulphide deposits and steam pipe corrosion.*   The action of H_2S on mild steel is to form a protective skin of iron sulphide which usually inhibits further attack. Thus, although a black deposit will usually be found on the inside walls of all steam pipes and pressure vessels, this is quite harmless and has enabled cheap mild steel to be used for all these plant components, which form a high proportion of the whole equipment. However, the curious attack experienced on Wairakei on the walls of the steam mains, described in Section 9.10, shows that the skin of iron sulphide does not always offer effective protection. This particular trouble was first detected 19 years after the pipelines had been put into service. The attack takes the form of crystalline deposits of magnetite of up to 3 mm thickness. At random intervals within this deposit large corrosion pits have been formed that have penetrated well into the pipe wall. The phenomenon seems to be due to the CO_2 and H_2S entering into solution with the pipe wall condensate. Silica inhibits attack, so it is only in the downstream parts of the steam pipes where scrubbing has removed nearly all the silica carried over from the wells that the trouble has been experienced. The cure, effected by deliberately injecting small quantities of bore water into the vulnerable lengths of pipe, must be applied with care lest an excess of chlorides should enter the turbines.

15.4 Corrosion and stress-corrosion

Despite the wide variation in the composition of the chemical impurities from field to field, the broad characteristics of geothermal fluids are similar in quality though they may differ in degree. Table 15.1 may be taken as broadly indicative of the resistance or vulnerability of the various materials likely to be used in component parts of a geothermal exploitation plant.

15.5 Turbine materials

As shown in Table 15.1, 13% chrome iron of the type normally used for turbine blades is, in the hardened or martensitic state, susceptible to stress-corrosion cracking in the presence of geothermal steam, but enjoys immunity if in the soft or ferritic state, provided that chlorides are not present except in minute quantities (not exceeding 10 ppm in the residual water droplets which may be present even in very small quantities in the steam). Turbine blades of softened and tempered stainless iron are therefore commonly used. However, softened metal is liable to suffer *erosion* in the presence of water droplets and, owing to the relatively high wetness of the steam at the lower pressure stages of a geothermal turbine fed with dry saturated steam at inlet, this could be a hazard. It is inadvisable to protect the blades by brazing on erosion shields on account of the risks of local hardening and the vulnerability of spelter. It is therefore customary to limit turbine blade tip speeds to a moderate figure (say 900 ft/s or 275 m/s) so as to reduce the erosion risks, which are proportional to the *square* of the tip speed. More recently tip speeds of up to 1000 ft/s have been found admissible if the blades are protected by stellite shields.

15.6 Condenser and gas extraction equipment

The most arduous conditions from the corrosion aspect occur in the condenser and gas extraction equipment of a geothermal turbine. Here, bore gas, water vapour and oxygen (released from solution out of the cooling water in jet condensers) are all present, forming a proper 'witch's brew'. The internal surfaces of jet condensers and ejector condensers should therefore be protected with various proprietary resins or with aluminium spraying. Austenitic stainless steels should be used for gas exhauster rotors where the gases are wet, and inter-cooler tubes are usually made of 99.5% pure aluminium, which is resistant to geothermal steam except at high temperatures. Care should be taken not to over-cool the gases in the exhauster system, as this is conducive to attack on exhauster rotors.

15.7 Expansion bellows pieces

The 'concertina' pieces of expansion bellows are made of stainless steel. This material is immune to corrosion from geothermal fluids except when simultaneously exposed to oxygen and chlorides. On the inside surfaces of the bellows, oxygen will normally be absent as the bore fluid is always oxygen-free and as the maintenance of internal pressure will exclude the possibility of air entering from outside. There may, however, be a certain amount of

Table 15.1 Resistance or vulnerability of materials to chemical attack in the presence of geothermal fluids.

Material	Remarks
Category I. Resistant to corrosion	
Good quality close-grained cast iron	
Ordinary mild steel or low alloy steel (in the unhardened, fully pearlitic condition)	H_2S forms a protective coat of FeS, but below 30° C (86° F) this is not so, and the material becomes vulnerable. Also, the presence of oxygen makes the material vulnerable
13% chrome iron in the soft, or ferritic state (but not in the martensitic state). Also the same material, if hardened and tempered at 750° C	This material is suitable for turbine blades. It is essential, however, that chlorides are either absent or present to a minute degree only
Ordinary 60/40 brass	
Admiralty brass (70% Cu, 29% Zn, 1% Sn)	Resistance is poor in the presence of oxygen
Aluminium brass	
Pure aluminium	Vulnerable only at very high temperatures
Hard chromium plate	
Stellites Nos. 1 and 6	
18/8/3 (austenitic) Cr/Ni/Mo steel	
18/8 (austenitic) Cr/Ni steel, columbium or titanium stabilized; (non-precipitation hardening)	
Category II. Vulnerable to corrosion	
Copper and copper-based alloys other than the brasses shown in Category I	
Nickel and nickel-based alloys	
Brazing spelter	Potentially dangerous
Zinc	Unreliable at higher temperatures
Aluminium alloys	*Pure* aluminium is reliable (see Category I)
Category III. Vulnerable to stress-corrosion	
Aluminium bronze	
13% chrome iron in the hardened, or martensitic state	
A.T.V. (an austenitic steel)	11% Cr, 35% Ni, 1.3% Mn, 0.5% Si, 0.3% C
1.5/0.3 Mn/Mo steel	
Hardened carbon and low alloy steels (incorporating martensitic or quasi-martensitic structure)	
Most, if not all, stainless steels if chlorides are present (aggravated if oxygen is also present)	External stress-corrosion has occurred on stainless steel bellows where bore water has leaked onto them
13% chrome iron, even in soft state, if chlorides present except minimally	Hence the need for flash steam scrubbers to protect turbine blades

chlorides present in the pockets of condensed steam that will collect in the underside corrugations – especially near the bores before the pipeline scrubbing effect has had time to act to any great extent. Only when a pipeline has been shut down for inspection and maintenance, and air admitted, should there by any danger from these pockets of slightly chloridic water. Before opening up a steam pipeline, it is therefore advisable to flush the pipe with clean water. Bellows pieces at the tops of expansion arches should be altogether removed and quickly flushed out when air has to be admitted into a line. On the *outside* of a bellows piece, the only danger likely to arise is if a leaky flange allows chloridic water to drip onto the corrugations exposed to the air. This has actually led to the failure of bellows pieces at Wairakei on one or two occasions. Great care should therefore be taken to ensure that expansion piece flanges are absolutely tight against leaks.

15.8 Standby corrosion

The internal parts of a geothermal turbine, provided the blading is made of the correct materials, are normally immune against attack from geothermal steam. However, when a machine is shut down, air may enter through drains or glands as the trapped steam condenses as a result of radiation heat losses. This air, in the presence of the non-condensible gases remaining in the turbine, could cause serious corrosion of both stator and rotor. Hence it is advisable to provide an electrically heated hot air blower that will quickly dry out and purge the remaining wetness, vapours and gases, so that all chemically hostile substances may be purged from inside the machine before the top half of the stator is removed if inspection or repairs are necessary. Suitable connecting and venting points must be provided on the turbines. The stored heat of the hot metal of the turbine will speed its drying out and purging effect.

15.9 Chemical attack on equipment from polluted atmosphere

In the neighbourhood of geothermal fields, even before exploitation, there is usually a rather strong smell of H_2S. Although natural selection may have ensured that local vegetation is unharmed by this gas, the same immunity will not extend to all man-made objects after exploitation. In Chapter 16 it will be explained how the source of this hazard – the very presence of H_2S – will be greatly reduced if not entirely eliminated in due course; but for the present it cannot be ignored. Unless special precautions are taken, exploitation will tend to increase the atmospheric concentration of H_2S. Aluminiun conductors in substations and on transmission lines will usually take on a protective coating of black sulphide which inhibits further attack. However instruments and relay contacts will almost certainly suffer if they feature exposed copper, as sealing is seldom perfect. Contacts and bare connectors of silver are advisable and, as an additional precaution, control rooms may sometimes be sealed with air-locks and ventilated with purified air. Exciter commutators of copper can be very troublesome, not only because the copper itself is attacked by H_2S but also because the sulphide film causes sparking at the brushes which wear away at an alarming rate. Self-excited generators are now therefore generally used.

15.10 Sulphate attack on cooling towers

Carry-over of bore gases into the condensate and cooling water can cause sulphate attack on the concrete structures of cooling towers. At Otake, Japan, alkaline river water has been introduced into the cooling system as a means of countering this hazard [153], while at the Geysers, California, a coal tar epoxy protection is provided on concrete surfaces [206].

15.11 Conclusion

The subject of chemical and metallurgical problems in geothermal installations is too vast to be adequately covered by a single chapter in a general book such as this. For further and more detailed information the interested reader is referred to [91, 219–222] in addition to the references that have appeared earlier in this chapter.

16 **Environmental problems**

16.1 Pollutive masochism

A century ago, except for a few murky industrial areas and slums, the world was on the whole a beautiful place. It is true that a few large cities suffered from choking fogs due to the excessive local burning of coal, but the countryside – whether dominated by untouched primeval features or by a happy blend of nature and human husbandry – was quiet and restful: an infinite variety of flora and fauna flourished as it had for aeons past. The steamship and the railway scarcely affected the Eden-like quality of most of the earth's open spaces. Then came the motor-car, the aeroplane, the industrial 'explosion', the centripetal migration of people from the country into the cities, the oil age, the proliferation and use of weapons capable of ever-increasing devastation; above all, the alarming increase of population (running now at an excess of new-born babies over deaths at a rate of some 160 human beings per minute). These teaming millions consume ever-increasing quantities of food and energy, and produce more and more indestructible wastes. The growing millions of motor-cars require more and larger roads to be carved out of our farming lands, and pour vast quantities of poisonous fumes into the atmosphere. In densely populated countries, ugly housing estates rob us of ever more agricultural land, while 'ribbon development' is tending to turn the greater parts of these countries into a uniform and soulless 'subtopia'. The human thirst for electricity is rendering the countryside hideous with forests of huge 'pylons'. Chemical wastes pollute the rivers, killing fish, and occasionally produce clouds of lethal vapours that have taken toll of human and animal life. Leviathan tankers are sometimes wrecked at sea, while many other ships callously and deliberately discharge oil waste into the oceans, spewing hundreds of thousands of tons of oil upon the surface of the waters, endangering bird life and fish and fouling the beaches. Sewage is being poured untreated into the sea, with horrible consequences, including the slow destruction of one of the world's most beautiful cities – Venice. Callous people throw their litter – much of which, like plastic and glass, is indestructible – all over the countryside, while the dismal sight of car cemeteries is becoming commoner every year. The annual combustion of thousands of millions of tons of coal, or the equivalent in other fuels, is pouring noxious fumes into the air and raising the CO_2 content of the atmosphere so rapidly that the climatologists are alarmed about the possible adverse effects on the world climate. In recent years the CO_2 content of the earth has been rising at a compound rate of 0.7% p.a., which implies doubling in a century. This gas has a 'greenhouse effect' in that it admits solar heat but inhibits its escape. The result of rising CO_2 content could be to make the earth hotter and, in the extreme, to approach the intolerable conditions that prevail on the planet Venus.

Predatory men are killing many animal species, such as the whale, the elephant and other wild game, at a rate that makes their early extinction (following the disappearance of the dodo and the moa) more than a possibility. All this has been accompanied by more and more noise, a horrific decline in the quality of the Arts and a growth of philistinism. In short, Man is very busily fouling his nest at an alarming rate.

It is true that there are few trends in a happier direction. The virtual death of the domestic coal grate has made the 'pea-soup' fogs of London a feature of the past. The heating of Reykjavík by means of earth heat has created a city of almost pure clean air. Many societies have been formed for the preservation of wild life and for the protection of places and buildings of outstanding beauty and interest. The quality of the water in the Thames has recently been so greatly improved that its fish life, not so many years ago on the verge of extinction, has almost returned to normal. However, these commendable trends, welcome though they be, amount only to a gentle zephyr in competition with the hurricane of self-inflicted pollution that threatens to engulf us.

Until about the middle of this century, despite the warnings of a tiny minority of thoughtful people, the world in general remained quite indifferent to what we were doing to ourselves. Those who thought about the problem at all were mostly apt to dismiss it with a cynical 'après nous le déluge'. During the last two or three decades, however, there have been welcome signs of a belated stirring of the public conscience concerning the growing threats to the quality of our environment.

16.2 How can geothermal development help solve the problem of pollution?

Geothermal enthusiasts have been rather too ready to claim that the exploitation of earth heat is an entirely 'clean' process. Certainly it is far less guilty of pollution than fuel combustion, but it must frankly be recognized that it is not entirely blameless, unless certain precautions are taken: even these precautions cannot always *totally* eliminate pollution, though they can reduce it to acceptable levels. The enthusiasts do not serve the cause of geothermal development if they fail to face the facts. Fortunately, the necessary precautions can mostly be, and in some cases are already being, taken with remarkable efficiency. The 'laissez faire' attitude that prevailed in the early days of geothermal development has now become very unwise in view of the greatly increased scale of development. It is far better to examine the environmental problems one by one and to consider how they may best be overcome. As to the *cost* of the necessary counter measures, this largely depends upon the zeal with which they are pursued. It is often possible to mitigate most of the pollutive consequences of geothermal development at quite moderate cost; but if idealism be carried to excess, the cost may be heavy. This question will be discussed in the next section. However, it is generally true that power generation from geothermal energy is much less pollutive than from the combustion of fuels, and that the use of earth heat for non-power purposes is virtually non-pollutive. Hence, even if anti-pollutive measures are taken with only moderate zeal the increased substitution of geothermal energy for fuel combustion will at least moderate the degree of environmental deterioration and may well lead to a marked amelioration.

16.3 American anti-pollution laws and the cost of their too rapid enforcement

In the USA very stringent anti-pollution laws have been enacted and are now being rigidly enforced at the power plants and steam fields in the Geysers area, not only at all the newer installations but also retrospectively at the older ones. Although these laws are on principle to be welcomed, it is arguable that they have been too suddenly and too rigidly applied. Full compliance with the new laws has had the effect of delaying construction by a year or more, and delay costs money. It has already been mentioned that 1 kW of geothermal base load can save about 2 tons of oil fuel annually. With oil fuel at US $200/ton the cost of a year's delay would be about $400/kW; and a glance at Table 14.2 will show that a delay of about a year would have the effect of approximately *doubling* the cost of a Geysers power station. It would seem that a more gradual introduction of the anti-pollution laws, with some permissible relaxation over a few years, would have been more in the national interest, especially as the marginal effects of geothermal development in California upon pollution would undoubtedly have been less than those caused by the combustion of the fuel that could have been saved by avoiding the delays consequent upon the rigid and immediate enforcement of the new laws.

This is not the end of the matter; for not only are the direct costs of conforming with the anti-pollution laws very heavy, but the *indirect* costs are even more burdensome. The insistence on retro-fitting the means of H_2S

abatement at the older power stations has resulted in a very marked decline in the annual plant factor at which the Geysers stations could operate over a few years, thus depriving geothermal power of one of its greatest assets – the ability to produce electrical energy at very high annual plant factor and therefore at low average cost per kWh.

16.4 Reinjection

Before considering, one by one, the various pollutive effects of geothermal development, it is well first to examine an operation that could well provide the antidote to several of these effects, namely reinjection of thermal fluids into the ground after use. The possibility of reinjection was considered many years ago, but was regarded with some timidity because there were many doubts as to the practicability of the process. One of the uncertainties was how the reinjected water could be persuaded to re-enter a pressurized aquifer. This invoked the time-honoured problem of the veterinary surgeon who is trying to administer a pill to a sick horse through a blow-pipe. Would the horse blow first? In more technical terms, would excessive pumping power be absorbed in forcing unwanted fluid into the ground? Another question was whether the introduction of cooler waters into the permeable substrata would interfere with the useful output of heat from the productive bores. Would the permeability of the aquifer be reduced or even totally eliminated by chemical deposition? Would reinjected waters outcrop else-where, thus simply transferring the pollution problem from one place to another? Would reinjection trigger off seismic shocks by inducing cooling stresses in the rock?

At the UN Geothermal Symposium held in Pisa in 1970 some discussion was devoted to reinjection, but there were many sceptics then. By the time the next UN Geothermal Symposium was held, in San Francisco in 1975, much useful empirical data had been gained and still more has been gathered since then. Although it would be unwise to be dogmatic about the pros and cons of reinjection, practical experience now offers promising evidence that it could often (but not necessarily always) be not only practicable but even thermally advantageous. For by reinjecting at some strategic point in a field, well removed from the producing bores, it could be possible to impose a warm barrier that delays the ultimate ingress of cold water from beyond the confines of a field. Reinjection could thus become a useful tool in field management, by recycling both water and heat, and could provide the equivalent of 'boiler feed' as an alternative to the wasteful rejection of heat at the surface. This concept of thermal conservation by means of reinjection is well illustrated by Einarsson et al. in Fig. 16.1 which shows the authors' interpretation of the heat balance in the Ahuachapan field in El Salvador, with and witout reinjection. It will be seen that for a given heat input to the

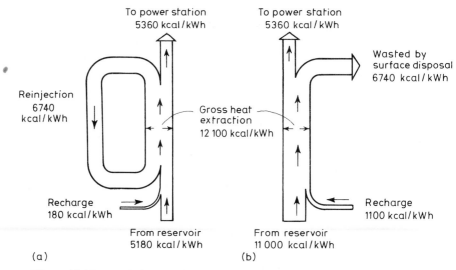

Figure 16.1 Energy balances of the reservoir for a water-dominated field, estimated for Ahuachapan, El Salvador. (After Fig. 11 of [223]) (a) Single-stage flashing with reinjection; (b) single-stage flashing with surface disposal.

power station reinjection could both reduce the amount of heat extracted from the reservoir and also the quantity of recharge.

If this thermal gain is to be won, great care must be taken, with the advice of a hydrogeologist, in choosing the point(s) of reinjection. These must not be close to the production bores lest there should be interference. The reinjected water should also enter the aquifer near the confines of the field in a zone from which the recharge fluid is expected to flow from outside for the replenishment of the fluid that is continuously being withdrawn from the aquifer through the production bores. It may sometimes be expected that the recharge fluid enters the aquifer from below, rather than from the edges of the field; in which case it may be necessary to sink very deep reinjection bores. Experiments in El Salvador [223] have been performed by entraining tritium tracer with the reinjected water. After two days a very small quantity of this tracer appeared in the discharge of another well about 400 m away, while after a few weeks much smaller traces were detected in two other wells at distances of about 400 m and 1050 m from the injection well. Less than 1% of the tritium reappeared. As in this case a production bore was used for reinjection experiments, whereas if reinjection were to be practised continuously a far more remote site would undoubtedly be chosen, this would seem to show that direct interference – at least in the Ahuachapan field – would not be a significant problem.

Experience has shown that in most cases a simple gravity feed suffices to force the injected fluid into the aquifer without recourse to pumping. During some reinjection experiments in New Zealand it was surprisingly discovered that the acceptance permeability of the aquifer actually *improved* with time [224]. This could be attributable to three possible factors: the lower temperature of the reinjected water could be fracturing the rock by thermal stresses; the negative buoyancy of the reinjected water may be ensuring its quick removal on entering the aquifer; or a two-phase zone may be locally converted into a single-phase zone, which would offer reduced flow resistance. It has also been found in New Zealand that if the reinjected bore water is denied contact with the air, and if the temperature is maintained at 150°C or higher, no deposition occurs in the reinjection bore, while the ready acceptance of the fluid by the formation suggests that no trouble need be expected in the rock. Moreover, if the reinjected bore water should be supersaturated with silica, it would soon be restored to a state of normal saturation after encountering the hot rock. In the Geysers field there has been a decline of 'acceptance' with some of the reinjection bores, and this has been attributed to the plugging of the formation with elemental sulphur. This trouble has been cured quite simply by shutting down the bore for a while to let it heat up. As the melting point of sulphur is 238°F and the reservoir temperature is 475°F, the sulphur just melts away and the bore can be reused until the trouble recurs [225].

No trouble has been reported from water-dominated fields of seismic shocks resulting from reinjection, though it cannot be stated categorically that this would never occur. In Italy some shocks were triggered off during some reinjection experiments in a steam-dominated field; but such fields are rare, and the need for reinjection is confined to far smaller fluid quantities than in wet fields, being only the surplus condensate from the cooling towers and the effluents from any chemical treatment plants.

Reinjection is now being practised in El Salvador [223], Japan [226], the Geysers field in California [225], the Imperial Valley, California, and in France [227]. In Mexico there has been experimentation, during which it has been found that if the rejected bore water is passed directly, under pressure, from the wellhead separators to the reinjection bore, no pumping is necessary. In New Zealand too there have been experiments with a view to halting the wasteful discharge into the river of 48 million tons p.a. of hot water that has been going on at Wairakei ever since 1958 [224]. Though there could probably be occasions where local conditions, such as excessive dissolved solids or low subterranean permeability, would preclude the adoption of reinjection it seems probable that the practice will become increasingly common and will thus greatly assist in the prevention of pollution. Although dissolved solids could often be precipitated before reinjection, it would seldom be practicable owing to the disposal problem caused by huge quantities of precipitate.

Reinjection, where practicable, can cure at a stroke the sources of pollution described in Sections 16.8 and 16.11 and can greatly mitigate those described in Sections 16.10–16.13 and 16.15–16.17. Experience at the Geysers shows that it is not expensive in steam-dominated fields. In 1980 it accounted for only 2.68% of the price of steam and only 1.61% of the total production cost of electricity. In wet fields, where large quantities of bore water would have to be handled, the costs would undoubtedly be greater, but not excessive in relation to the benefits derived.

16.5 Hydrogen sulphide

H_2S is almost invariably present in geothermal fields. Its properties and dangers have already been mentioned in Chapter 13.4.3. Rare fatalities have occurred in the vicinities of fumaroles, but fortunately no incidents have yet been reported from this hazard to geothermal exploitation plants. This can probably be attributed to the provision of adequate ventilation in cellars and basements – a most necessary precaution. H_2S also attacks electrical equipment (see Section 15.9), and it may have adverse effects on crops and on river life.

At geothermal power plants H_2S occurs in high concentration at the gas ejector discharge points and to a lesser extent in the cooling towers where jet condensers are used. Until the mid-1970s it had been deemed sufficient to place gas ejector discharges at high level and to rely on the high temperature of the gases to produce the necessary buoyancy and thus ensure their wide dispersal. Apart from this precaution, and the provision of good ventilation in buildings, the tendency had hitherto been more or less to ignore H_2S on the argument that in any case it occurs naturally in thermal fields. However, this overlooked the fact that exploitation concentrates and increases the release of this gas into the air; and the scale of development has now grown so rapidly in certain fields that greater precautions are now necessary. It has been estimated that if suitable precautions had not been taken, no less than 50 tons/day of H_2S would have been emitted in the Geysers field in 1980. At Cerro Prieto about 110 tons/day of H_2S are being discharged into the air with the present 150 MW installation, and by 1984, if the present extension plans are realized, this figure will have risen to something over 400 tons/day unless something is done about it. Such quantities cannot be ignored indefinitely. American legislation now insists on the reduction of the H_2S content of the air almost to within the threshold of odour, so that if it can be smelled, the law is perhaps being broken. This has led to a curious *reductio ad absurdum* in that Nature herself is guilty of breaking the law in several places such as the Yellowstone National Park! The stringency of the American anti-pollution laws together with the rapid development of the Geysers field compelled energetic attempts to tackle the problem of H_2S abatements in that field.

The method first tried was a modified form of the Claus process which converts the H_2S into elemental sulphur $(2H_2S+SO_2 = 3S+2H_2O)$. The sulphur dioxide is obtained by burning part of the ejector discharge gases $(2H_2S+3O_2 = 2SO_2+2H_2O)$, and the combustion products with the residual H_2S are scrubbed into the cooling water system. The addition of an iron catalyst encouraged the dissolved oxygen in the cooling water to aid the process of H_2S removal $(2H_2S+O_2 = 2S+2H_2O)$ [228, 229]. This method had only limited success, as the presence of moisture and CO_2 were detrimental to the process. Moreover the removal of the sulphur was both troublesome and expensive: the recovered sulphur was impure and unmarketable. A treatment involving the use of iron sulphate and sulphuric acid was tried at units 3 and 4, with caustic soda injected into the condensers and hydrogen peroxide into the cooling tower sumps. This achieved about 90% removal of the H_2S but gave rise to even worse troubles. Heavy corrosive sludge formed in the cooling towers, reducing their efficiency and even sometimes causing the fill material to collapse with the weight; and contaminated spray from the towers deposited filth on nearby structures and vegetation.

Plate 38 Power unit No. 15 at the Geysers, California, showing the Stretford H_2S abatement plant to the left of the turbine house. (Note that the space occupied by the abatement plant is comparable with that of the turbine house.) (By courtesy of The Pacific Gas and Electric Company)

The next method tried was the more complicated and expensive Stretford process [230, 231] which necessitates the substitution of the costlier surface condenser in place of the far cheaper jet, or direct contact, condenser so as to reduce drastically the quantity of water that comes into direct contact with the H_2S carried through with the bore steam. The Stretford process is very efficient in an acid environment; but unfortunately the ammonia present in the Geysers steam causes the pH value to be about 8.5 at the turbine exhaust, and this reduces the process efficiency by about 30%, which is unacceptable. As a result it became necessary to supplement the Stretford process with a secondary abatement process involving dosage with hydrogen peroxide and acetic acid, with the use of an iron catalyst. This secondary abatement process is too costly to be used as the sole means of treatment. The cynics say that the efforts to conform with the Californian pollution laws have caused the Geysers installations to become chemical factories with ancillary power plants appended! (Plate 38.)

In 1980 complete H_2S abatement had still not yet been achieved at the Geysers, and the Californian Government were threatening to license no new power plants until the law had been fully complied with. There were, however, certain options that were still being examined, two of which were upstream processes and one which involved the acid dosing of the steam before entering the condenser so that the Stretford process might become more effective. There is little doubt that complete success will ultimately be achieved, but at what cost! In Sections 16.2 and 16.3 the wisdom of excessive zeal in H_2S abatement has been questioned, and it is perhaps worth quoting the following figures:

Maximum permissible ambient concentration of H_2S according to Californian law	30 ppb
Typical concentration of H_2S at which eye irritation begins	10000 ppb
Typical concentration of H_2S at which lung irritation begins	20000 ppb
Typical concentration of H_2S which is fatal after 30 minutes	600000 ppb

Thus the legally permissible concentration is about 1/300th of that at which physical discomfort begins to be felt and about 1/20000th part of what is generally regarded as a fatal dose. Do not these figures suggest 'excessive zeal'?

Although most of the efforts towards H_2S abatement have hitherto been made in California, other countries developing geothermal energy will almost certainly have to take active steps in this direction very soon. It is understood that at Cerro Prieto, Mexico, power station staff have to wear gas masks for certain tasks in windless weather; and it may be that by the time that this book is in print proper gas abatement equipment will have been installed there.

Even if effective H_2S abatement is practised at geothermal power plants, there is always likely to be intermittent releases of the gas through 'untamed' escape paths such as the rare 'rogue' bore, wells undergoing test or bypassed under emergency conditions, or even from vent valves when exercising their control function (see Chapter 13).

It is most advisable that at all geothermal installations there should be an adequacy of gas detectors and alarms to give warning of dangerous conditions at key points in enclosed areas for the protection of personnel and electrical equipment.

Where river cooling is adopted and where jet condensers are used, as at Wairakei, some H_2S will enter the river in solution with the condenser cooling water. If the river flow is very large, as in the case of the Waikato River at Wairakei, this may be unimportant (though this is debatable); but with small rivers fisheries and weed growth could be affected. Matters are made far worse if unwanted bore waters are also discharged into the river. It has been suggested [232] that river pollution is in fact by no means negligible at Wairakei from both these sources despite the normally high rate of river flow. It could sometimes be necessary to use surface condensers, or adopt cooling towers, and to reinject unwanted bore waters, if river flow rates are inadequate.

16.6 Carbon dioxide

By far the greater part of the incondensible gases from a thermal field consist of CO_2. This can escape into the air or into local water courses. The fact that fuel combustion usually produces far greater quantities of this gas in terms of the heat consumed than does geothermal exploitation has generally been regarded as an excuse for inaction, particularly as the gas is not toxic. However, in certain exceptional fields, such as those in the Monte Amiata region of Italy, the CO_2 discharged to the air may be far greater than from fuel-fired plants of comparable size and duty. As stated in Section 16.1, concern about the rising CO_2 content in the atmosphere is already worrying the climatologists, and it is known that high CO_2 content in river waters can aggravate weed growth. It is most undesirable that geothermal exploitation, especially if this is to bear a growing share of the world's energy burden in future, should contribute towards these ill effects even though it may be far less guilty, on average, than fuel combustion. Ideas for the commercial extraction of CO_2 from effluents are being closely studied – e.g. the manufacture of dry ice, carbonic acid for beverages, the production of methyl alcohol – but have not yet proved economic. (Moreover, in each of these cases the CO_2 would be returned to the atmosphere after use!) Meanwhile the emission of CO_2 to the atmosphere seems to be inevitable until some permanent method of chemical fixation can be found. The problem has not yet become seriously urgent, but if geothermal development grows dramatically, as well it soon may, it should be tackled seriously before long.

16.7 Land erosion

At the Geysers field, where heavy rains and steep slopes of incompetent rock often cause natural landslides and high erosion rates, the artificial levelling of the ground for the accommodation of field works, roads and power plants has sometimes aggravated these conditions by creating very steep local gradients (Plate 22) and by the removal of local vegetation which normally protects the land from erosion by the binding action of its roots. Closer control, replanting of shrubs and trees, more careful site selection and improved construction methods are helping to solve this problem [233]. The close spacing of several deviated wells within a single levelled area (see Fig. 7.4) could sometimes be helpful.

16.8 Water-borne poisons

The water phase in wet fields sometimes contains toxic ingredients such as boron, arsenic, ammonia and mercury which, if discharged into water courses, could contaminate downstream waters used for farming, fisheries or human water supplies [234, 235]. Although not strictly 'poisonous', very saline bore waters too can be harmful. Possible solutions to the problem of water-borne poisons include reinjection, disposal into the sea (if not too remote) through ducts or channels, the use of evaporator ponds as at Cerro Prieto [146, 236], storage during the dry season (in 'monsoon' countries) with subsequent release into rivers when in spate during the wet season when dilution would render the offending substances more or less innocuous, or even chemical treatment [234].

16.9 Air-borne poisons

From ejector exhausts, from the upward vaporous effluents from cooling towers, from silencers, drains and traps, from discharging bores under test, from 'wild' bores and also from control vent-valves, various harmful substances (beside H_2S) may sometimes escape into the air at geothermal sites. These may include mercury and arsenic compounds and radioactive elements. Certain noxious, though not poisonous, emissions such as rock dust and silica-laden spray may also be air-borne. At Wairakei a certain amount of forest trees were damaged by silica deposition, while in El Salvador the contamination of coffee plantations was a problem. At Cerro Prieto salt deposition on buildings and agricultural lands caused some trouble [236] while at Wairakei the build-up of hard silica on car windscreens, from the spray emitted from discharging bores and silencers, has caused nuisance. Horizontal or inclined well discharge in a controlled direction, where there are no vulnerable 'targets' can sometimes offer a solution to this problem, but this is clearly no more than a makeshift solution that would seldom be acceptable. Fortunately, air-borne toxicants are seldom present in harmful concentrations (other than silica, which is not truly a 'toxicant' but rather a nuisance); but systematic monitoring is advisable to enable a careful watch to be kept on possible future dangers.

16.10 Noise

The nuisance caused by noise can cause a serious health hazard unless workers on new well sites and at test sites wear ear-plugs or muffs, without which their hearing can be permanently damaged. Though the noise from regular escape paths of fluids can be reduced by means of silencers, there are occasions when noisy discharges cannot be avoided; e.g. bores newly 'blown' or those undergoing test or maintenance. The erection of temporary sound barriers can sometime be helpful in such cases. Reinjection can eliminate the noise arising from flashing steam where hot water is now being discharged to waste (as at Wairakei). Drilling operations can be somewhat noisy, but no worse than road repairs, and they do not persist for very long. Noise also occurs in and near power plant buildings, from the whirr of machinery, but this is no worse than in conventional thermal power plants and is very difficult to control in machinery halls or near outdoor auxiliaries such as ventilating fans. Control rooms and offices can be sound-proofed, so that most of the attendant personnel are protected. Legislation against noise has been enacted in the USA and other countries, and although its strict enforcement may sometimes be difficult its very existence acts as a powerful incentive to designers to overcome the nuisance.

16.11 Heat pollution

This can sometimes be very troublesome. Apart from being pollutive, it is very wasteful and so every effort should be made to avoid its occurrence. The necessary adoption of relatively low temperatures for geothermal power production must result in low efficiencies and the emission of huge quantities of waste heat in far greater proportions than in conventional thermal power plants of the same capacities. Where cooling towers are used, this waste heat escapes into the air and by way of the surplus condensate which, in the latter case, may be reinjected. With direct river cooling the waste heat simply raises the temperature of the river. This of course is a wastage that differs only in degree from that which occurs in conventional thermal power stations, with the added factor that higher condenser back-pressures are usually adopted in geothermal power plants, so that the water is a good deal warmer. What is far more serious is when the hot water phase in a wet field is discarded to waste, for even if power is generated from the flash steam from this water, the temperature of the water at discharge is likely to be at or near to the atmospheric boiling point. The discharge of huge amounts of such hot water into rivers and streams (as is being done at Wairakei) can damage fisheries and encourage the growth of unwanted water weeds. If, as at Cerro Prieto, the hot water is simply discharged into lagoons, from which it can evaporate, the heat escapes as vapour into the air. The escape of heat and vapour from cooling towers and 'dump' lagoons may affect the local climate by the formation of fog and/or ice [237], although in certain climates the increased humidity could have a beneficial effect. The most damaging form of heat pollution is undoubtedly the discharge of very hot waters into rivers or streams. Possible remedies include reinjection of unwanted hot bore water, and also of surplus cooling tower water; the generation of additional power by means of binary fluid cycles; or, best of all where possible, the establishment of dual or multi-purpose projects which usefully extract low grade heat from turbine exhausts and/or from rejected hot bore waters. Where rivers of high flow rates are used to receive very hot discharged waste waters, as in the case of the Waikato River at Wairakei, a zone of heated water is apt to be formed along one bank, extending downstream for perhaps half a kilometre until mixing has become complete. Apart from a few casualties, most of the fish learn to avoid the very hot zone. At Wairakei, during normal river flow, the average temperature rise of the river, allowing both for the condensers and the hot bore discharge, is about 3°F after mixing, and about 5°F at times of low flow. (At a time of drought in 1974, a temperature rise approaching 11°F was actually observed.) It has been claimed that heat pollution has had an adverse effect on the fish in the Waikato River [232]. With smaller rivers or larger power installations, the effects could of course be worse.

The question is sometimes asked whether heat pollution from our growing energy consumption could ultimately affect the world climate. In Chapter 4 it was mentioned that the *natural* outward flow of earth heat amounts only to about 1/40th of 1% of the heat received by the earth from the sun. At present the annual world primary energy consumption is about 285 million TJ, equivalent to a continuous steady power consumption of about 9×10^9 kW, compared with a natural outward flow of earth heat of about 3×10^{10} kW (thermal) – see Section 3.5. Thus the present world energy consumption is only about 0.3 of 1/40th of 1% of the solar energy falling continuously upon the earth – i.e. about 0.0075%. Practically all of our energy consumption reappears as heat surrendered to the atmosphere and seas, but this is at present such a microscopic, proportion of the solar energy we receive that we could clearly afford to multiply our energy consumption by quite a large factor without imposing any *thermal* threat to our climate. The question of CO_2 affecting the upper atmosphere (see Section 16.6) is another matter.

16.12 Silica

Silica has already been mentioned under Section 16.9 as a relatively minor form of air-borne pollution, but silica can be extremely troublesome with district heating systems [218]. Dilution with fresh water has proved to be fairly successful in Iceland, but temperatures will not always be such as to make this possible. Silica could also perhaps foul up the permeability of the aquifer where reinjecion is practised, though little has been heard of this form of trouble. Some extensive research has been carried out at the Ohaki (Broadlands) field in New Zealand into the problems of reinjecting silica-laden bore waters. Chemical treatment was tried there by dosing the bore water with slaked lime so as to precipitate calcium silicate. It was found that in the presence of air, less lime is needed if the bore water is allowed to 'age' for a few hours before dosage; so aging tanks were provided to hold up, or delay, the bore water before treatment, and settling tanks were provided for the collection of the precipitated silicate. Although this treatment was effective in removing the silica it was rather expensive in itself, and it furthermore created a new problem in the disposal of huge quantities of silicate sludge, for which no commercial market could be readily found. The experience of New Zealand suggested that immediate hot reinjection was preferable to chemical treatment, if the conditions mentioned in Section 16.4 are maintained. The phenomenon of improved acceptance per-meability, also reported in that subsection, suggests that at Ohaki at least no trouble need be expected from the fouling up of the aquifer by reinjected silica-laden water. Experimentation continues in New Zealand, and it would be rash to conclude that other fields would necessarily behave in the same way as Ohaki; for the chemistry of bore waters can differ widely from one field to another. The chemistry of treatment for silica has been discussed by Axtmann [238] and by Rothbaum [234]. No trouble has been reported of silica-plugging of the aquifer near reinjection wells at the Geysers field, but of course the quantities of water to be reinjected in dry fields are very much less than in wet. Experience in this respect in Japan and El Salvador is not known.

16.13 Subsidence

Although ground movements have been observed in unexploited fields [239] and were detected at Wairakei between 1950 and 1956, during which time only a few bores had been sunk [240] there can be no doubt that the withdrawal of huge quantities of subterranean water from a wet field can cause substantial ground subsidence which could result in the tilting and stressing of pipelines and surface structures, perhaps with serious or even disastrous consequences. Vertical movements had been observed at certain points in the Wairakei field of up to 4.5 m after 10 years of exploitation [241], together with smaller, but appreciable horizontal movements also. By the end of 1978 the maximum vertical settlement had reached 7.6 m and was still increasing at about 0.4 m p.a. [107]. The area of subsidence by then covered about 260 km^2 and the volume of settlement was of the order of 38 million m^3. This volume could be related to, though not equated with, the total volume of fluid withdrawn from the field less the replenishment inflow during the same period. Fortunately the main area of subsidence at Wairakei lay mostly to one side of the pipelines and was clear of the power station buildings. Nevertheless there was some pipe movement that had to be corrected.

Reinjection of the bore fluid would almost certainly mitigate the problem of subsidence, though it would probably not cure it completely as the volume of reinjectable bore water would be less than that of the withdrawn fluid by the amount flashed. Natural recharge would help, and it is possible that the injection of additional river water, where available, at strategic points could effect a complete cure. It is reported that in Venezuela, where the extraction of oil caused subsidence, the injection of an extraneous supply of water fully cured the trouble.

16.14 Seismicity

Fears have sometimes been expressed that prolonged geothermal exploitation of a field could trigger off earthquakes, especially if reinjection is practised in zones of high sheer stress where fairly large temperature differentials could occur. These fears arise because all existing exploitations of hyperthermal fields are in naturally seismic areas where ground instability is self-evident, and any interference with nature might precipitate earthquake shocks [237]. At the present state of geothermal development these risks do not seem to be great, but if very large scale geothermal exploitation should be undertaken in future by exploiting hot dry rock the problem could perhaps become a serious one. Fortunately it is already the subject of study, so that reliable information and means of forestalling the danger should be ready in good time.

No reports have been received of seismic shocks caused by the exploitation of water-dominated fields, even where reinjection is practised. In Italy, however, some shocks were triggered off during some reinjection experiments. This suggests that steam-dominated fields may be more sensitive in this respect. However, steam-dominated fields are comparatively rare, and the quantities of fluid to be reinjected in such fields are far less than in water-dominated fields. Vigilance is clearly required, but from experience to date it seems unlikely that any serious hazards will arise from seismicity during the exploitation of hydrothermal fields.

16.15 Escaping steam

The large quantities of waste heat escaping from the tops of cooling towers at geothermal power plants give rise to immense volumes of water vapour. In dry climates this vapour is usually absorbed quickly into the atmosphere (Plate 23) but, in colder and more humid climates, huge clouds of billowing condensing vapour can and do persist long enough to create serious local fog hazards (Plate 24), and in some cases cause ice precipitation. Worse than cooling towers can be the clouds of condensing vapour issuing from the tops of silencers in wet fields such as Wairakei (Plate 11), where vast quantities of superheated flashing hot water are being discharged to waste. At Cerro Prieto, Mexico, discharged bore water is dumped into an open lagoon, from the surface of which a great deal of vapour rises into the air. Fortunately the climate at Cerro Prieto is very dry, so no trouble is caused by this, but if dump lagoons were used for hot bore water disposal in cool and humid climates fog formation could be a nuisance. Bore steam escaping from control vent-valves can also be troublesome, but if properly adjusted this can be kept to small proportions except on the rare occasions when an unexpected change of plant conditions (such as a tripped-off turbine) gives rise to a large escape of steam through the vent valves. The drifting of large clouds of condensing steam across roadways can give rise to traffic hazards. Warning signs and diversionary routes can of course mitigate the trouble, but the best remedies for dealing with the nuisance arising from discharged hot water are either to use the inherent heat productively for some industry or other application, or to reinject. These steps would not of course cure the trouble from cooling tower vapour or from vent-valve discharges, or from bores newly blown or under test. In practice, no really serious inconvenience has arisen from this purely local and relatively minor nuisance, with which one must learn to live, after taking whatever precautions as may be practicable.

16.16 Scenery spoliation

It is virtually impossible to be entirely objective in matters of aesthetics, but it cannot be denied that thermal areas often occur in areas of outstanding beauty, highly prized by the local population and frequented by tourists. Conservationists will often oppose geothermal development on the ground that scenic amenities will be destroyed. Though it cannot be denied that man-made engineering works can seldom compete aesthetically with natural scenic beauty, it must be conceded that the power installations in the Geysers field, California, have been most tastefully camouflaged. The pipe-lines have been painted to blend with the background; scarcely a puff of steam is visible except on the rare occasions when vent-valve operation is unavoidable or when a well is newly blown or under test; the power plants themselves are relatively inconspicuous; and the normally dry climate quickly absorbs the plumes of water vapour arising from the cooling towers. A serious setback to the aesthetic aspect of the Geysers field occurred when the iron sulphate treatment was used at units 3 and 4 (see Section 16.5) and filth was scattered around the vicinity; but it is hoped that this will have been a temporary phase only, pending on the development of a better and cleaner method of H_2S abatement.

At Wairakei, New Zealand, where the scarred ground surface has been rehabilitated by careful 'landscaping' and damaged trees have been removed so that only the healthy forest can be seen, visitors flock to see the geothermal development in greater numbers than those who frequented the area before exploitation. It could even be claimed that the Wairakei scene has a certain majesty of its own, where the billowing steam from the wellhead silencers – itself a form of 'pollution – contributes a dramatic aesthetic quality (Plate 11).

It is true that geothermal exploitation can interfere with natural surface manifestations. For instance, the activity of the geysers and boiling pools in the once famous Geysers Valley, close to the exploited area of Wairakei, has virtually ceased. The question of scenic amenities is one that can only be judged subjectively on a balance of considerations, by weighing the value of the energy won against that of touristic attractions and national heritage, a balance that must be largely emotional and cannot be strictly quantitative. The declaration of an area such as Yellowstone Park as a zone of outstanding beauty and interest, not to be exploited for the winning of energy, would be a value judgment having its protagonists and its opponents: there can be no absolute standards in such matters. There is a certain schizophrenic streak in human nature. We may hate the sight of electricity transmission lines extending over the countryside, but we would bitterly complain if our homes were deprived of electricity. We become demented by the noise of aircraft, forgetting the joys of travel that have been brought to us by the aeroplane; we object to the depressing sight of industrial areas and mining works, yet we enjoy their products. In so far as geothermal development is concerned, it is only fair to point out that we cannot do without energy, and that a geothermal power plant, which produces no smoke, has no unsightly chimney stacks, no ungainly coal- or ash-handling equipment, no coal storage yard or oil storage tanks, and no boiler house can be far less displeasing to the eye than a fuel-fired plant of the same capacity. For similar reasons, industrial establishment using geothermal heat are likely to be far less obtrusive than those which rely upon fuels. On balance, it may justly be claimed that geothermal exploitation is far less guilty of scenery spoliation than fuel combustion.

16.17 Ecology

This is a subject that has only recently begun to attract well-deserved attention. The discharge of chemicals into the air and into streams and rivers and thence into the ground water; small, but appreciable local changes of temperature and humidity; noise; a degree of deforestation; all these factors could, and possibly do, disturb the natural balance of nature prevailing in a thermal area before exploitation. The effects of geothermal exploitation upon local fauna and flora are already being studied in New Zealand, California and very possibly elsewhere. Protectors of wild life and of fisheries would do well to encourage more intensive research in this direction, although it is only fair to say that on present evidence there seems to be little cause for alarm. However, it is always possible that the development of certain fields hitherto untouched might have effects not yet observed elsewhere.

Action of a positive nature has recently been taken in California, where hybrid pine seedlings have been planted in the Geysers field. These hybrids are capable of surviving on steep slopes in soil of poor quality lacking good irrigation, and in an atmosphere having an appreciable content of H_2S.

16.18 Conclusion

The comparatively recent upsurge of interest in geothermal development has roughly coincided with the genesis of a public sensitivity to the dangers threatening our environment. This is probably a happy coincidence, for the latter has acted as a timely check to a certain 'laissez faire' attitude that accompanied the earlier geothermal exploitation projects. This attitude, although relatively harmless when the exploitation of earth heat was in its infancy, could later become a serious hazard were it allowed to persist until this form of energy bears a major share – as it undoubtedly will – of the world's energy burden. A clear change of mood can now be discerned in geothermal circles from one of unreasoning optimism to one of sober understanding of the environmental aspects of geothermal exploitation. Gone is the pious belief that earth heat exploitation is entirely 'clean' and wholly guiltless of infecting the environment. Nevertheless, it is an undoubted fact that geothermal exploitation is far less culpable than fuel combustion of fouling the human nest, and that an antidote can be found to nearly every potential source of geothermal pollution. The more rapidly the use of earth heat takes over the energy burden from fuels, the better will it be for out human environment. Timely legislation in certain countries has enforced attention to this very important matter, even though it could perhaps have been introduced at a rather less drastic pace. The advances made in environmental studies in the 1970s have been impressive, and the good work may be expected to continue.

(Also of relevance to this Chapter are [242–244].)

17 Miscellaneous points

17.1 Systematic approach

Exploration for geothermal fields can be a costly affair, perhaps involving a few millions of US dollars, depending partly on luck and partly on the skills of the exploration team. To minimize the cost it is absolutely essential that development should be set about systematically. So often on past occasions countries have squandered their limited resources in looking for exploitable earth heat in an unsystematic way. They may have flitted from area to area, sinking a few holes here and a few holes there, in places chosen solely on the superficial evidence of surface manifestations which, as explained in Section 6.3, are not always a reliable guide to the exact location of a field and may even be entirely absent from an area overlying a valable field. Such hit-or-miss methods are followed in the pious hope of making a lucky strike. International 'experts' may have been invited to pay short visits to thermal areas in the expectation that they will be able to assess the geothermal possibilities from a superficial inspection lasting only a day or two. Such unmethodical activities are not necessarily wholly wasted, for every hole sunk adds to the cumulative knowledge of an area, and every visiting expert may contribute some useful suggestions as to the next steps to be taken. It must be admitted that successes have been sometimes achieved by such haphazard methods. Nevertheless, without a systematic approach there will always be a risk that much money may be needlessly wasted. If exploration costs are to be kept to a minimum, and if good results are to be obtained quickly, assuming that a useful field *does* exist, it is vitally important that a logical system be followed from the outset. Even if no field exists at all, at least that negative fact can be established more quickly and at less cost by proceeding systematically.

A suggested methodology has been proposed [245] and is schematically illustrated in the process diagram shown in Fig. 17.1 which is more or less self-explanatory. Its main purpose is to concentrate first on the less expensive investigations and to postpone drilling operations, which absorb by far

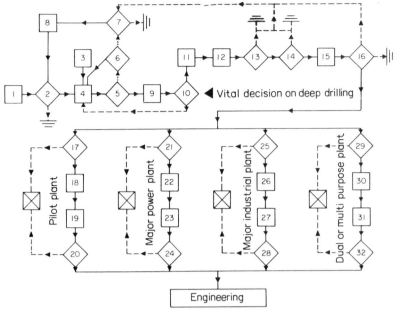

Vital decision on deep drilling

Pilot plant · Major power plant · Major industrial plant · Dual or multi purpose plant

Engineering

1. Inventory of alternative energy costs
2. Is geothermal energy likely to be competitive?
3. Collection and review of preliminary field data
4. Collection of more extensive field data
5. Are data enough for conjectural field model?
6. Are data sufficiently promising to proceed?
7. Is there an alternative area worth investigating?
8. Selection of alternative area
9. Conjecture of tentative working model
10. Are data sufficient to justify deep exploration drilling?
11. Minimum economic yield of bores
12. Deep exploratory drilling and borehole measurements
13. Are production bores likely to yield economic minimum?
14. Are corrosion problems soluble?
15. Estimation of energy potential of field
16. Are other engineering problems soluble?
17. Is a pilot plant required?
18. Preliminary design of pilot plant, with estimates
19. Re-appraisal of economics of pilot plant
20. Is pilot plant still economic?
21. Is field potential enough for major power plant (> 20 MW)?
22. Preliminary design of major power plant, with estimates
23. Re-appraisal of economics of major power plant
24. Is major power plant still economic?

25. Is field potential enough for major industrial plant?
26. Preliminary design of major industrial plant, with estimates
27. Re-appraisal of economics of major industrial plant
28. Is major industrial plant still economic?
29. Is there a demand for waste heat from power plant?
30. Preliminary design of dual or multi-purpose plant, with estimates
31. Economic study of dual or multi-purpose plant
32. Is dual or multi-purpose plant economic?

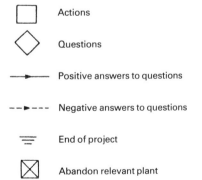

Actions

Questions

Positive answers to questions

Negative answers to questions

End of project

Abandon relevant plant

Figure 17.1 Suggested process diagram for systematic approach to geothermal development. (From [245])

the greatest part of exploration costs, until a reasonably probable field model has been conceived which offers good chances of a high success ratio in the subsequent drillings. Actual exploration techniques are not described or specified either in Fig. 17.1 or in [245] because these are covered by the available literature: it is the *procedure* that is emphasized.

17.2 Legal questions

Before any country attempts to develop its geothermal resources it is most advisable that the legal position be fully clarified, and that suitable legislation be enacted as quickly as possible if the situation is not clear. Is the exploitation of earth heat to be a Government monopoly, or is it to be open to private enterprise? Is earth heat to be treated, legally, as a mineral, as ground water, or as neither? What are the rights of a land-owner under whose property lies an exploitable field, even though the field may be tapped (perhaps by taking advantage of subterranean flow patterns or perhaps by using directional bores, as in Fig. 7.4) from land owned by another party? Has the Government the right of compulsory acquisition, at a fair price, of land from which earth heat can supposedly be exploited? What of the rights to discharge unwanted fluids into water courses, or even to reinject them into the substrata if the quality of the ground water could thereby be effected? What are fair rates for leasing rights? Are permissible pollution levels clearly defined, or the subject of total veto in certain cases? What redress has a land owner – e.g. of coffee plantations, of other crops or grazing lands – against pollution damage consequential to exploitation nearby (e.g. crop destruction from silica-laden spray)? What are the access rights across the land of third parties? Are drilling activities to be restricted to Government agencies, or may they be pursued by private enterprise under licence subject to certain safety precautions? May a private individual exploit earth heat under his own land; and if so will a special licence be needed? The questions are almost endless, but they must be comprehensively listed and answered with legal backing. Where geothermal legislation exists, as in California, existing acts can no doubt form a useful starting point for the guidance of other countries intent upon drafting their own legislation; but there could well be special problems peculiar to individual countries.

17.3 Training

Geothermal development calls for the services of many specialists versed in various disciplines – geology, geochemistry, geophysics, hydrogeology, engineering, economics, industry, district heating, husbandry, etc. Graduates and technologists in many of these fields are regularly being turned out by the universities and technical colleges of the world; but if geothermal development is to be fostered at an accelerating rate, these young people must be 'slanted' towards the geothermal application of their chosen discipline, and this requires at least some partial understanding of related disciplines as well as a deep understanding of their own. The purpose of the UNESCO geothermal review [4] was to encourage this process, but so much has happened since that was written that the breadth of necessary training has increased in recent years. Until the 1950s there were very few geothermal specialists in the world; but the proliferation of conferences, symposia, government reports, papers read before learned societies and articles in the technical press bears witness to the accelerating awareness of the importance of geothermics and to the ever-growing number of such specialists. It is becoming important not only that sufficient people shall be trained in the necessary disciplines at a rate to keep pace with the development of the world's geothermal resources, but also that enough trained nationals of the developing countries possessing these resources are available for their exploitation without too much reliance on foreign aid.

In 1973 the United Nations University came into being, armed with a charter unanimously approved by the UN General Assembly [246]. The purpose of this new university was to mobilize the international academic community for the solution of the many pressing problems relating to human survival, development and welfare: its facilities virtually amount to the sum total of those of all the other universities of the world. The UNU is an autonomous academic institution under the sponsorship of the UN and UNESCO, having its own endowment fund, rector and governing council. One of the functions of the UNU is to provide specialized training courses in selected subjects of international urgency or importance. Fellowships are awarded to enable trainees, nominated by their national governments, to attend these courses which are held in appropriate locations chosen for the availability of specialized teaching staff and of research facilities. The activities of the UNU embrace such subjects as food, health, cultural activities and natural resources. It is under this last category that geothermal training programmes have been established: courses are now periodically held in Iceland, New Zealand, Italy and Japan. Fellows attend these courses for a period of some months, during which they hear lectures by geothermal specialists and participate in field research. All the chosen training centres are close to geothermally active regions. From a fairly modest beginning it is hoped that a large international organization may grow, capable of feeding the world with a steady flow of well trained personnel who may return to their home countries and help them to develop their geothermal resources. In this way the necessary legions of specialists should be available for the expected 'geothermal renaissance' that probably lies shortly ahead.

17.4 Publication of data

As with so many sciences, geothermics suffers rather from a plethora than from a shortage of information. The would-be student of the subject is apt to be bewildered by the sheer volume of literature available to him and one of the purposes of this book is to ease his approach to the subject. The UN and UNESCO alone had, by 1980, been responsible for publishing between 600 and 700 papers and rapporteurs' summaries, mostly produced at various international conferences and symposia; but this is only a fraction of what has been written and recorded on the subject of geothermal energy since the 1950s. A few periodicals, such as *Geothermics* (published by the Istituto Internazionale per le Ricerche Geotermiche, Pisa, Italy) and the *Geothermal Energy Magazine* (published by the Geothermal World Corporation, 18014 Sherman Way, Suite 169, Reseda, California 91335) and others mentioned below are devoted solely to the geothermal sciences, while many other technical periodicals often contain articles of geothermal interest. The proceedings of learned societies in many countries, and of national and international seminars abound in articles and discussions on earth heat, while even the popular press do not entirely neglect the subject. Several complete geothermal textbooks have also been published, most devoted to some particular facet of the subject. The USA are particularly prolific in the production of learned reports and treatises on every aspect of geothermal research and development. A positive flood of informative documents issue from Government Departments such as the US Department of Energy, and the US Geological Survey, while the output from the various American national laboratories such as the Los Alamos Scientific Laboratory, the Lawrence Berkeley Laboratory, the Lawrence Livermore Laboratory and the Sandia National Laboratories is equally impressive. Other countries and regions too publish various reports and newsletters devoted to geothermal affairs – e.g. the Japan Geothermal Energy Association's reports and the Commission of the European Communities' *Geothermal News*. It would be no exaggeration to say that the world's geothermal literature by 1980 could be numbered in thousands, rather than hundreds, of documents.

In 1974 an attempt was made to gather and codify the growing flood of geothermal information by the establishment of the International Geothermal Information Exchange Programme for the prompt exchange and dissemination of all published articles and data, including test results from various sites. Two computerized centres were established – one in Pisa, using the combined facilities of the International Insititute of Geothermal Research (CNR) and the Italian National Research Council (CNUCE), and the other in the USA using the facilities of the Lawrence Berkeley Laboratory, California, and the US Geological Survey in Reston, Virginia. In 1974, the coordinating group of IGIEP initiated the following activities:

(i) GRID a computerized bibliography of geothermal literature at the Lawrence Berkeley Laboratory, California.

(ii) GEOTHERM a computerized data file on geothermal fields, wells, and other geothermal topics compiled by the US Geological Survey.
(iii) Implementation of GEOTHERM at the computer centre of the International Centre for Geothermal Research, Pisa, and at the Lawrence Berkeley Laboratory.
(iv) Coordination of activities between the United States, New Zealand, Italy and other countries, and between the US Geological Survey and the Lawrence Berkeley Laboratory.

The agreement between the Italian and American centres operated for five years until 1979, but was not renewed thereafter. There is no longer a single world centre for published geothermal data, so the student of geothermal energy undoubtedly has a problem in keeping up-to-date with the latest developments. For concise reporting on worldwide geothermal developments the following publications are of considerable value:

The monthly bulletins of the Geothermal Research Council (P.O. Box 98, Davis, California 95617, USA).
The quarterly bulletins of the Geo-Heat Centre (Oregon Institute of Technology, Klamath Falls, Oregon 97601, USA).
The Geothermal Progress Monitor Reports (published from time to time and available from the National Technical Information Service, US Department of Commerce, 5285 Port Royal Road, Springfield, Virginia 22161, UA).

The first of these is worldwide in scope, and covers geothermal developments of all kinds. The second is devoted to direct applications of earth heat only, and is also worldwide in scope. The third is a valuable survey, reviewed periodically, of geothermal developments in the USA only.

17.5 Field management [84, 242]

The art of field management may be described as that of extracting a maximum of heat of the required grade, in accordance with availability, in such a manner as to eke out the life of a field over a sufficiently long period to avoid large changes of pressure and temperature during the life of the exploitation plant – all as efficiently as possible and with a minimum of wastage. Over-exploitation of a field – that is, excessive fluid withdrawal – may pay an immediate but spurious dividend in the production of huge quantities of heat which can be used for power generation or other purposes, but it will probably result in pressures falling so rapidly that a plant can no longer operate at its full initial rating after only a fraction of its potentially useful life has passed. This would mean that for the latter part of the plant's life it would have to operate at reduced capacity, which implies a waste of capital resources. By drawing on the field at a slower rate, a smaller plant might serve its full useful life of, say 25 years, with no decline – or at least with only minimal decline – of output during the latter part of its service. A second generation (and even a third or more) of plant could then follow, perhaps designed to operate at rather lower pressures than its predecessors. In this way a field could possibly be run down over a century or more, during which it will have extracted far more useful energy that if the field had been over-exploited in the first place. Clearly the withdrawal of steam only, where a field is dry, is likely to run down a field less rapidly than the withdrawal of flashing hot water owing to the much higher enthalpy of the former and, where power generation is involved, its greater power potential.

The choice of an optimum wellhead pressure (see Section 8.8) is of great importance in the good management of a field, as is the location of reinjection bores where the bore water of a wet field is returned to the aquifer. The depth of reinjection bores, in relation to that of the production bores, will also have considerable influence on field life, as it is desirable that the moderately hot reinjected water should sweep as much as possible of the virgin aquifer water ahead of it, while recovering a maximum of the heat retained in the hot rocks formerly impregnated by the original hotter water. From time to time it could well be advisable to change the sites of reinjection bores (even if not forced to do so for reasons of chemical deposition) in order to avoid over-cooling part of a field while other parts of that field still remain much hotter. The 'model' of the field must constantly be kept under review, and the pattern of recharge water flows from beyond the confines of the field should be constantly studied, so that the approach of relatively cool waters may be 'headed off' by changing the points of reinjection as may be necessary.

Constant logging of well characteristics, of down-hole temperatures and pressures under static conditions; observations of geophysical and geochemical variations, of ground level changes and of the degree of activity of natural surface phenomena in the neighbourhood; and a continuous record of the quantities of fluids withdrawn and reinjected into the field; will all inform the enlightened earth scientist of what is happening in the aquifer below ground [242].

17.6 Geothermal energy prospects in countries lying outside the seismic belt

A reference to Fig. 5.1 will show that the chances of finding a hyperthermal field outside the seismic belt are virtually nil. Nevertheless it is always possible that either semithermal fields, or low-grade aquifers in non-thermal areas, may yet be discovered in almost any country. In Hungary semithermal fields are already being used to great advantage for farming and space-heating, and in the USSR it is claimed that vast semithermal fields exist within that country's boundaries. The occurrence of low-grade aquifers in non-thermal areas is probably even commoner than that of semithermal fields, particularly as they are less easily detected. No country should therefore despair of finding some useful natural heat within its borders, even if of low grade only.

The existence of *semi-thermal areas* in many countries far from the seismic belt is a known fact, and it seems probable that these areas betray, in many cases, the presence of hot rock at not excessive depths, but lacking a permeable aquifer. The granites of south-west and north-east England are cases in point, where fairly high temperature gradients have been observed. Many other countries have similar 'warm spots' beneath which hot rock is believed to exist within commercial range of the conventional drill. The development of these semithermal areas awaits only the successful mastery of rock-fracturing techniques, in which promising progress has already been made at Los Alamos in New Mexico and in Cornwall, England.

For the achievement of the ultimate goal, the recovery of very deep seated heat even in non-thermal areas, we must await not only the successful achievement of rock-fracturing but also new and cheaper methods of penetration to great depths – methods that will economically supersede the performance of conventional drilling by a large factor (see Chapters 7 and 18). There may, however, be an earlier and small-scale partial development towards this goal brought about by the use of deep oil or gas drillings that have either failed to make a strike or that have become obsolete through the depletion of the fuel deposits wich they once conducted to the surface. There must be many abortive or disused oil drillings to great depth in various parts of the world, capable of exploitation. The proposal has already been made to use the North Sea oil and gas drillings for geothermal purposes after the oil and gas have run out [92].

17.7 Geothermal developments in Europe

Until the 1970s Europe, apart from Italy, Iceland, Hungary and to some extent, the USSR, was generally regarded as more or less worthless geothermally, except by a few far-sighted people; but the highly successful French developments of low-grade aquifers in the Paris region and the rapid inflation of fuel costs encouraged the Commission of the European Communities to take an active interest in the geothermal resources of all member countries. An international committee of experts was twice convened in Milan in 1974, and their recommendations led to the First European Geothermal Research and Development Programme (1975–79), for which a budget of 13 million ECU (European Currency Units) was devoted. The programme was directed to the collection of data, the improvement of exploration methods, the encouragement of hot fluid recovery and the feasibility of hot dry rock exploitation. The results were considered sufficiently promising to justify a second programme (1979–83) with a budget of 18 million ECU. It is hoped that during the four years of this programme the installed power capacity of geothermal power plant in the EEC countries will be raised from 423 to 550 MWe, and of direct applications from 200 to 3000 MWt.

18 The future

18.1 The broadening horizon

Until 1973 it was unlikely that even the most optimistic of geothermal enthusiasts would ever have ventured to suggest that earth heat could one day supplant fossil fuels as the world's main source of energy. Even today, probably only a handful of people (including the author) believe in such a possibility. As recently as 1975 Worthington [247] expressed the opinion that geothermal energy was most unlikely ever to supply more than 10% of the energy needs of the United States – a country very richly endowed with natural geothermal resources.

The reason for the soft-pedalling of the potentialities of earth heat was that Nature appeared to have strictly rationed the supplies of this form of energy by making it available only in rather few places at low or moderate temperatures and in still fewer places at reasonably high temperatures. By far the greater part of the world's land surfaces appeared to be both non-thermal and devoid of permeable aquifers beneath. Although it was well known that gigantic quantities of heat lay beneath the outer crust at all points in the world, it seems that this rich store of energy was locked away in a safe deposit, beyond the reach of man. The fuel crisis of 1973 turned a few men's thoughts to the possibilities of safe-breaking! Attention suddenly became focused upon all unorthodox forms of energy, and the apparent absurdity of really large scale geothermal development was called into question.

Meanwhile, the Americans had been devoting much thought and research effort to the problem of whether the availability of earth heat was truly limited, as had hitherto been tacitly assumed, to geothermal fields and to low-grade aquifers in non-thermal areas. The fact that thermal *areas*, which through lack of permeability could not be regarded as *fields*, were known to exist in plenty led them to examine the possibilities of creating artificial fields in these areas. If natural fields are a rarity, then clearly our task must be to create artificial fields; and where better could they be created than in the hot

347

dry impermeable rock which is known to exist in many places at depths within commercial reach to the drill?

As soon as the oil crisis of 1973 occurred, the author started to consider an even more ambitious goal, namely the recovery of very deep seated heat from *non-thermal* areas, so that immense reserves of heat could be tapped and made universally available at any point of our choice on the surface of the globe. His initial thoughts on this possibility were somewhat crude, as he then lacked the necessary knowledge of rock mechanics. He was also ignorant at that time of the valuable work being done by the University of California at Los Alamos, and at the Sandia National Laboratories, both in New Mexico, towards finding the two principal missing keys to the 'safe deposit' of deep seated earth heat – rock fracturing and new penetration methods.

In the space of a few years the geothermal horizon suddenly became greatly widened from the mere search for natural thermal fields and low-grade aquifers to the creation of artificial fields in hot rock at moderate depth and even to the ultimate extraction of very deep seated heat at any point in the world. Obviously it is the hot rock at moderate depths that would be the first to be exploited when the necessary technology has been mastered; and the energy content of these is gigantic. In has been estimated that in the USA alone the crustal heat content in areas having thermal gradients exceeding $40°$ C/km is about 4×10^{20} Btu to a depth of 10 km – well over 1000 times the world consumption of primary energy in 1980 (disregarding grade). This new goal of heat recovery from hot dry rock may logically be described as HEAT MINING – a process that differs from the more conventional methods of geothermal exploitation in that natural subterranean fluids are not directly involved. Heat mining is at present in the experimental stage: as a commercial activity it belongs to the future – or so it is believed by an increasing number of people. For that reason, and because of the complexity of the subject, only the rudimentary principles of heat mining will be described in this chapter. For a more detailed treatment of the subject a separate new book is in the course of preparation*.

18.2 The search for more natural fields and low-grade aquifers

Although heat mining operations are now being conceived for the recovery of heat from hot dry rock and although the author is confident of ultimate success in this task, he is fully conscious of the fact that success will not be achieved overnight. Meanwhile, the energy situation is steadily deteriorating (even though there may be a short deceptive easement in the 1980s when the flow of North Sea oil is at its peak and while the world suffers from an economic recession), and we must not relax in our search for more of

Heat Mining by H. Christopher H. Armstead and Jefferson W. Tester, to be published by E. and F.N. Spon of London.

those natural geothermal resources that Nature has rather grudgingly provided. First we must thoroughly examine all available worldwide records that have been collected by the earth scientists, and at the same time must extend our inventory of relevant knowledge by undertaking in likely zones as much exploratory field work as we can afford, in order to locate promising areas where geothermal resources could perhaps be found. This is a task both for national governments and for the United Nations. After locating promising areas, we should then undertake systematic and detailed exploration of all of them as rapidly as funds, trained personnel and equipment permit. As previously emphasized, *time* is the essence of geothermal development. Doubtless many commercially exploitable resources will thus come to light quite rapidly, so that the curve of Fig. 0.1 will rise ever more steeply for at least two or three decades, and direct applications of earth heat will grow in number and variety, so enabling geothermal energy to take over a much more significant share of the world's energy burden.

The process of exploiting natural fields and low-grade aquifers, however vigorously pursued, could never provide the full answer to our energy problems, although it could well win for us a valuable time-gaining palliative to tide us over the comparatively few years during which we can discover and develop ways of fulfilling our more ambitious aims of winning earth heat in far greater quantities by entirely new methods.

18.3 Rock-fracturing

It has already been emphasized that the fundamental difference between a thermal *area* and a thermal *field* is one of permeability. A thermal area is caused by the presence of large masses of hot dry rock at differing depths below the surface. To convert such an area into a *field* it is necessary to create permeability where none exists, and this can only be done by shattering, or at least fracturing, the hot dry rock so as to create voids and/or fissures through which water may be circulated to pick up the heat content of the rock and bring it to the surface for exploitation. Even in non-thermal areas there is plenty of heat at greater depths, and the same principles may – at least in theory – be applied to them also.

Rocks are very poor thermal conductors, so if useful quantities of heat are to be collected at a useful rate it is necessary that the artificially created voids or fractures should have the following properties:

(i) The water/rock interface must have the greatest possible contact area.
(ii) The volume of voids and fissures must be large enough to ensure that the circulating water passes at low velocity over the hot rock surfaces so as to extract a maximum of heat during its passage.
(iii) The configuration of voids and fissures must be such as to offer a minimum resistance to fluid flow.

Ideally, the rock should be shattered into a rubble formation, which would meet all three of these *desiderata*. However, a mass of rubble, containing a high volumetric ratio of voids to solids, can only occur within a confined space of greater volume than that of the sum total of the rubble pieces – e.g. as with lumps of rock or coal contained in a hopper. How could such a condition be brought about in a solid mass of more or less homogeneous rock? It has been possible to do this ever since the advent of the nuclear bomb.

18.3.1 *Underground nuclear explosions.*   If a nuclear bomb is exploded at depth in competent rock, the resulting underground formation will take the form of a tall rubble-filled 'chimney' about six times as high as its diameter. The theory is that the intensely hot vaporized fission products expand within a split second of detonation to form a spheroidal cavity of such a size that the pressure within the cavity balances that of the overburden plus an excess pressure depending upon the forces of shear and friction within the overburden. The cavity is oblate, being wider horizontally than its vertical height, because the vertical lithostatic pressure exceeds the lateral rock pressure by a factor depending upon Poisson's ratio. Temperatures within the cavity momentarily rise to millions of degrees, and pressures to several millions of atmosphere. Within seconds of the explosion, and persisting for minutes or even hours, the roof of the cavity starts to collapse; and the process continues until the rubble so formed rises to such a height that the roof is sufficiently supported thereby to prevent further rock-falls. The force of the explosion has created an 'empty' volume of voids, made possible partly by slight surface deformation over a wide area and partly by the squeezing out of existence of other voids that had existed in the surrounding formation before the explosion.

Most of the radioactive fission products of long half-life would be trapped within a vitreous layer of molten rock which flows to the bottom of the chimney and there solidifies. Obviously the bomb would have to be placed at sufficient depth to ensure adequate cover above the top of the chimney, lest the explosion should break the surface.

Nuclear explosions are a relatively cheap form of 'potted' energy, particularly if a hydrogen bomb is used. (Conventional explosives would be prohibitively expensive.) Another great economic attraction of the nuclear explosion is that the price of the bomb rises only very slowly with the kiloton capacity, so that the use of a big bomb would be far more economical in terms of heat gained or of power potential won, than for a small bomb. For example, in 1966 a 10 kiloton H-bomb cost US $350000 whereas a 2000 kiloton bomb (200 times as powerful) cost less than US $600000 [248]. Doubtless these costs have been hopelessly overtaken by inflation since then, but the proportions could well be valid still. The trouble with large bombs, however, is the resulting shock waves that could damage buildings and other surface structures. In densely populated areas their use would be ruled out, but in deserts or very thinly populated areas they might be practicable within electrical transmission range of the nearest population centre. It would be safe to detonate a 45 kiloton bomb at a distance of 19 km, or a 125 kiloton bomb at 48 km from the nearest unexpendable structure. The method could thus be practicable in some of the deserts of the western USA and doubtless in other parts of the world where cities are contiguous with deserts. It has also been pointed out [249] that off-shore nuclear explosions could be safely let off beneath the continental shelf – e.g. in the North Sea – at a safe distance from the coast, so it is not inconceivable that we may one day have off-shore power stations cable-connected to the land.

The possibility of shattering hot rocks in this way, and of extracting their heat for power generation purposes was analysed by Professor George Kennedy of the University of California, Los Angeles, as long ago as 1964 [250], no less than 9 years before the oil crisis of 1973. He pointed out that the heat energy of the bomb would account for about 20% of the total recoverable heat, the remaining 80% being truly geothermal, and he made out a good economic case for the suggested project. The idea has never been followed up, probably because, as with all matters concerning nuclear energy, opposition is encountered both on political grounds and from fear that despite the alleged sealing off of the more dangerous radioactive products in the vitreous layer at the base of the chimney, there could be a risk of radioactive contamination. Nevertheless, it is understood that the idea is by no means 'dead', and that the US Atomic Energy Commission is still interested. The project may yet be revived.

18.3.2 *Hydro-fracturing.* Hydro-fracturing, or the underground splitting of rocks by means of hydraulic pressure, has been practised for many years in the petroleum industry as a means of increasing the permeability of a formation in order to enhance the yield of oil or gas from a well. The application of a high water pressure to the uncased section at the base of a bore, steel-cased to within a short distance from the bottom, will cause the formation of a more or less lenticular fracture in the rock. In homogeneous rock this fracture will tend to lie in a more or less vertical plane unless very high unrelieved tectonic stresses are present. Given sufficient pressure, such fractures may be of some hundreds of metres in diameter, though of only very few millimetres width at the centre and tapering to zero at the periphery: they will be positioned more or less symmetrically about the point at which the pressure is applied.

Lenticular cracks of this type bear little resemblance to the ideal 'labyrinth' that we are seeking. Nevertheless, it is believed that they would enable heat to be extracted from the rock at commercially acceptable cost. A great deal of experimentation has been performed at Los Alamos, New Mexico, USA, by the Los Alamos Scientific Laboratory of the University of California since the early 1970s. The method used there is illustrated in the simplest terms in Fig. 18.1.

A vertical hole is sunk to the required depth into a mass of hot dry rock, and a lenticular fracture is formed by applying a high hydraulic pressure at

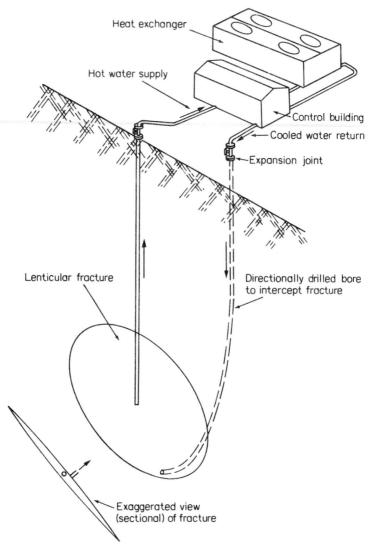

Fig. 18.1 Conceptual diagram of man-made lenticular fracture in hot dry rock with circulating system. (After Fig. 1 of [251]).

the *un*cased base of the bore which is cased almost to the bottom. A second hole is then drilled *directionally* from a short distance away, aimed to strike the lower part of the lenticular crack. Water is then circulated in the manner shown – if possible by thermo-syphon action, otherwise by pumping. It is necessary to pressurize the circulating fluid so as to avoid the formation of steam, to which the crack pattern would offer excessive flow resistance owing to the high specific volume of steam and its consequent high velocities. Moreover, water has a better thermal capacity for heat absorption than steam.

For the experimental plant at Los Alamos the hot water rising from below ground dissipates its heat to the atmosphere in heat exchangers, but for a working plant the heat would be usefully applied to some practical purpose. It is assumed that the cool water enters the bottom of the crack, is convected round it in a swirling motion, and ultimately returns to the surface up the original straight bore. It is possible that better results would be obtained if the lower part of the vertical pipe were blocked off and an opening cut in it nearer to the top: this is a matter for empirical testing.

After some preliminary experiments stable conditions had been established at Fenton Hill, Los Alamos, by early 1980, with a loop circuit like that illustrated in Fig. 18.1. The vertical bore extended to a depth of 2990 m, where a temperature of approximately 200°C was encountered. This implied a thermal gradient of about 62°C/km, or rather more than double the world average. The diameter of the fracture disc was believed to be not less than 600 m and the average width about 2 mm. Circulation was maintained at more than 16 kg/s for some months, during which time the flow impedance fell by a factor of more than 5. By March 1980 more than 7.5 MWt were being extracted continuously from the fissure. A reinjection pressure of 1540 psi was being maintained and a pressure of 150 psi was observed at the top of the vertical bore. Reinjection from the heat-exchangers was at 15 to 16°C while the upriser water was 158°C at the bottom and 131°C at the top. The temperature drop in the upriser pipe – 27°C – was then slowly falling as the surrounding rock was gradually warming up, but the down-hole temperature seemed to have stabilized. Water losses from the loop into the surrounding granite were less than 2% and the total dissolved solids in the circulated water had stabilized at about 2000 ppm.

Later in 1980 a small 60 kWe binary cycle power unit was activated by this same loop, using Freon 114. The ability of heat mining to extract heat from the earth's crust both for direct applications and for power generation has thus been proved, though the economic aspects have yet to be established.

A second loop was initiated in 1980 at Fenton Hill, in which the vertical bore was sunk to a depth of 4423 m and from which the extraction of about 50 MWt is hoped for. This heat will probably be put to practical use. The initial

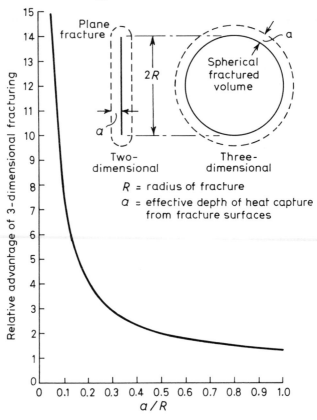

Figure 18.2 Relative advantage in heat capture of three-dimensional versus two-dimensional rock fracturing.

rock temperature will be about 290° C. Owing to the greater degree of rock compaction under the higher lithostatic pressure the water losses are expected to be substantially less than the 2% experienced with the first loop. The Fenton Hill experiments continue.

Professor R.W. Rex has suggested [252] that instead of continuous uni-directional water circulation there might be advantages in establishing a 'huff-puff' system of alternate injection of cold water and withdrawal of hot. This proposal has yet to be substantiated.

There are various secondary effects that could or would favour the Los Alamos system of extracting heat from a planar crack:

(i) *Thermal stresses* resulting from the contact of relatively cool water with the hotter rock would cause the walls of the crack to shrink. Owing to the fact that the horizontal compressive stresses at depth are less than the vertical lithostatic pressure, the effect of this shrinkage could be to create vertical grooves in the walls of the main fissure. This is because the transition from compression to tension would occur in vertical planes earlier than in horizontal planes. The effect of these vertical grooves should be to reduce the flow resistance to the circulating fluid. The fact that the flow resistance in the first experimental loop at Fenton Hill did in fact fall so greatly within a short time supports the belief that secondary thermal cracking is actually occurring.

(ii) *Viscosity/temperature effects.* It might be expected that flow from the inlet to the outlet bore would tend to be confined to the shortest path, with the result that this path would be rapidly cooled before any substantial quantities of heat had been removed from the more circuitous paths which, in consequence, would remain hot. However, the viscosity of water decreases rapidly at higher temperatures, so that flow resistance would tend to fall in the longer, hotter flow paths; so there would be a tendency for the flow to be shared fairly evenly between the infinite number of possible paths.

(iii) *Wedge action.* The thermal stresses mentioned under (i) above would be very marked along the lower part of the periphery of the original fracture, where the rock temperature is highest. Owing to the very narrow taper of the fracture, the effect of this would be like the tearing of a piece of cloth after making an incision with scissors to concentrate the stress at one point. It seems probable that fractures would tend to be self-propagating when water is circulated through them – especially downwards into hotter rock.

(iv) *Propping.* If, in the process of fracturing, some rock fragments should break off from the walls of the newly formed crack while the high fracturing pressure is applied, they would fall to a depth at which their size corresponds to the width of the crack. On reducing the pressure these fragments, though they would be partly crushed by the converging crack walls, would nevertheless prevent the complete closure of the crack, wedging it slightly open. It might even be practicable to inject gravel or hard steel pellets when applying the maximum fracturing pressure so as to sustain dilation and thus reduce the flow resistance to the circulating water. Where residual tectonic shear stresses are present in the rock, perfect re-closure after fracturing is never complete. This has a similar effect to propping.

(v) *Chemical leaching.* It is hoped that by circulating suitable solvents through the system, large volumes of solubles may be removed from the fracture, thus increasing the volume of contained fluid and effectively widening the flow passages to a small but useful extent.

18.3.3 *Three-dimensional fracturing.* Although these Los Alamos experiments have attained a fair measure of success, there is one obvious weakness in the concept of using a plane fracture for the purpose of extracting heat from hot rock: it is *two-dimensional*. At best, a plane fracture would only extract heat from a coin-shaped slab of rock, with decreasing efficiency as the width of the 'coin' increases. What is really needed is a *three-dimensional* method of heat extraction whereby a web of radiating fractures extends outwards in all directions from a central point so as to embrace a heat capture zone of approximately spherical shape. The relative gain in heat capture volume will depend upon the ratio of the extraction depth (i.e. the distance from a fracture at which heat extraction is reasonably efficient) to the radius of the fracture. Obviously the 'extraction depth' is not a precise concept, as the effectiveness of extraction will attenuate rapidly within a short distance of the fracture, but will theoretically never fall to zero. Neverthless for practical purposes the extraction depth will be a small fraction of the fracture radius. The relative gain from perfect three-dimensional heat extraction by comparison with two-dimensional extraction is shown in Fig. 18.2. If the heat capture depth were 1/20th of the fracture radius – e.g. 15 m in a fracture of 300 m radius – the relative gain from three-dimensional extraction would be almost 15 times.

At the Camborne School of Mines in Cornwall, England, experiments are in progress for the production of three-dimensional fractures in granite. Instead of using a steady hydraulic pressure to create rock fractures a conventional explosion is detonated within a small confined space at the base of a bore. Instantaneous pressures of 1 or 2 million psi are thus generated which cause multiple radial fractures to form in all directions (Fig. 18.3(a), because the inertia of the rock substance prevents it from moving sufficiently rapidly to provide an easy escape path through a single fracture for the high-pressure gases from the explosion. Hence, other fractures will also form during the split second while the high pressures persist. Multiple radial fractures will in fact form and continue to grow until sufficient volume

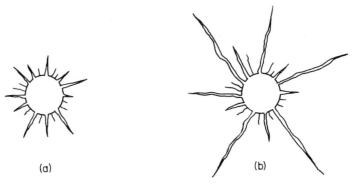

(a)   (b)

Figure 18.3 Formation of multiple radial fractures by means of explosions at the base of bores and their subsequent selective extension by hydro-fracturing.

has been created to reduce the gas pressure to a value below that required for further fissure propagation. An advantage of this procedure is that small fragments of rock are forced violently into the radiating cracks at the moment of their formation, thus providing useful propping against subsequent collapse. After this initial 'star' formation has been created, the hydro-fracture technique may then be applied, and it is found that instead of a single crack being formed in one plane, as with straight hydro-fracturing, the applied pressure will extend *several* of the weakest cracks to considerable distances (Fig. 18.3(b)). Moreover it is hoped that the radiating 'star' fractures will communicate with relatively weakly sealed natural rock joints and open them up so as to establish a complex system of fissures that could form a 'reservoir'. Not only would three-dimensional crack systems offer hydraulic access to far larger effective masses of hot rock: they would also provide good chances of fissures intersecting one another, thus forming the basis of a true labyrinth with multiple flow paths between adjacent bores. A group of bores in hexagonal formation could offer an alternating arrangement of downward and upward fluid flows with wide dispersal through the labyrinth fractures. Such a group could easily be extended outwards, preserving the hexagonal pattern and alternating flow directions, with improved ratio of heat capture volume to the number of bores and fracture systems.

Another approach to the three-dimensional ideal is to create a series of parallel lenticular cracks spaced fairly close to one another. If the spacing does not exceed twice the 'extraction depth', as loosely defined above, then such a pattern of parallel cracks would effectively be equivalent to a three-dimensional cylindrical heat-capture zone of hot rock. The formation of parallel cracks by hydro-fracturing should not be difficult, for no natural mass of hot rock is entirely free from tectonic stresses, whether active or passive. The effect of a horizontal tectonic stress over a wide area is to impart the equivalent of a 'grain' to the rock, so that if a group of fairly closely spaced bores were all subjected to adequate hydraulic pressures it is very probable that the fissures so formed would lie parallel to one another. Secondary thermal stress cracks radiating outwards from each primary fissure would tend to extend the 'extraction depths' towards one another, and might even establish flow communication from one primary crack to the next if two or more secondary fissures were to intersect.

Although steadily applied hydraulic pressures and the use of explosives have here been mentioned as means of generating three-dimensional fractures in hot dry rock, there are other methods that could perhaps attain the same ends. Proposals for dynamic impact and for induced water-hammer were advanced by the author and described in the first edition of this book, but were received with scepticism. Others have mooted different ideas. One proposal is to set up a hydraulic pressure resonance within a bore, whereby very high pressures could momentarily be attained at the point where fracturing is to be initiated: another is to subject the rock to very high frequency pressure vibrations that would ultimately induce fatigue in the rock material and cause it to fracture. These ideas, however, are at present only in the stage of theoretical concepts. Yet another thought would be to inject liquid air into the base of a bore where the rock is hot, thereby inducing sudden shrinkage and local cracking; the cracks being subsequently extended by means of hydraulic pressure.

An advantage of heat mining by extracting heat from hot dry rock is that exploration costs are relatively cheap. All that has to be established is the nature of the lithography and the existence of a good thermal gradient: the discovery of a permeable aquifer is not needed. Thus the success ratio of drilling should be 100% once the presence of a semithermal field has been found.

The theory and practice of rock-fracturing form far too wide a subject to be adequately covered in a single subsection of a single chapter; but it is hoped that the broad principles have here been adequately explained. The art is still in its infancy; but there are already good grounds for believing that well before the end of the 20th century heat mining will be practised on a commercial scale. If so, it seems likely that earth heat will soon be contributing an impressive share of the world's energy needs; for it is believed that far larger quantities of useful heat are contained in hot dry rock within commercial range of the conventional drill than in all the hydrothermal fields and low-grade aquifers in the world.

18.4 Improved penetration methods

The economic limitations of conventional rotary drilling have been stressed in Sections 7.15 and 14.10, while novel methods of penetration have been broadly mentioned in Section 7.17. Percussion drilling has certain marked advantages over rotary drilling in dry rocks, principally because the bit life can be extended to twelve or more times as long; but unfortunately the method cannot be used in wet holes nor at great depths.

Owing to the rapidity with which rotary drilling costs rise with increasing depth it follows that with any given temperature gradient there will always be a depth at which the ratio of heat potential to drilling costs will be a maximum. This depth will not necessarily be the economic optimum depth, because the cost of the heat won would also depend on the rate of extraction, and this would be determined by the thermal capacity and conductivity of the rock and on the degree of fracturization. It would also depend on the rate of interest on the investment. The temperature of the heated water could be adjusted, up to a maximum somewhat short of the rock temperature, by varying the rate of fluid circulation; so that we would have a choice between extracting a large quantity of low-grade heat or a smaller quantity of higher-grade heat: the choice would depend upon the intended use.

Clearly any gain in our ability to penetrate to greater depths would bring rapidly increasing rewards – not only in the volume of rock to which access would thereby be gained but also in the grade of the recoverable heat per cubic metre of rock. The various improved methods of drilling described in Section 7.17 will all tend to benefit the economic prospects of heat mining by rock fracturing; but the dream of *universal* heat mining (see Section 18.6) can only be realized by some completely novel method of penetration with a totally different cost/depth relationship from that of conventional rotary drilling or of its refinements.

In the mid-1970s great hopes were placed on the 'subterrene' – a melt-penetration device [253, 254]. By applying intense heat to a small local area, the rock may be fused so that the hot penetrator can advance downwards, melting and displacing the fused rock outwards in the process. Most igneous rocks melt at temperatures of about 1500 K – approximately the same as steel; so the melting penetrators must be made of refractory metals such as molybdenum or tungsten which melt at 2880 and 3650 K respectively, and which also have low creep rates at the rock-melting temperatures. The fused rock forms a dense, vitreous lining to the hole, intimately bonded with the surrounding rock (Fig. 18.4) having very great compressive strength capable of supporting the bore walls against collapse. Thus, no steel casings are required, with their attendant problems of jointing and grouting. The energy per metre of penetration in most igneous rocks, is substantially more than that required for conventional rotary drilling, but at great depths this could be more than offset by the savings in casings and cementation and, more important, by the very great time savings in the avoidance of bit

Figure 18.4 Vitreous bore lining formed by melt-drilling intimately bonded with the surrounding rock.

renewal, restringing and in correcting circulation losses. Moreover, as the penetrator enters hotter ground with increasing depth, less additional heat is needed per metre of penetration, so that the energy differential by comparison with conventional drilling improves. The result would be that in place of the quasi-exponential cost/depth relationship shown in Fig. 14.3 the average cost per metre of penetration should actually *decline* at great depths. Another feature of melt-drilling is that the penetration energy varies only slightly between one type of rock and another, so that the relative advantages of the method are far greater with very hard rocks such as granites and other igneous rocks likely to be found at great depths. The problem of increasing bit wear with rising temperature and increasing depth would be entirely eliminated; no mud or mud cooling system would be necessary; and directional drilling could be easily effected by controlling the temperature circumferentially around the penetrator. In that it does not rotate, the penetrator would be free from high torque and shock loads in deep wells. No large ground surface area would be required for all the stores, mud-cooling and other ancillary equipment needed for rotary drilling. Finally, vitreous encased continuous core can be extruded from the entire length of the hole. Hot fluids from any aquifer penetrated by melt-drilling could be captured by perforating the vitreous walls of the bore by means of explosive charges or other means of locally opening up the glass-like casing, but the method is more likely to be of real value in terrains where no natural aquifers exist, and where the object is to reach very deep-seated hot dry rock.

Hitherto, electrically heated penetrators only have been used, but the ultimate intention was to use nuclear-heated heads which would be expendable, advancing at an increasing rate through the crust and finally entering the mantle and disappearing within the cores of the earth. In 1971 it was prophesied [253] that within 10 or 15 years the nuclear subterrene would be developed so as to be capable of producing holes of several metres diameter and several tens of kilometres in length or depth – either for purposes of tunnelling or of geothermal heat recovery.

In 1976 these great hopes were suddenly shattered by the unexpected abandonment of the American subterrene experimental programme at Los Alamos. The reasons for this unhappy decision are not known, but evidently the early and apparently successful experiments that gave ground for the optimism expressed by several eminent American scientists [253] were not supported by the results of later and more ambitious experiments. Had they been, the abandonment of the subterrene programme would be hard to explain. It is conceivable that the forces of lithostatic compression at great depth became unmanageably large, or that heat was conducted too rapidly away from the point of rock fusion for the melting action to be sustained with sufficient intensity, or that cooling facilities for projecting the main body of the subterrene were incapable of withstanding the intense heat generated only a short distance away. There are several possible explanations, but they are not known to the author. Whatever the reasons, they must have been well founded. As will be shown in Section 18.6, the real break-through in geothermal development that would enable it ultimately to compete with, and even surpass, fossil fuels as a principal (if not *the* principal) source of world energy, must largely depend upon the satisfactory development of a relatively cheap method of very deep penetration. Hence the abandonment of the subterrene programme can only be described as a disastrous setback. It seems probable that sheer necessity will sooner or later compel the readoption of the subterrene experiments with concentration upon those factors that led to their temporary abandonment in 1976; and it is to be hoped that if the financial burden is too great to be borne by the USA every effort will be made for the enterprise to be supported internationally by a concerted effort, since it would scarcely be an exaggeration to say that the future prosperity of the human race is at stake. Meanwhile, reliance must be placed on the several improvements to conventional rotary drilling that have been mentioned in Section 7.17.

18.5 Direct tapping of magma

The direct tapping of earth heat from active volcanoes has, in recent years, been the subject of much discussion, but not very much seems yet to have been achieved in the way of tangible results. Between 300 and 400 active volcanoes are known to exist in the world; and since many inactive ones may merely be dormant, the total number of volcanoes that *could* at some time become active may perhaps be measured in thousands. Volcanoes have the advantage over hydrothermal fields in that their existence and location are self-evident, so that they do not have to be 'discovered', even though their hidden underground structure may have to be divined by exploration methods. A joint USA – Japanese seminar was held in Hawaii in February 1974, at which the possibilities of direct volcanic exploitation were discussed. It has also been reported that the Russians plan to develop about 300 MW of power by boring nearly 10000 ft – say 3 km – into the heart of the Avachinskaya Spoka volcano in Kamchatka. At the base of this volcano is a mass of hot magma at 1000–1200°C, at a depth of 3 to 5 km, and having a diameter of about 6 km. The volume of this magma has been estimated at about 100 km^3 [38]. The heat content of this magmatic mass would be something of the order of 1% of the world's total measured fossil fuel reserves, which represents a gigantic concentration of energy within the comparatively small confines of a single volcano.

Apart from volcanoes there are known to exist many pockets of magma lying at various depths within the more tectonically active parts of the earth's crust – both near rifts and subduction zones. These pockets vary in temperature from about 600 to 1300°C and may be found at depths of 4 or 5 km or less, and of volumes measuring up to several tens of km³. They represent immense concentrations of heat of a grade far exceeding that of anything else that lies in the earth's crust within commercial range of the conventional drill. Many such pockets almost certainly exist in the western USA, Hawaii, Kamchatka, Iceland, the Azores, Sicily and elsewhere. In Hawaii and Iceland they lie close to the surface, and even break the surface from time to time to form new volcanoes. It is the larger magma pockets that have the best chance of surviving until they have travelled all the way up from the mantle to the surface: many of the smaller ones cool off during their upward journey, but in their death throes they may serve to heat an overlying aquifer to form a hydrothermal field, or provide a semithermal area with a high thermal gradient and a large mass of exploitable hot dry rock.

How could very hot magma pockets be usefully tapped? This is a question that is still under active consideration. The difficulties of penetrating the magma pockets by means of rotary drilling, even at shallow depth, are formidable. In the first place the very high temperatures encountered as the pockets are approached could be too much for conventional bits to contend with: secondly a temperature may ultimately be reached at which there is danger of the drill stem welding itself into the magma and seizing itself up. Assuming for the present that some means will be found for penetrating the rock right up to the edge of the magma pockets, exploitation could be effected in two ways. A refractory heat-exchanger could be forced into the mass of molten viscous magma. A chilled crust of solidified magma would form on the outer surface of this exchanger while a circulated fluid would remove the heat transmitted to the inner surfaces. Thermal equilibrium would be established when the heat conducted through the crust balances the convective heat flow in the surrounding molten magma. Another method would be to inject water into hot basaltic magma to produce hydrogen by dissociation. If organic matter were added to the injected water, it would be possible to produce carbon monoxide and methane also.

18.6 The ultimate goal: universal heat mining

The ultimate geothermal goal is, of course, to win earth heat wherever we wish – even in non-thermal areas. The temperature distribution with depth in these areas would not simply depend upon the thermal gradient at the surface, but also on the local distribution of radioactive matter throughout the thickness of the crust, the actual thickness of the crust, and the local composition and distribution of the rocks forming the crust. However, it seems likely that in non-thermal areas it would be necessary to penetrate about 2.5 to 3.5 km in order to encounter low-grade heat that could be used for district heating or farming; about 6.5 to 9 km for power generation at thermal efficiencies of 15 to 20%; and much deeper if power is to be generated at efficiencies anywhere approaching those attainable with modern conventional thermal power plants. Even now we can penetrate to depths of about 15 km with conventional rotary drilling, but at costs that would be quite unacceptable in terms of the temperatures that could be reached in non-thermal areas. For the art of universal heat mining to become commercially possible, three prerequisites must be satisfied:

(i) The successful development of the subterrene or of some other relatively cheap method of very deep penetration (see Section 18.4).
(ii) The successful mastery of rock-fracturing techniques – preferably three-dimensional (see Section 18.3.2).
(iii) The maintenance of excess pressure to dilate the very deep artificial cracks so as to permit the circulation of water through them with acceptable power expenditure.

The first two of these prerequisites have already been discussed at some length. The third is relatively simple. As is well known, fractured rocks and rubble can exist in a state of stability to moderate depths with a fairly high proportion of voids or permeability. As the depth increases, the rising lithostatic pressure tends to squeeze such voids out of existence, thus at first reducing the percentage of voids and ultimately – at perhaps 3 or 4 km or thereabouts and at greater depths – rendering the rock totally impermeable. It is this that accounts for the presence of bedrock beneath aquifers in hydrothermal fields, as shown in Fig. 5.2 (Section 5.7). Hence, even after we have mastered the art of rock-fracturing to the required pattern, the man-made fractures would instantly reclose on the removal of the fracturing pressure. To recover earth heat at very great depths it will therefore be necessary not only to fracture the rock but also to keep the fractures permanently open by means of a dilating pressure so that water circulation may take place with a minimum of resistence to flow. Fortunately simple hydrostatic pressure would provide a very large share of the requisite dilation pressure, and the balance (to be applied by pumping) would be less than needed to fracture the rock in the first place. At Los Alamos a dilating pressure of 88 atü was applied for the first loop experiment.

It may be asked 'If voids are squeezed out of existence at great depths, how can a bore be sunk to these depths without itself collapsing?' This is because even a plain uncased vertical cylindrical hole could survive to far greater depths than a flat or amorphous fissure, owing to the 'arch' effect of the hole walls which are in compression. If the hole were supported from within by hydrostatic pressure and also strengthened with a tough vitreous lining, it could doubtless resist collapse to immense depths. Even if water could not be admitted during the actual penetration process, high pressure inert gases or even steam could be admitted to give internal support.

19 *Epilogue: The frame of the geothermal picture*

19.1 Rationale

The broad theme of this book is that the development of geothermal energy is desirable and that it should be encouraged to expand as rapidly as possible. That is the picture, and the book has to some extent analysed the composition and structure of that picture. Still, what of the frame into which the picture is to be set? Surely that frame must be the overall world energy problem, a brief study of which will provide a suitable setting for the geothermal picture. This final chapter, in fact, seeks to show *why* there is a pressing need to develop the use of earth heat with a sense of great urgency. Geothermal development is not being advocated just for the sake of growth, as such, or because it is an interesting and novel idea: it is being urged as a way of trying to avert a human disaster of terrifying proportions.

19.2 Energy hunger

About 175 years ago Malthus foretold that the human race would founder on the twin rocks of population growth and food scarcity. Basically, he was one of the first men to foresee the obvious consequences of living on a finite planet. He prophesied that widespread war, famine, pestilence, misery and vice would inevitably be the lot of man before the end of the nineteenth century. The fact that human prosperity had broadly improved, rather than diminished, by the early twentieth century caused Malthus to be largely discredited as an alarmist prophet. Nevertheless, there was little wrong with his general thesis: it was only his time scale that proved to be unduly pessimistic. He should not be blamed for failing to foresee, in the early days of the Regency, the astonishing effects of mechanical transport and colonialization which so immensely increased the areas of the world from which food and raw materials could be drawn in the nineteenth century. In short, the arguments of Malthus were sound, but the early universal disasters which he foretold were mercifully postponed by the expected discovery of vast new resources.

364

Ever since the dawn of the Industrial Revolution, man's appetite for *energy* has grown far more rapidly than his appetite for food, and there is no lack of neo-Malthusians who foretell the early coming of a world energy famine. The gloomy prognostications of these prophets of doom are understandable when one pauses to note the following facts:

(i) From 1900 to 1950 the world's annual energy demands rose by a factor of about 3.6, equivalent to an average annual growth rate of rather more than 2.5%. From 1950 to 1960, in only one-fifth of that time, they rose at an average annual growth rate of about 4%; and from 1960 to 1970 by about 5.5% p.a. (compound). These growth rates are hyper-exponential, and give cause for alarm.

(ii) As late as 1980 (the latest year for which published world statistics were available when this book went to press) 97% of the world's energy needs were supplied by fossil fuels. The remaining 3% came mainly from hydro-power, a small amount from nuclear fission and a minuscule quantity from geothermal energy. We are still almost wholly dependent upon fossil fuels for our energy.

(iii) The World Energy Conference Survey of Energy Resources, 1980, quoted a figure of *proved recoverable reserves* of fossil fuels (excluding nuclear resources) equivalent to 131 years supply at the world consumption level at 1972 (the last year before the pattern of world energy growth became disturbed by the Middle East oil crisis and the sympathetic price rises of other fossil fuels that followed).

(iv) Now 131 years may sound reassuring to those who pursue the philosophies of 'sufficient unto the day is the evil thereof', but at best that is only about five generations from 1973, and could carry us only into the early part of the 22nd century: in practice our reprieve would be far less, as will now be shown.

(v) If *growth* were to continue at anything approaching the rates to which we have become accustomed, the 131 years would shrink in a spectacular fashion, as Table 19.1 will show.

Table 19.1 Life of (1980) proved recoverable reserves of energy for various growth rates.

Growth rate p.a. (%)	Life of proved recoverable reserves (as from 1972; in years)	
nil	131	
1	84	
2	65	to the nearest
3	54	whole number
4	47	
5	41	
6	37	

These figures show that if the growth rate of the 1960s – 5.5% p.a. – had persisted after 1972 the (1980) proved recoverable reserves of fossil fuels would have become totally exhausted after 40 years – i.e. by the year 2012. However, this presupposes that no further recoverable reserves of fossil fuels are proved in the meanwhile – a needlessly pessimistic assumption. Nevertheless, it is clear that time could be running out unless growth rates are greatly curbed or unless we become less reliant upon fossil fuels.

(vi)   This is not all. As stocks become depleted, prices will soar and wars could perhaps break out in a free-for-all scramble for what remains. There could well be chaos before the end of this century if we are to continue to rely upon fossil fuels as our main sources of energy.

(vii)   Although 'proved recoverable reserves' are probably a very cautious figure which grossly under-estimates the ultimately available fossil fuel resources, the benefits to be gained from higher figures are largely offset by the evils of continued growth unless the growth be very small, zero, or even negative, as can be seen from Table 19.2.

(viii)   Yet another argument used, with good reason, by the pessimists, is that not all fossil fuels will be burnt: much will be needed as feed stock for industry.

(ix)   In certain countries the coal miners seem to be pricing themselves out of the market. Certainly the winning of coal has failed to keep pace with the rising energy demand, and in some countries the rate of coal production has actually fallen.

(x)   The continued growth of fossil fuel combustion may so pollute the earth to spoil the environment, menace public health, and perhaps even make the world uninhabitable.

Table 19.2 Life of possible reserves of energy for various growth rates.

Annual growth rate (compound) p.a. (%)	Life in years reckoned from 1972, assuming that ultimate fossil fuel resources are:		
	1980 proved recoverable reserves	10 times as much	100 times as much
nil	131	1310	13 100
1	84	266	491
2	65	167	281
3	54	125	202
4	47	101	160
5	41	86	133
6	37	75	114

(xi) Whereas for several decades we were discovering new deposits of fossil fuels more rapidly than we were consuming them, the reverse is now true: we are consuming them more rapidly than we are discovering new sources.

These eleven points broadly reflect the arguments of the pessimists – or as some would call them, the scaremongers. Even if they do not emulate the precision of Bishop Ussher's extrapolations into the past, the pessimists make a rough stab at the early part of the next century as the likely time when doom will catch up with us.

19.3 The need for timely action

Now what have the optimists to say? How do they propose to act? Their first duty is to recognize that the arguments of the pessimists cannot be lightly dismissed, for they contain plenty of sound sense. Disaster will not be averted by Micawberism: we must *act,* and act vigorously and urgently; otherwise the pessimists may well prove to be right. Complacency is a very common human failing. Too often have we drifted into devastating wars through failure to grasp a nettle in time. Too often do we read in the press letters and reports of speeches – sometimes from eminent men of science who should know better – giving assurance that there is plenty of energy available and that it always will be so, and accusing those who say otherwise of being scaremongers.

Two main lessons are to be learned from the points enumerated in Section 19.2:

(a) We just cannot afford to sustain growth at anything approaching the levels to which we had become accustomed in the 1960s. There is no way of avoiding, sooner or later (and it had better be sooner) a reduction in growth rates. This reduction must be judiciously applied. Too drastic reduction could impose intolerable social stresses: too slow reduction could fail to avert the disasters foretold by the pessimists. Depending upon the outcome of (b) which follows, we may perhaps have to reduce the growth rate to zero within the foreseeable future, or even reverse it to negative values. Certainly anything approaching *exponential* growth, even at a very low rate, cannot for long be tolerated; for such growth is a divergent series that expands at an ever-increasing rate towards infinity, whereas we live on a planet with large, but *finite* resources.

(b) We cannot for much longer rely almost entirely upon fossil fuels for our main sources of energy. Herein lies our best hope of mitigating the horrors that would otherwise follow from (a) if tackled too rapidly; for at long last we are becoming aware of the importance of 'non-fuel' energy sources.

It cannot be over-emphasized that the principal enemy is not so much energy *demand* as energy *growth,* as can be seen from Tables 19.1 and 19.2. The author makes no apology for repeating the well-known legend of the Chinese traveller who, some centuries ago, introduced to the Moghul Emperor in Delhi the game of chess. The Emperor was so delighted with the game that he invited the Chinese to name his own reward. The Chinese said that he would gratefully accept one grain of rice for the first square of the chessboard, two for the second, four for the third, and so on – doubling each time. The Emperor thought him mad to demand such a trifling reward, until he discovered that the whole world did not contain enough rice to satisfy the request. The effects of exponential growth – and even more, of *hyper-*exponential growth such as has occurred in the first 70 years of this century – are not unlike those grains of rice.

Politicians and economists are constantly preaching the need for growth as a necessary adjunct to prosperity. This may well be true for developing nations who have not yet attained a high standard of living, but for affluent countries it could prove disastrous in the long term, however attractive its short term benefits may appear; for it amounts to the mortgaging of a comfortable present against a painful future. The effects of demand growth on the life of natural resources can be both surprising and alarming [255].

It is a curious fact that until 1972 no difficulties, other than temporary and local ones, were experienced in matching energy supplies to the rapidly growing demand. A belief in quasi-exponential energy growth had become almost an Article of Faith. Prognostications of electricity demands (which incidentally had been growing more rapidly than total energy demands), based on simple exponential growth modified by a small 'regressive factor', proved to be astonishingly accurate. By 1972, thanks to the general availability of abundant and cheap energy, the human way of life had so accustomed itself to hyper-exponential growth that dislocation, or even disaster, seemed to threaten us if our voracious energy appetite could not be fully satisfied. Then came the events of late 1973, which shattered our complacency and showed that dislocation could indeed occur almost overnight at the crack of the Arab whip. The days of cheap energy had apparently gone forever. Astronomical rises in the price of oil were soon followed by sympathetic rises in the prices of other energy sources. National balances of payment were thrown into disorder and many countries were beset by serious economic troubles. It is perhaps still too early to gain a proper perspective of the results of the 1973 fuel crisis (it takes about 3 years for world energy statistics to be compiled and a few more years before short term 'wobbles' in the demand pattern can become smoothed out). Nevertheless, it is clear that this demand pattern was undoubtedly disturbed by the events of 1973 (Fig. 19.2) although it seems equally clear that the broad energy demand is still upward despite two brief falls. A decline in oil consumption has been more than offset by increased uses of other energy forms.

If our principal enemy is growth, it is equally clear that our potential allies are new forms of energy. To stave off disaster we must weaken the enemy (growth) and rally our potential allies (new energy resources). Those are our two urgent tasks that will brook no delay.

19.4 Curtailment of growth

It is easy to talk of growth curtailment, but far from easy to bring it about. There is of course great scope for effecting economies: enormous quantities of energy are at present being needlessly wasted. Profligacy is a luxury no longer to be afforded. Apart from the avoidance of waste, which need hurt no one, it must be recognized that curtailment of energy growth could be a very painful process; for man is an industrial animal, and industrial growth has been achieved only at the cost of energy growth. A too sudden reduction in energy growth rate could well have serious social consequences – even to the extent of political revolutions (which of course would not solve the dilemma but are apt to serve as an emotional vent to people subjected to unaccustomed restraints). Democracy, as we know it, could well be an early casualty of too rapid growth reduction. The brakes should be applied as gently as we can afford, so that people can become accustomed to the necessary changes to their habits that must be an inevitable corollary to a declining energy growth rate. In the last quarter century, what was formerly regarded as a luxury has come to be regarded as an essential, or even as a right. That attitude must gradually be reversed, as has been successfully done in times of war, and a more proper distinction must be restored as between necessities and luxuries. It is not inconceivable that if this process is not too rapidly applied it could somewhat paradoxically improve the quality of life; for there is little to show that the peoples of those countries that enjoy a high material standard of life are any happier than their forefathers whose demands were far simpler.

Another theoretical partial solution to the energy problem lies in reducing, halting, or even reversing population growth. However, to effect this painlessly would take several generations of intense propaganda. Deeply ingrained religious and social attitudes would have to be surmounted, and the process would inevitably be slow. We lack the time to rely on this solution alone, although every effort in this direction could be helpful.

19.5 Development of new energy sources and alternatives to fossil fuels

Our greatest source of hope of averting disaster is that just as Malthus fell down in his prophesies of early impending chaos by failing to take into account new developments which, to do him justice, he could not have been expected to foresee, so may there well be new developments in the near future *which can already be discerned* (so we lack the excuse of being unable to foresee them) that will drive away the spectre of an energy famine. It is clear that however successful we may be in curtailing growth, there can be no *permanent* solution in that direction. The only ultimate key to the energy problem lies in the development of new energy sources on a scale hitherto unimagined. It is therefore useful and instructive to take stock of the options apparently available to us.

19.5.1 *Hydro-power.*   Of all conceivable alternative energy sources to fossil fuels, the first that naturally comes to mind is hydro-power, which is clean and renewable and often brings secondary benefits like irrigation and flood control. According to the World Energy Conference Survey of Energy Resources 1980 only about one-sixth of the total world hydro-potential had been developed by that year; and according to the UN Energy Statistics hydro-power accounted for 2.3% of the total world primary energy production in 1980. It might therefore be thought that the (five-sixths) undeveloped hydro-resources could add to the world's energy production an amount equal to 5×2.3, or 11.5% of the 1980 total energy production. This, however, would be misleading, because although the UN treat hydro-power as primary energy and equate it to the heat equivalent of the electricity generated it is in fact *secondary* energy. The undeveloped hydro-potential could in fact provide a quantity of electricity that would otherwise have to be generated thermally at, say about 30% average efficiency. Thus the true value of the undeveloped hydro-potential should more logically be assessed at something like 11.3/0.3, or about 38% of the total world primary energy produced in 1980. Clearly hydro-power could make an immensely valuable contribution to the world's energy needs, and for that reason should be developed as rapidly as possible. Nevertheless, even this very substantial resource could not be regarded as the ultimate solution to the world's energy problem. It could be a time-winning palliative of immense value, but if the rate of world energy consumption continues to grow – and it seems likely to for some time at least – 38% of the 1980 world energy production would account for a steadily shrinking percentage of future annual production figures. Moreover, since hydro-power is anchored to location, and much of it is available at very remote and sparsely populated places, the economic cost of development could be prohibitive for quite a large proportion of the total undeveloped hydro-resources – particularly as hydro-power can be very capital-intensive. Much could doubtless be done by siting certain energy-intensive industries close to the larger remote hydro-sites, but there is a limit to this.

19.5.2 *Other renewable energy sources.*   Wave, wind, solar and tidal energy are now receiving much deserved attention and may yet contribute quite large shares of our total energy needs. However, it is doubtful whether, even collectively, they will ever be able to offer more than a very useful palliative. Moreover, with the exception of tidal power, they are likely to involve the use of obstructive, ungainly and unsightly structures that would be aesthetically offensive.

19.5.3 *Nuclear fission.*   Here of course we have a very valuable potential source of energy which, with the use of breeder reactors and of thorium, could perhaps treble our energy reserves. This could give us an immensely valuable breathing space. However, from Table 19.2 it will be seen that even a trebling of resources would not gain us a great deal of time unless we first succeed in reducing the growth rate to a very low annual percentage. Moreover, there is a very powerful anti-nuclear lobby, and also a certain reluctance in high places to make decisions. These together are raising serious doubts as to the extent to which this energy source will ultimately be used. That there are possible environmental, and even terrorist, hazards cannot be denied, but these are perhaps not much greater than those associated with off-shore oil rigs: in any case, little that is worthwhile has ever been accomplished without some element of calculated risk.

19.5.4 *Nuclear fusion.*   Controlled thermo-nuclear fusion would indeed open the floodgates of available energy, so that we could afford to relax somewhat in our otherwise vitally urgent task of reducing the energy growth rate. So far the mastery of this vast energy source has eluded us, and although it is devoutly to be hoped that our efforts will ultimately be crowned with success, we cannot afford to place all our eggs in one basket.

19.5.5 *Geothermal energy.*   This really is our last remaining option. Great strides have been made in recent years, both in exploration and in exploitation, while valuable research work is in progress that will almost certainly extend enormously the accessibility to earth heat. Herein would seem to lie our best hopes.

19.6 Mutual inter-dependence of growth curtailment and the development of new energy sources

The more rapidly we can develop new energy sources, the less urgent becomes the need to reduce the energy growth rate, and *vice versa*. The abundant availability of new energy sources within the near future would greatly help to absorb the shock of reducing the growth rate; while conversely, the more rapidly we can reduce the growth rate, the more time will be available within which we must develop new energy sources. Both remedies should be tackled simultaneously, and success in either will reduce the urgency of the other. What we cannot possibly afford is to fail in both endeavours; for that could lead only to inescapable disaster – the end of Industrial Man and perhaps a reversion to barbarism, if not to the total extinction of Mankind.

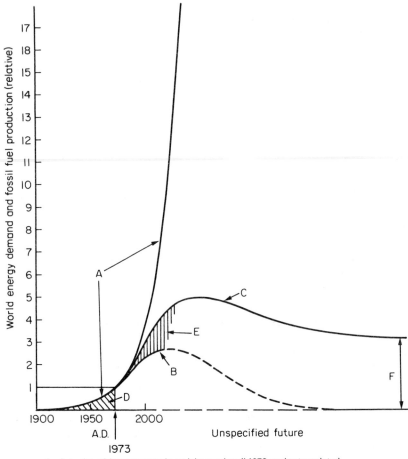

A, Actual world energy supply and demand until 1973, and extrapolated future demand assuming 5% annual exponential growth.
B, Probable *shape* (not to scale) of future fossil fuel production.
C, Probable *shape* (not to scale) of future world energy demand if disaster is to be avoided.
D, Total world energy consumption from 1900 to 1973.
E, Energy deficit between world energy demand and fossil fuel availability, to be met from new energy sources and economies.
F, Ultimate stability when world energy demand can be met from renewable sources alone.

Figure 19.1 Probable pattern of future world energy supply and demand (see [256]).
Note: The vertical scale is in multiples of the world energy demand in 1973 – the last year to be scarcely affected by the Middle East oil crisis.

19.7 World energy: the probable shape of things to come

Figure 19.1 illustrates the sort of pattern we may expect with world energy development. Except for curve A, which is precise, it is purely qualitative and makes no claim to accuracy as to the future time scale: it is the *shapes* of curves B and C that are of significance.

If we take as the unit of the vertical scale the world energy consumption (8130 million tons coal equivalent) in 1973 – the last year before the Middle East oil crisis really started to take effect – curve A represents the *actual* annual world energy consumption from the beginning of the century until 1973, and the *extrapolated* world energy consumption thereafter at an assumed exponential growth rate of 5% p.a. This figure of 5% is arbitrary, and has been chosen simply by way of illustration. It has already been pointed out that the mean annual growth rate in the 1960s was 5.5%. It is true that from 1970 to 1973 there was a slight recession to rather less than 5% p.a. growth, but this could have been merely due to inevitable short term fluctuations lacking long term significance. The chosen figure of 5% p.a. is slightly less than the average growth rate from 1960 to 1973, and has been taken in order to show the utter impossibility of maintaining for long growth rates anywhere approaching those to which we had become accustomed by 1973.

Curve B shows the actual growth in the supply of fossil fuels from 1900 to 1973. It lies so close to curve A that it is almost indistinguishable from it, as these fuels supplied nearly all the world's energy needs during the 73 years under consideration. The effect of the huge increases in fossil fuel prices since 1973 are not likely to cause a sustained *fall* in fuel production for some time owing to the pressures of still rising demands, but they will almost certainly lead to a falling off of the growth rate of production. Moreover, it is the avowed policy of some of the oil-producing countries to keep more of their reserves in the ground as an insurance against future contingencies. Against these retarding factors, new oil and gas fields are now coming into production, and high prices are likely to stimulate the exploitation of other more exotic fuels such as oil shales and tar sands. Before very long fossil fuels will inevitably begin to show signs of approaching exhaustion after one or two short-lived 'bonanzas' have been played out. Also, as new energy forms are developed to an increasing extent, market forces will start to retard the growth in demand for fossil fuels. On the whole, it seems likely that the pattern of fossil fuel production will follow some such shape as the extrapolated part of curve B after 1973. Production could continue to rise at a declining pace to reach a peak – perhaps in the early part of the next century – and then gradually decline to approach zero almost asymptotically at some remote and unspecified future date when all the accessible stocks of fossil fuels have become exhausted. Against the background of history, the bulk of the world's fossil fuels will have been consumed within perhaps a couple of centuries – a mere instant in eternity – with very attenuated production over longer periods on either side of the 'hump'. The area D approximately represents the total fuel consumption during the first 73 years of this century – which will not differ greatly from the total fuel consumed from the dawn of history until 1973. It represents about 17% of the proved recoverable reserves as estimated in 1980 (excluding nuclear resources), but probably a far smaller percentage of the fossil fuel stocks ultimately recoverable. In fact according to the World Energy Conference 1980 estimates of world energy resources the total believed stock of fossil fuels, proved *and* inferred, is about 12 times the proved recoverable reserves; so the shaded area D of Fig. 19.2 could perhaps represent only about 1.5% of the ultimately recoverable fossil fuel resources of the world. If the WEC 1980 estimates of total recoverable non-nuclear fuel stocks, proved and inferred, should be correct, then they would last for rather longer periods

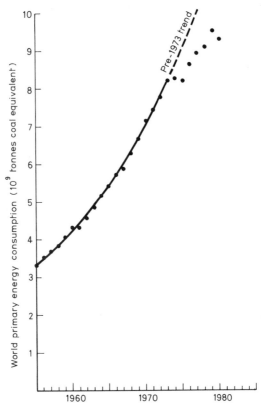

Figure 19.2 World consumption of primary energy, year by year, since 1955, showing the clear impact of the 1973 Middle East crisis.

than those shown in the second column of Table 19.2 as the various growth rates shown.

Now it is quite obvious that we could not possibly hope to fill the gap between curves A and B for more than a very short time, since the two curves will soon diverge from one another so rapidly that the gap between them would become unmanageable, with curve A soaring away merrily towards infinity and curve B bending further away from it. Even if we could *discover* new energy sources at a rate undreamed of, there must be a limit to the rate at which we could construct the necessary equipment for harnessing them and converting them into the form required. For instance, in 50 years time the energy demand at a sustained exponential growth of 5% p.a. would have increased about 11.5 times whereas the population, which for quite a long time has been growing at a fairly steady rate of 2% p.a. would have increased only about 2.7 times. Hence the production of exploitation equipment *per capita* would have to increase by about 4.5 times. After 100 years this factor would have risen to more than 18 times. Clearly such a trend could not possibly be maintained. Moreover, it is often overlooked that all artifacts, including energy exploitation equipment, themselves consume energy in the process of manufacture – sometimes on an immense scale. There could be a danger that the energy absorbed in making exploitation equipment would be so great that the net gain of energy usable for other purposes might be negligible.

Thus it is abundantly clear that curve A, or anything approaching it is doomed as totally impracticable. The demand curve must be forcibly 'bent' into some shape such as curve C in Fig. 19.2. Curve C could be permitted to rise above curve B only to the extent that we can supplement fossil fuels with new forms of energy. The shaded area E represents these new energy sources, and *it is within this area that the importance of geothermal energy will lie.* This clearly shows the inter-dependence between growth curtailment and the development of new energy sources (see Section 19.6). As drawn in Fig. 19.2, it will be seen that curve C need depart from curve A only fairly slowly at first, provided that we can develop area E fairly rapidly. If we cannot do this, curve C can lie only slightly above curve B, and this could be very painful. Obviously, if E can be expanded rapidly it will give us much more time in which to adapt ourselves to a new, and drastically re-shaped, demand curve C; whereas if curve C can be bent over quickly, smaller values of E would be acceptable.

It is possible that the maximum height to which curve C can be allowed to grow will be limited not so much by our ingenuity in winning new energy sources as by an approach towards the ultimate of environmental acceptance. Certainly geothermal energy would permit a far higher peak value to curve C than would fuel combustion.

Ultimately, in the very remote future, we shall have to learn to live on renewable energy sources only – F in Fig. 19.2. When we have learned to do that, and to limit our population accordingly, there will no longer be a world energy problem.

Figure 19.2 suggests that there is already clear evidence to support the belief that curve C of Fig. 19.1 is starting to bend away from the extrapolated growth trend of the 1960s and early 1970s. It is perhaps worth mentioning that Fig. 19.1 was first drawn [256] when the latest available world energy statistics were for 1972 only.

Our future lies largely in our own hands [256]. Mankind has shown himself to be particularly adept in failing to learn from the lessons of history and in blundering from crisis to crisis through his inability to see ahead. The purpose of this chapter is to show that to some extent we *can* see ahead, and that by taking timely action we may avert the disasters that may otherwise be inevitable. The obstacles that lie in the path of a brilliant energy future are probably no greater than those that were successfully overcome within a few years in attaining mastery over nuclear fission, space travel and micro-chip technology. All that is needed is a sense of urgency, and our problems will vanish. Despite the evidence that this sense of urgency is now being felt by a growing number of people, we must be on our guard against complacency which may arise in the 1980s when a temporary glut of energy could engender a false sense of security, as North Sea oil flows in spate. The future of energy supplies is the problem of Mankind, and not only of a small group of countries.

This chapter is intended to be no more than a warning: it makes no claim to being a precise prophecy. As already remarked in Chapter 4, long term prognostications are a vain pursuit. At all costs we should not attempt to extrapolate the future into the realms of absurdity. Who can possibly foresee how Mankind will have developed in tens of thousands of years? He may by then have learned to harness energy from sources as yet not dreamed of. He may have learned so to modify his way of life as to be able to do with far less energy – *per capita* and *in toto* – than at present. There may have occurred tectonic or solar cataclysms that will have destroyed or drastically reduced human and animal life. He may even (not altogether improbably) have destroyed himself by continuing to follow his foolish policies, perhaps at the same time annihilating all other forms of life on this planet, or perhaps leaving other species to survive, thus putting back the evolutionary clock by aeons.

One thing is certain: the next few decades will be critical for the human race. We may perhaps be faced with a Morton's Fork of discomfort on the one hand and disaster on the other; but by timely action – and there are encouraging signs of an awareness of the need for such action – we may weather the storm successfully. In any case, babies already born will see incredible changes in the human way of life before they are elderly, or even middle-aged.

References

1 *Proceedings of the United Nations Conference on New Sources of Energy, Rome, August 1961,* Vols **2** and **3** (1963). United Nations, Geneva. E/CONF 35/3 and E/CONF 35/4. Document sales Nos 63.1.36 and 63.1.37.

2 *Proceedings of the United Nations Symposium on the Development and Utilization Geothermal Resources, Pisa, Sept/Oct 1970.* Published in *Geothermics* special issue, Vols **1** and **2** (1974).

3 *Proceedings of the Second United Nations Symposium on the Development and Use of Geothermal Resources, San Francisco, USA May 1975,* Vols **1, 2** and **3** (1976). Lawrence Berkeley Laboratory, University of California.

4 *Geothermal Energy, a review of research and development.* Earth Sciences **12** (1973). UNESCO, Paris.

5 Circular 790 (1978). US Geological Survey, Reston, Virginia.

6 Einarsson, S.S. (1961). Proposed 15 MW geothermal power station at Hveragerdi, Iceland. *Paper G/9,* **3,** 354 of [1].

7 Reynolds, G. (1970). Cooling with geothermal heat, **2,** Part 2, 1658 of [2].

8 Einarsson, S.S. (1975). Geothermal space heating and cooling, **3,** 2117 of [3].

9 Líndal, B. (1973). Industrial and other applications of geothermal energy, 135 of [4].

10 Kunze, J.F., Richardson, A.S., Hollenbaugh, K.M., Nichols, C.R. and Mink, L.L. (1975). Non-electric utilization project, Boisé, Idaho, **3,** 2141 of [3].

11 Kunze, J.F., Miller, L.G. and Whitbeck, J.F. (1975). Moderate temperature utilization project in the Raft River valley, **3,** 2021 of [3].

12 Lund, J. (1981). Direct use: the emerging giant. Geothermal Resources Council Bulletin, May, 1981, p. 9.

13 Líndal, B. (1973). Industrial and other applications of geothermal energy, 146 of [4].

14 Special Report No. 7 (1979). Geothermal Resources Council, Davis, California, p. 4-26.

15 Reistad, G.M. (1975). Potential for non-electrical applications of geothermal energy and their place in the national economy, **3,** 2155 of [3].

16 Tamrazian, G.P. (1970). Continental drift and thermal fields, **2,** Part 2, 1212 of [2].

377

17   Bullard, Sir Edwards, FRS. (1973). Basic theories, 19 of [4].

18   Szádeczky-Kardoss, E. (1975). Geothermal energy and plate tectonics, 1977, *Atti Accad. Naz. dei Lincei*, 30, Roma, p. 123.

19   Bullard, Sir Edward, FRS. (1969). The origin of the oceans. *Scientific American*, **221,** 68.

20   Bullard, Sir Edward, FRS. (1973). Basic theories, 20 of [4].

21   Map of world seismicity 1961–9. (1970). National Earthquake Centre, Washington DC.

22   Bullard, Sir Edward, FRS. (1969). The origin of the oceans. *Scientific American*, **221,** 69.

23   Boldiszár, T. (1975). Terrestrial heat and geothermal energy production, 1977. *Atti Accad. Naz. dei Lincei*, 30, Roma, p. 141.

24   World Energy Resources: 1985–2020. (1980). World Energy Conference, London.

25   Boldiszár, T. (1975). **1,** 297 of [3].

26   Bodvarsson, G. (1975). **1,** 35 of [3].

27   Status of geothermal development in 1980. *UN University Report No. 1982–2.* Orkustofnun, Iceland, p. 27.

28   Status of geothermal development in 1980. *UN University Report No. 1982–2.* Orkustofnun, Iceland, p. 29.

29   Takashima, I. (1980). *Present status of geothermal research and development in Japan.* UN Conference on New and Renewable Sources of Energy, Nairobi, 1981. United Nations, Geneva, pp. 2–4.

30   Takashima, I. (1980). *Present status of geothermal development in Japan.* UN Conference on New and Renewable Sources of Energy, Nairobi, 1981. United Nations, Geneva, p. 17.

31   Alonso, H. (1975). **1,** 23 of [3].

32   Bolton, R.S. (1975). **1,** 38 of [3].

33   Circular 790 (1978). US Geological Survey, Reston, Virginia, p. 41.

34   Fomin, V.M., Mavritsky, L.F., Polubotko, L.F. and Makarenko, F.A. (1975). **1,** 129 of [3].

35   White, D.E. (1970). Geochemistry applied to the discovery, evaluation and exploitation of geothermal energy resources. **1,** 58 of [2].

36   Boldiszár, T. (1970). Geothermal energy production from porous sediments in Hungary, **2,** Part 1, 101 of [2].

37   Calami, A. and Ceron, P. (1970). Air convection within Montaña del Fuego, Lanzarote Island, Canary Archipelago, **2,** Part 1, 611 of [2].

38   Tikhonov, A.N. and Dvorov, I.M. (1970). Development of research and utilization of geothermal resources in the USSR, **2,** Part 2, 1072 of [2].

39   Facca, G. and Tonani, F. (1967). The self-sealing geothermal field. *Bull. Volcanologique,* **30,** 271.

40   Haldane, T.G.N. and Armstead, H.C.H. (1962). The geothermal power development at Wairakei, New Zealand. *Proc. Instn Mech. Engrs, London,* **176,** No. 23, 603.

41   Comisión Federal de Electricidad, México (1971). Official brochure on the Cerro Prieto geothermal development.

42   Facca, G. (1973). The structure and behaviour of geothermal fields, 61 of [4].

43   Stefánsson, V. (1979). The Krafla geothermal field, North-east Iceland. National Energy Authority, Geothermal Division, Reykjavík.

44   Sheinbaum, T. (1978). Geothermal well stimulation with a secondary fluid. *Geothermal Energy Mag.* **6,** No. 1, 33.

45   Chierici, A. (1961). Planning of a geothermoelectric power plant: technical and economic principles. *Paper G/62,* **3,** 299 of [1].

46   Hunt, T.M. (1970). Net mass loss from the Wairakei geothermal field, New Zealand, **2,** Part 1, 487 of [2].

47   House, P.A., Johnson, P.M. and Towse, D.F. (1975). Potential power generation and gas production from Gulf Coast geopressured reservoirs, **3,** 2001 of [3].

48   Boldiszár, T. (1976). Private communication.

49   Banwell, C.J. (1961). Geothermal drillholes: physical investigations. *Paper G/53,* **2,** 60 of [1].

50   McNitt, J.R. (1973). The rôle of geology and hydrology in geothermal exploration, 33 of [4].

51   Grindlay, G.W. (1961). Geology of New Zealand geothermal steam fields. *Paper G/34,* **2,** 242 of [1].

52   Garnish, J.D. (1976). Geothermal energy: the case for research in the United Kingdom. *Energy paper No. 9.* Report prepared for the Department of Energy by the Energy Technology Support Unit, Harwell. HMSO, London.

53   McNitt, J.R. (1970). The geologic environment of geothermal fields as a guide to exploration, **1,** 24 of [2].

54   Craig, H.. Boato, G. and White, D. (1956). The isotopic geochemistry of thermal waters. *National Research Council on Nuclear Science Service Report No. 19,* 29–44.

55   Hulston, J.R. (1961). Isotope geology in the hydrothermal areas of New Zealand. *Paper G/31,* **2,** 259 of [1].

56   Banwell, C.J. (1973). Geophysical methods in geothermal exploration, 41 of [4].

57   Banwell, C.J. (1970). Geophysical techniques in geothermal exploration, **1,** 32 of [2].

58   Dobrin, M.B. (1960). *Introduction to Geophysical Prospecting,* McGraw Hill, New York.

59   Sestini, G. (1970). Heat flow measurement in non-homogeneous terrains. Its application to geothermal areas, **2,** Part 1, 424 of [2].

60   Lee, W.H.K. and Uyeda, S. (1965). Terrestrial heat flow. *Geophysical Monograph,* **8,** American Geophysical Union, Washington D.C.

61   Burgassi, R., Battini, F., and Mouton, J. (1961). Prospection géothermique pour la recherche des forces endogènes. *Paper G/61,* **2,** 134 of [1].

62   Meidav, T. (1970). Application of electrical resistivity and gravimetry in deep geothermal exploration, **2,** Part 1, 303 of [2].

63   Gorhan, H.L. (1971). Geophysics *versus* drilling in engineering geology and hydrogeology. *Queensland Government Mining Journal,* September 1971.

64   Lumb, J.T. and Macdonald W.J.P. (1970). Near-surface resistivity surveys of geothermal areas using the electromagnetic method. **2,** Part 1, 311 of [2].

65   Clacy, G.R.T. (1973). Geothermal ground noise, amplitude and frequency spectra in the New Zealand volcanic region. *J. Geophys. Res.,* 5377.

66   Brune, J.N. and Allen, C.R. (1967). A micro-earthquake survey of the San Andreas fault system in southern California. *Bull. Seismol. Soc. Am.*, **57**, 277.

67   Ward, P.L., Pálmason, G., Drake, C. and Oliver, J. (1968). The microseismicity of Iceland and its relation to the regional tectonics. *49th Annual Meeting of the American Geophysical Union.*

68   Pálmason, G., Friedman, J.D., Williams, R.S.Jr., Jonnsson, J. and Saemundsson, K. (1970). Aerial infra-red surveys of Reykjanes and Torfajokull thermal areas, with a section on cost of exploration surveys, **2**, Part 1, 399 of [2].

69   Meidav, T. (1979). Trends in geothermal exploration technology. UNITAR Conference long-term energy resources, Montreal.

70   Schirra, W.Jr. (1976). National Geothermal Conference, Palm Springs. *Geothermal Energy Mag.*, **4**, No. 6, 32.

71   Barbier, E. and Fanelli, M. (1975). Relationships as shown in ERTS satellite images between main fractures and geothermal manifestations in Italy, **2**, 883 of [3].

72   Sigvaldason, G.E. (1973). Geochemical methods in geothermal exploration, 49 of [4].

73   Craig, S.B. (1961). Geothermal drilling practices at Wairakei, New Zealand. *Paper G/14*, **3**, 121 of [1].

74   Matsuo, K. (1973). Drilling for geothermal steam and hot water, 73 of [4].

75   Cigni, U. and Giovannoni, A. (1970). Planning methods in geothermal drilling, **2**, Part 1, 725 of [2].

76   Cigni, U. (1970). Machinery and equipment for harnessing of endogenous fluid, **2**, Part 1, 704 of [2].

77   Giovannoni, A. (1970). Drilling Technology, **1**, 81 of [2].

78   Woods, D.I. (1961). Drilling mud in geothermal drilling. *Paper G/21*, **3**, 270 of [1].

79   Fabbri, F. and Vidali, M. (1970). Drilling mud in geothermal wells, **2**, Part 1, 735 of [2].

80   Contini, R. and Cigni, U. (1961). Air drilling in geothermal bores. *Paper G/70*, **3**, 89 of [1].

81   Dench, N.D. (1970). Casing string design for geothermal wells, **2**, Part 2, 1485 of [2].

82   Fabbri, F. and Giovannoni, A. (1970). Cements and cementation in geothermal well drilling, **2**, Part 1, 742 of [2].

83   Bolton, R.S. (1961). Blowout prevention and other aspects of safety in geothermal steam drilling. *Paper G/43*, **3**, 78 of [1].

84   Innes, I.A. (1961). Management, in relation to measurements, and bore maintenance of an operating geothermal steam field. *Paper G/15*, **3**, 208 of [1].

85   Stilwell, W.B. (1970). Drilling practices and equipment in use at Wairakei, **2**, Part 1, 714 of [2].

86   Brunetti, V. and Mezzetti, E. (1970). On some troubles most frequently occurring in geothermal drilling, **2**, Part 1, 751 of [2].

87   Smith, J.H. (1961). Casing failures in geothermal bores at Wairakei. *Paper G/44*, **3**, 254 of [1].

88   Special Report No. 7 (1979). Geothermal Resources Council, Davis, California. pp. 3–7 to 3–9.

89 Matsuo, K. (1973). Drilling for geothermal steam and hot water, 75 of [4].

90 New Zealand Government (1970). Wairakei. Power from the Earth. The story of the Wairakei geothermal project. Official brochure.

91 Tolivia, M.E. (1970). Corrosion measurements in a geothermal environment, **2**, Part 2, 1596 of [2].

92 Garnish, J.D. (1976) Geothermal energy: the case for research in the United Kingdom. *Energy paper No. 9.* Report prepared for the Department of Energy by the Energy Technology Support Unit, Harwell. HMSO, London, p. 35.

93 Kestin, J. (ed.) (1980). *Sourcebook on the Production of Electricity from Geothermal Energy,* US Department of Energy, p. 145.

94 Budd, C.F. (1973). Producing geothermal steam at the Geysers field. In *Geothermal Energy Resources, Production and Stimulation* (eds P. Kruger and C. Otte), Stanford University Press.

95 Maurer, W.C. (1975). Geothermal drilling technology, **2**, 1509 of [3].

96 Hunt, A.M. (1961). The measurement of borehole discharges, downhole temperatures and pressures, and surface heat flows at Wairakei. *Paper G/19,* **3**, 196 of [1].

97 James, R. (1970). Factors controlling borehole performance, **2**, Part 2, 1502 of [2].

98 James, R. (1975). Rapid estimation of electric power potential of discharging geothermal wells, **3**, 1685 of [3].

99 James, R. (1967). Optimum wellhead pressure for geothermal power. *N. Zealand Engng,* **22**, No. 6, 221–8.

100 Iga, H. (1976). Personal communication.

101 Armstead, H.C.H. and Shaw, J.R. (1970). The control and safety of geothermal installations, **2**, Part 1, 848 of [2].

102 Contini, R. (1961). Methods of exploitation of geothermal energy and the equipment required. *Paper G/71,* **3**, 111 of [1].

103 Bangma, P. (1961). The development and performance of a steam-water separator for use on geothermal bores. *Paper G/13,* **3**, 60 of [1].

104 Dench, N.D. (1961). Silencers for geothermal bore discharge. *Paper G/18,* **3**, 134 [1].

105 Pollastri, G. (1970). Design and construction of steam pipelines, **2**, Part 1, 780 of [2].

106 Armstead, H.C.H. (1968). The extraction of power from hot water. Seventh World Power Conference, Moscow, August, 1968, Paper 174, Section C4.

107 Thain, I.A. (1979). Wairakei – the first 20 years. New Zealand Electricity Department document.

108 James, R. (1968). Pipeline transmission of steam–water mixtures for geothermal power. *New Zealand Engng,* **23**, 55–61.

109 Takahashi, Y., Hayashida, T., Soezima, S., Aramaki, S. and Soda, M. (1970). An experiment on pipeline transportation of steam–water mixtures at Otake geothermal field, **2**, Part 1, 882 of [2].

110 Ryley, D.J. (1980). In *Sourcebook on the Production of Electricity from Geothermal Energy* (ed. J. Kestin), Section 2.6.5.1 *et seq.,* US Department of Energy.

111   Aikawa, K. and Soda, M. (1975). Advanced design in Hatchobaru geothermal power station, **3,** 1881 of [3].

112   Eliásson, E.T., Björnsson, G., Matthíasson, M., Maask, R., Sigfússon, S. and Jónsson, V.K. (1982). *Collection and transmission of two-phase geofluid at the Krafla geothermal power plant, Iceland.* British Hydromechanics Research Association, Geothermal Conference in Florence, Italy. BHRA, Bedford, England.

113   James, R. (1973) Control orifices replace steam traps on overland transmission pipelines, **3,** 1699 of [3].

114   Maslan, F., Gordon, T.J. and Stover, J. (1975). **3,** 2414 of [3].

115   Definition Report (1975). ERDA, Washington DC, Document ERDA-86, p. L-12.

116   Third Annual Report of Geothermal Energy Research, Development and Demonstration Program (1979). DOE, Washington DC, Document DOE/ET-00900 IGCC-4, p. 14.

117   Moskvicheva, V.N. and Popov, A.E. (1970). Geothermal power plant on the Paratunka River, **2,** Part 2, 1567 of [2].

118   Rollet, A. (1957). Centrale géothermique de Kiabukwa: leçons tirées de quatre années d'exploitation. *Bull Acad. R. Sci. Col., Bruxelles,* **3,** 1246.

119   Hansen, A. (1961). Thermal cycles for geothermal sites and turbine installation at the Geysers power plant, California. *Paper G/41,* **3,** 365 of [1].

120   Harrowell, R.V. (1975). Private communication.

121   Wigley, D.M. (1970). Recovery of flash steam from hot bore water, **2,** Part 2, 1588 of [2].

122   Wood, B. (1973). Geothermal power, 109 of [4].

123   Armstead, H.C.H. (1970). Utilization of steam and high enthalpy water for electric power generation and other purposes, **1,** 106 of [2].

124   Dal Secco, A. (1970). Turbo-compressors for geothermal plants, **2,** Part 1, 819 of [2].

125   Dal Secco, A. (1975). Geothermal plants: gas removal from jet condensers, **3,** 1943 of [3].

126   James, R. (1970). Power station strategy, **2,** Part 2, 1676 of [2].

127   Armstead, H.C.H. (1970). Geothermal power for non-base load purposes, **2,** Part 1, 936 of [2].

128   Ciapica, I. (1970). Present development of turbines for geothermal applications, **2,** Part 1, 834 of [2].

129   Ricci, G. and Viviani, G. (1970). Maintenance operations in geothermal power plants, **2,** Part 1, 839 of [2].

130   Matthew, P. (1975). Geothermal operating experience, Geysers power plant, **3,** 2049 of [3].

131   Geothermal Resources Council Bulletin, May, 1980, p. 4. Geothermal Resources Council, Davis, California.

132   Palmer, T.D., Howard, J.H. and Lande, D.P. (1975). Geothermal development of the Salton Trough, California and Mexico. Lawrence Livermore Laboratory Document, UCRL-51775.

133   Kestin, J. (ed.) (1980). *Sourcebook on the Production of Electricity from Geothermal Energy.* US Department of Energy, p. 296.

134 US Department of Energy brochure (1979). *Geopressurized Geothermal Resources: an unconventional energy source.*

135 HPC Brochure (1974). Hydrothermal Power Co., Pasadena, California.

136 Austin, A.L. (1975) Prospects for advances in energy conversion technologies for geothermal energy development, **3**, 1925 of [3].

137 Kestin, J. (ed.) (1980). *Sourcebook on the Production of Electricity from Geothermal Energy.* US Department of Energy, pp. 512, 513.

138 Armstead, H.C.H. (1975). Basic design of a cheap wellhead non-condensing turbine, **3**, 1905 of [3].

139 Pessina, S., Rumi, O., Silvestri, M. and Sotgia, G. (1970). Gravimetric loop for the generation of electrical power from low temperature water, **2**, Part 1, 901 of [2].

140 Daman, E.L. (1966). *Electrogasdynamic power generation,* Gourdine Systems, Inc., New Jersey.

141 Naymanov, O.S. (1970). A pilot geothermoelectric power station in Pauzhetka, Kamchatka, **2**, Part 2, 1560 of [2].

142 Alger, T.W. (1975). Performance of two-phase nozzles for total-flow geothermal impulse turbines, **3**, 1889 of [3].

143 Armstead, H.C.H. (1975). Some unusual ways of developing power from a geothermal field, **3**, 1897 of [3].

144 Charropin, P., Despois, J., Fauconnier, J-.C. and Nougarède, F. (1976). Paper presented at the first International Total Energy Congress, Copenhagen, *Total Energy,* Miller Freeman Publications, San Francisco, p. 460.

145 Meyer, C.F. and Hausz, W. (1978). Energy management objectives and economics of heat storage wells. Total energy storage in aquifers workshop, Berkeley, California, 1978. Tempo GEC Centre for advanced studies, Santa Barbara, California, USA.

146 Guiza, J.L. (1975). Power generation at Cerro Prieto geothermal field, **3**, 1976 of [3].

147 Barton, D.B. (1970). Current status of geothermal power plants at the Geysers, Sonoma County, California, **2**, Part 2, 1552 of [2].

148 Finney, J.P., Miller, F.J. and Mills, D.B. (1972). Geothermal power project of Pacific Gas and Electric Co. at the Geysers, California, presented at the summer meeting of the IEEE Power Enginering Society, July, 1972.

149 Gudmundsson, J.S., Thórhallsson, S. and Ragnars, K. (1981). Geothermal electric power in Iceland. Orkustofnun, Geothermal Division, Iceland.

150 Villa, F.P. (1975). Geothermal plants in Italy: their evolution and problems, **3**, 2061 of [3].

151 Ente Nazionale per l'energia elettrica (ENEL). Larderello and Monte Amiata: electric power by endogenous steam. Official brochure.

152 Sato, H. (1970). On Matsukawa geothermal power plant, **2**, Part 2, 1546 of [2].

153 Usui, T. and Aikawa, K. (1970). Engineering and design features of the Otake geothermal power plant, **2**, Part 2, 1533 of [2].

154 New Zealand Government (1970). Wairakei: power from the earth. The story of the Wairakei geothermal project. Official brochure.

155 Howard, J.H. (1975). Principal conclusions of the committee on the Challenges of Modern Society Non-electrical Applications Project, **3**, 2127 of [3].

156   World survey of low temperature geothermal energy utilization (1981). Document OS81005/JHD02, Orkustafnun, Iceland.

157   Boldiszár, T. (1976). Hungary cuts back oil imports by increasing geothermal energy production. *Geothermal Energy Mag.*, August 1976, p. 5.

158   Boldiszár, T. (1970). Geothermal energy production from porous sediments in Hungary, **2**, Part 1, p. 104 of [2].

159   Geothermal Progress Monitor No. 5: progress report, June, 1981. US Department of Energy, Division of Geothermal Energy.

160   Arnórsson, S., Ragnars, K., Benediktsson, S., Gislason, G. and Thórhallsson, S., with Björnsson, S., Grönvold, K. and Líndal, B. Exploitation of saline high temperature water for space heating, **3**, 2077 of [3].

161   Dvorov, I.M. and Ledentsova, N.A. (1975). Utilization of geothermal water for domestic heating and hot water supply, **3**, 2109 of [3].

162   Buachidse, I.M., Buachidse, G.I. and Shaorshadse, M.P. (1970). Thermal waters of Georgia, **2**, Part 2, 1092 of [2].

163   Kremnjov, O.A., Zhuravlenko, V.J. and Shurtshkov, A.V. (1970). Technical–economic estimation of geothermal resources, **2**, Part 2, 1688 of [2].

164   Balogh, J. (1975). Results achieved in Hungary in the utilization of geothermal energy, **1**, 29 of [3].

165   Garnish, J.D. (1977) Geothermal energy usage in the Paris Basin. Report ETSU N2/77. Energy Technology Support Unit, Harwell.

166   Coulbois, P. and Hérault, J-P. (1975). Conditions for the competitive use of geothermal energy in home heating, **3**, 2104 of [3].

167   Mashiko, Y. and Hirano, Y. (1970). New supply systems of thermal waters to a wide area in Japan, **2**, Part 2, 1592 of [2].

168   Behl, S.C., Jegadeesan, K. and Reddy, D.S. (1975). Some aspects of the utilization of geothermal fluids in the North-West Himalayas, **3**, 2083 of [3].

169   Shannon, R.J. (1975). Geothermal heating of Government buildings in Rotorua, **3**, 2165 of [3].

170   Einarsson, S.S. (1973). Geothermal district heating, 123 of [4].

171   Takashima, I. (1980). Present status of geothermal research and development in Japan. UN Conference on New and Renewable Sources of Energy: Nairobi, 1981. United Nations, Geneva.

172   Bodvarsson, G. (1961). Utilization of geothermal energy for heating purposes and combined schemes involving power generation, heating, and/or by-products. *Paper GR/5(G)*, **3**, 429 of [1].

173   Gutman, P.W. (1975). Geothermal hydroponics, **3**, 2217 of [3].

174   Lokchine, B.A. and Dvorov, I.M. (1970). Applications expérimentales et industrielles de l'energie géothermique en URSS, **2**, Part 2, 1079 of [2].

175   Japan Geothermal Energy Association (1974). Geothermal energy utilization in Japan, 1974. Official brochure.

176   Komagata, S., Iga, H., Nakamura, H. and Minohara, Y. (1970). The status of geothermal utilization in Japan, **2**, Part 1, 185 of [2].

177   Einarsson, S.S. (1970). Utilization of low enthalpy water for space heating, industrial, agricultural and other uses, **1**, 112 of [2].

178   Líndal, B. (1961). Greenhouses by geothermal heating in Iceland. *Paper G/32*, **3**, 476 of [1].

179   Hallsson, S. (1970). Drying seaweeds and grass by geothermal energy. *Timarit Verkfraed. Islands*, **4.**

180   Kerr, R.N., Bangma, R., Cooke, W.L., Furness, F.G. and Vamos, G. (1961). Recent developments in New Zealand in the utilization of geothermal energy for heating purposes. *Paper G/52*, **3,** 456 of [1].

181   Thiagarajan, B. (1967). Report of a UNIDO mission on the manufacture of chemicals from seawater in Iceland, United Nations Industrial Development Organisation, Vienna.

182   Líndal, B. (1961). The extraction of salt from sea water by multiple-effect evaporators using natural steam. *Paper G/27*, **3,** 479 of [1].

183   Mizutani, Y. (1961). Salt production by geothermal energy in Japan. *Paper G/7*, **3,** 483 of [1].

184   Líndal, B. (1970). The production of chemicals from brine and seawater, using geothermal energy, **2,** Part 1, 910 of [2].

185   Kennedy, A.M. (1961). The recovery of lithium and other minerals from geothemal water at Wairakei. *Paper G/56*, **3,** 502 of [1].

186   Werner, H.H. (1970). Contribution to the mineral extraction from supersaturated geothermal brines, Salton Sea area, California, **2,** Part 2, 1651 of [2].

187   Mazzoni, A. (1948). The steam vents of Tuscany and the Larderello plant. Amonina Arts Grafiche, Bologna, pp. 59–75.

188   Lenzi, D. (1961). Utilization de l'énergie géothermique pour la production de l'acide borique et des sous-produits contenus dans les 'soffioni' de Larderello. *Paper G/39*, **3,** 512 of [1].

189   Valfells, A. (1970). Heavy water production with geothermal steam, **2,** Part 1, 896 of [2].

190   Bottomley, J. (1980). Utilization of geothermal energy in a biomass-ethanol plant. Geo-heat Centre quarterly bulletin, Dec. 1980, p. 14.

191   Shreve, R.N. (1956). *Chemical process industries*, 2nd ed., McGraw Hill, New York.

192   Ludviksson, V. (1970). Nýting jardhitans (The application of natural heat). *National Research Council Report 70–3*, Reykjavík.

193   Burrows, W. (1970). Geothermal energy resources for heating and associated applications in Rotorua and surrounding areas, **2,** Part 2, 1662 of [2].

194   Armstead, H.C.H. and Rhodes, C. (1970). Desalination by geothermal means. *Proc. 3rd International Symposium on Fresh Water from the Sea, Dubrovnik*, **3,** 451–9. European Federation of Chemical Engineering, Athens.

195   De Anda, L.F., Reyes, S.C. and Tolivia, M.E. (1970). Production of fresh water from the endogenous steam of Cerro Prieto geothermal field, **2,** Part 2, 1632 of [2].

196   Chiostri, E. (1975). Geothermal resources for heat treatment, **3,** 2094 of [3].

197   Minohara, Y. and Sekioka, M. (1975). Geothermal utilization in the Atagawa tropical garden and alligator farm: an example of successful geothermal utilization, **3,** 2237 of [3].

198   Cooke, W.L. (1970). Some methods of dealing with low enthalpy water in the Rotorua area of New Zealand, **2,** Part 2, 1670 of [2].

199   Lund, J.W., Culver, G.G. and Svanevik, L.S. (1975). Utilization of intermediate temperature geothermal water in Klamath Falls, Oregon, **3**, 2147 of [3].

200   Thórhallsson, S. (1979). Combined generation of heat and electricity from a geothermal brine at Svartsengi in S.W. Iceland. Geothermal Resources Council Transactions, **3**, Sept. 1979.

201   Wong, C.M. (1970). Geothermal energy and desalination: partners in progress, **2**, Part 1, 892 of [2].

202   Palmer, H.D., Forns, J.M. and Green, J. (1975). Exploitation of sea-floor geothermal resources: multiple use concept, **3**, 2241 of [3].

203   Kunze, J.F., Miller, L.G. and Whitbeck, J.F. (1975). Moderate temperature utilization project in the Raft River Valley, **3**, 2021 of [3].

204   Verkajakrishnan, R. (1980). Artist's impression of a 'total geothermal community', embodying power generation, industrial, space heating, agricultural, aquacultural and recreational uses. Geo-heat quarterly bulletin, Sept. 1980, Cover.

205   Sekioko M. and Fujitomi, M. (1981). New national projects on the direct utilization of geothermal resources in Japan. Geo-heat quarterly bulletin, June 1981, p. 11.

206   Bruce, A.W. (1970). Engineering aspects of a geothermal power plant, **2**, Part 2, 1516 of [2].

207   Leardini, T. (1970), Economie de l'energie géothermique, **2**, Part 1, 958 of [2].

208   Armstead, H.C.H. (1973). Geothermal economics, 161 of [4].

209   Bodvarsson, G. and Zoëga, J. (1961). Production and distribution of natural heat for domestic and industrial heating in Iceland. *Paper G/37*, **3**, 449 of [1].

210   Special Report No. 7 (1979). Geothermal Resources Council, p. 3-21.

211   Robinson, E.S., Rowley, J.C., Potter, R.M., Armstrong, D.E., McInteer, B.B., Mills, R.L. and Smith, M.C. (ed.) (1971). A preliminary study of the nuclear subterrene. Los Alamos Scientific Laboratory Document LA-4547. UC-38.

212   Carson, C.C. and Lin, Y.T. (undated and un-numbered). Geothermal well costs and their sensitivities in drilling and completion operations. Sandia National Laboratories, New Mexico, p. 8–14.

213   Bowen, R.G. and Groh, E.A. (1971). Geothermal–Earth's primordial energy. *Technology Rev.*, 42–48.

214   Facca, G. and Ten Dam, A. (1964). *Geothermal power economics*. Worldwide Geothermal Exploration Co., Los Angeles, California.

215   Leardini, T. (1976). Private communication.

216   Leardini, T. (1974). Geothermal power. *Phil. Trans. R. Soc., London*, **A276**, 500.

217   Ragnars, K., Saemundsson, K., Benediktsson, S. and Einarsson, S.S. (1970). Development of the Námafjall area, Northern Iceland, **2**, Part 1, 925 of [2].

218   Thórhallsson, S., Ragnars, K., Arnórsson, S. and Kristmannsdóttir, H. (1975). Rapid scaling of silica in two district heating sytsems, **2**, 1445 of [3].

219   Allegrini, G. and Benvenuti, G. (1970). Corrosion characteristics and geothermal power plant protection; (collateral process of abrasion, erosion and scaling.), **2**, Part 1, 865 of [2].

220   Hermannsson, S. (1970). Corrosion of metals and the forming of a protective coating on the inside of pipes carrying thermal waters used by the Reykjavík Municipal District Heating Service, **2**, Part 2 1602 of [2].

221   Marshall, T. and Braithwaite, W.R. (1973). Corrosion control in geothermal systems, p. 151 of [4].

222   Ozawa, T. and Fujii, Y. (1970). A phenomenon of scaling in production wells and the geothermal power plant in the Matsukawa area, **2**, Part 2, 1613 of [2].

223   Einarsson, S.S., Vides, R.A. and Cuéllar, G. (1975). Disposal of geothermal waste water by re-injection, **2**, 1349 of [3].

224   New Zealand Ministry of Works and Development (1978). Report on the injection of geothermal waste water into the Broadlands field.

225   Chasteen, A.J. (1975). Geothermal steam condensate re-injection. **2**, 1335 of [3].

226   Kubota, K. and Aosaki, K. (1975). Re-injection of geothermal hot water at the Otake geothermal field, **2**, 1379 of [3].

227   Gringarten, A.C. and Sauty, J.P. (1975). The effect of re-injection on the temperature of a geothermal reservoir used for urban heating, **2**, 1370 of [3].

228   Hartley, R.P. and Di Pippo, R. (1980). Claus Process. In *Sourcebook on the Production of Electricity from Geothermal Energy* (ed. J. Kestin), Section 9.4.6, US Department of Energy.

229   Allen, G.W. and McCluer H.K. (1975). Abatement of hydrogen sulphide emissions from the Geysers geothermal power plant, **2**, 1313 of [3].

230   Hartley, R.P. and Di Pippo, R. (1980). Stretford Process. In *Sourcebook on the Production of Electricity from Geothermal Energy* (ed. J. Kestin), Section 9.4.1, US Department of Energy.

231   Laszlo, J. (1978). Application of the Stretford process for H_2S abatement at the Geysers geothermal power plant. Pacific Gas and Electric Co. publication.

232   Axtmann, R.C. (1975). Environmental impact of a geothermal power plant. *Science*, **187**, No. 4179.

233   Reed, M.J. and Campbell, G. (1975). Environmental impact of development in the Geysers geothermal field, USA, **2**, 1399 of [3].

234   Rothbaum, H.P. and Anderton, B.H. (1975). Removal of silica and arsenic from geothermal discharge waters by precipitation of useful calcium silicates, **2**, 1417 of [3].

235   Anderson, S.O. (1975). Environmental impacts of geothermal resource development on commercial agriculture: a case study of land use conflict, **2**, 1317 of [3].

236   Mercado, G.S. (1975) Cerro Prieto geothermoelectric project: pollution and basic protection, **2**, 1394 of [3].

237   Swanberg, C.A. (1975). Physical aspects of pollution related to geothermal energy development, **2**, 1435 of [3].

238   Axtmann, R.C. (1975). Chemical aspects of the environmental impact of geothermal power, **2**, 1323 of [3].

239   Smith, J.H. (1961). General report. *Paper GR/4(G)*, p. 21 of [1].

240   Hatton, J.W. (1970). Ground subsidence of a geothermal field during exploitation, **2**, Part 2, 1294 of [2].

241   Stilwell, W.B., Hall, W.K. and Tawhai, J. (1975). Ground movement in New Zealand geothermal fields, **2,** 1427 of [3].

242   Bolton, R.S. (1973). Management of a geothermal field, 175 of [4].

243   Armstead, H.C.H. (1975). Environmental factors and waste disposal, **1,** p. lxxxvii of [3].

244   Cuéllar, G. (1975). Behaviour of silica in geothermal waste waters, **2,** 1343 of [3].

245   Armstead, H.C.H., Gorhan, H.L. and Müller, H. (1974). Systematic approach to geothermal development. *Geothermics,* **3,** No. 2.

246   Fridleifsson, I.B. (1980). A UNU international geothermal energy training course in Iceland. World Energy Conference, Sept. 1980.

247   Worthington, J.D. (1975). Geothermal development in Chapter 9 of *Status Report–Energy Resources and Technology,* a report of the *ad hoc* Committee on Energy Resources and Technology, Atomic Industrial Form, Inc.

248   Witherspoon, P.A. (1966). Economics of nuclear explosives in developing underground gas storage. University of California, Lawrence Radiation Laboratory, Livermore, California, Document UCRL-14877.

249   Parker, K. (1974). Personal communication.

250   Kennedy, G.C. (1964). A proposal for a nuclear power programme. 3rd Ploughshare Symposium, University of California, Davis, Cal., April, 1964.

251   Pettitt, R.A. (1976). Environmental monitoring for the hot dry rock geothermal energy development project. Los Alamos Scientific Laboratory Document LA-6504-SR Status Report. UC-11 and UC-66e.

252   Harlow, F.H. and Pracht, W.E. (1972). A theoretical study of geothermal energy extraction. *J. Geophys. Res.,* **77,** 7041.

253   Robinson, E.S., Rowley, J.C., Potter, R.M., Armstrong, D.E., McInteer, B.B., Mills, R.L. and Smith, M.C. (ed.) (1971). A preliminary study of the nuclear subterrene. Los Alamos Scientific Laboratory Document LA-4547. UC-38.

254   Altseimer, J.H. (1975). Geothermal well technology and potential applications of subterrene devices – a status review, **2,** 1453 of [3].

255   Armstead, H.C.H. (1980). Effects of growth on natural resource life. *Energy International,* **17,** No. 4, April 1980.

256   Armstead, H.C.H. (1975). World energy: the shape of things to come. *Energy International,* **12,** No. 1, 13.

257   Armstead, H.C.H. (1982). A proposal for accelerating geothermal power development: especially for small systems. *GRC Bulletin,* August 1982.

Index

Page numbers given in ordinary type indicate textual matter or tables. Page numbers given in bold type indicate illustrations. Page numbers in italic indicate references (pp. 377–387).

389